ZEW Economic Studies

Publication Series of the Centre for
European Economic Research (ZEW),
Mannheim, Germany

ZEW Economic Studies

Vol. 1: O. Hohmeyer, K. Rennings (Eds.)
Man-Made Climate Change
Economic Aspects and Policy Options
1999. VIII, 401 pp. ISBN 3-7908-1146-7

Vol. 2: Th. Büttner
Agglomeration, Growth, and Adjustment
A Theoretical and Empirical Study
of Regional Labor Markets in Germany
1999. XI, 206 pp. ISBN 3-7908-1160-2

P. Capros · L. Mantzos · P. Criqui · N. Kouvaritakis
A. Soria Ramirez · L. Schrattenholzer
E.L. Vouyoukas

Climate Technology Strategies 1

Controlling Greenhouse Gases. Policy and Technology Options

With 80 Figures
and 95 Tables

Springer-Verlag
Berlin Heidelberg GmbH

ZEW
Zentrum für Europäische
Wirtschaftsforschung GmbH
Centre for European
Economic Research

Series Editor
Prof. Dr. Wolfgang Franz

Authors

Prof. Pantelis Capros
Dr. Leonidas Mantzos
National Technical University of Athens
Department of Electrical &
Computer Engineering
42 Patission Street
10682 Athens, Greece

Dr. Patrick Criqui
Institute of Energy
Policy and Economics – CNRS
BP 47
38049 Grenoble Cedex, France

Nikolaos Kouvaritakis
43 Quarry High Street
Headington
Oxford, UK

Dr. Antonio Soria Ramirez
Institute for Prospective Technological
Studies
Joint Research Centre
World Trade Center Building
Isla de la Cartuja s/n
41092 Sevilla, Spain

Dr. Leo Schrattenholzer
ECS Project
International Institute for
Applied Systems Analysis (IIASA)
Schlossplatz 1
2361 Laxenburg, Austria

E. Lakis Vouyoukas (Editorial Work)
Kodrou 13
10558 Athens, Greece

ISBN 978-3-7908-1229-9

Cataloging-in-Publication Data applied for
Die Deutsche Bibliothek – CIP-Einheitsaufnahme
Climate technology strategies / ZEW, Centre for European Economic Research. P. Capros ... – Heidelberg; New York: Physica-Verl.
Vol. 1. Controlling greenhouse gases: policy and technology options; with 95 tables. – 1999
(ZEW economic studies; Vol. 3)
ISBN 978-3-7908-1229-9 ISBN 978-3-642-58679-8 (eBook)
DOI 10.1007/978-3-642-58679-8

Originally published by Physica-Verlag Heidelberg in 1999

Cover design: Erich Dichiser, ZEW, Mannheim

SPIN 10733396 88/2202-5 4 3 2 1 0 – Printed on acid-free paper

Preface

This book provides a synthesis of the energy-related results of the large scale project entitled "Climate Technology Strategy within Competitive Energy Markets" partly financed by the European Commission under the Non Nuclear Energy (JOULE III) EU-RTD programme (Contract JOS3-CT95-0008). Research in this project involved several organisations from almost all European Union member states, including: ICCS/NTUA (co-ordinator), CES/KULeuven, GRETA Associati, University of Mannheim/ZEW, ESD Ltd., IEPE/CNRS, IDEI, IPTS/JRC, UCL/CORE, SSE/EFI, KFA/STE, ENSPM-FI/CEG, University of Strathclyde, IER/University of Stuttgart, VTT, ETSU, IIASA/ECS, ERASME/CCIP, ECOSIM Cons. Ltd., OME, BPB, CRES.

In relation of the "pre-and post-Kyoto" context for Climate Change and also with respect to the preparation of the Vth EU-RTD Framework Programme, a large exploitation of the results is being made.

Huguette Laval of the European Commission, DGXII, was the project scientific officer. With Pierre Valette, Head of the Unit in DGXII, they have given fruitful advices for the project. Thanks are given to them.[1]

---

[1] The views expressed in this volume are those of the authors and do not reflect the views of the Commission or its officials.

Table of Contents

Executive Summary

The global warming threat has become an important source of concern in recent years. The identification of energy-related emissions as the most important contributor to anthropogenic emissions contributing to global warming has highlighted the possibility that the problem may be especially difficult to solve. Energy consumption and economic growth have been closely linked for a very long time. Consequently, if alleviating the possibility of global warming implies the reduction in the use of energy the question arises whether this can be achieved without endangering the prospects of economic growth. The question is especially pertinent in a period of high unemployment in developed economies (this seems to be no longer true in the US) and of a high priority for rapid development among developing countries.

However, carbon emissions are not the first environmental problem that has worried energy planners. Indeed, the problem of finding compatible solutions to fulfil the energy supply requirements and limiting sulphur and nitrogen oxide emissions was of extreme importance during the seventies and the eighties. The solution of the acid rain problem came basically through technological solutions that were found after an aggressive environmental regulatory framework. The global warming issue, being linked to the use of carbon itself (although also depending on the way the carbon is combusted), seems a problem whose solution may require changes in human behaviour, an appropriate policy framework, as well as a large input from technological progress.

Technology is likely to be called upon to play a very important role in dealing with the greenhouse gas issue. Of course, "technology" must be defined sufficiently broadly to include both supply-side technologies and demand-side technologies. Amongst the first, energy transformation (and, particularly, electric power generation) is a very important subset, since it accounts for a remarkable share of the carbon emissions around the world. On the demand-side, energy efficient improvements in domestic and industrial appliances are fundamental in contributing to the greenhouse gas emissions reduction. Special mention should be made of the potential contribution of technology to the improvement of the efficiency in the transportation sector, one of the fastest growing energy intensive human activities.

It is some aspects of this contribution of technology that this volume aims to examine. The volume is divided into three parts. The first sets the stage of the analysis by presenting the European and global economic and energy outlook. The second examines the policy options available for controlling greenhouse gases in the period to 2010. Over such a relatively short period, the role of technology lies mostly in making possible the more efficient use of energy. Finally, the third part of the volume examines the longer term role of technology in the context of a case study for the power generation sector.

Economic, Energy and Emissions Outlook

The world economy is projected to increase by 3.3% pa in the period to 2030. Over the same period, the EU is likely to experience growth of close to 2% pa. Perhaps the most important aspect of the outlook is the further rapid increase in the weight of developing regions in the world economy and energy system. It is likely that most of the increase in energy consumption in the next forty years will come from regions outside the OECD, which account for only a fraction of current world energy consumption. By 2010, they will contribute to half the world GDP and almost half the world energy consumption.

International energy markets will be affected by this shift, as oil demand will largely grow along the development of transport and car equipment in these regions. In terms of energy, transportation remains the fastest growing sector in the EU. Although oil resources are large and might be extended by technological progress, non conventional fuels will need to be used after 2010. Oil prices are projected to increase in order both to satisfy the growing demand and to encourage the development of non conventional resources. Natural gas prices may also increase, as gas demand will be especially strong in every region of the world.

Overall global energy demand is likely to grow by slightly over 2% pa maintaining an improvement in enegry intensity of close to 1% pa. EU energy is projected to increase by less than 1% pa, leading to an improvement in energy intensity of close to 1.5% pa. This improvement in intensity is rather optimistic when compared to historical trends. However, the most important development, at least over the next 15 years, in global energy is the penetration of natural gas. The EU as well as the global energy system continue to be dominated by fossil fuels, at least until 2010. Even by 2030, new technologies based on renewables are likely to only slightly dent this dominance.

Consequent on the growth in world energy demand based on fossil fuels, world CO_2 emissions will be on an upward trend. Reducing emissions by 2010, as developed regions have undertaken to do, seems a very difficult task. Even European emissions are projected to increase by 8% in the absence of any specific CO_2 mitigation measures. The expected increase in developing country emissions is very large and stems directly from the strong dynamics in energy demand, particularly for transport and electricity. For the latter, coal use in power generation in Asia, and especially in China and India, is projected to be one of the main sources of incremental emissions in the medium term as well as in the period to 2030. Asian economies are likely to account for 40% of global emissions by 2030, compared to 18% in 1990.

Medium Term Issues: the Role of Policy

Over the past couple of years there has been a great deal of debate on how to reduce emissions in the medium term. This debate reached a turning point in the Kyoto Conference on climate change where all developed regions of the world undertook firm commitments and set targets for reductions in CO_2 emissions by 2010. The European Union agreed to decrease the CO_2 emissions in 2010 by 8% compared to the level of emissions in 1990. This would involve a reduction of nearly 16% when compared to the level of emission projected for that year under baseline assumptions.

The task of meeting the Kyoto targets should not be underestimated. While in most developed economies there is a tendency for energy demand to grow somewhat less than economic growth, there is a very strong tendency for any sizeable economic expansion to be associated with an increase in energy demand of about 1% less than GDP. Furthermore, energy patterns are almost certain to continue to be dominated by CO_2 emitting fossil fuels in the period to 2010. Hence, reducing CO_2 emissions for the EU as a whole over the next 15 years would amount to no less than breaking the historical link between economic growth and energy use.

These difficulties are not insurmountable and, even by 2010, there is a great deal that could be done to limit emissions. Indeed, two plausible scenarios are constructed showing some of the means by which the Kyoto target might be reached, at least for the EU. In one scenario, the prime driver of changes in energy use patterns is regulation, mostly on energy efficiency, while in the second scenario change is effected through carbon taxes. Technology plays a key role in both scenarios. However, most of its contribution will be in the form of integrating faster existing energy efficient technologies or accelerating the introduction of known advanced technologies. Much of the reduction in emissions by 2010 can only be reached through improving energy efficiency.

Both for the EU and for the rest of the world, there are large differences between the average and marginal costs of meeting any emission target. Furthermore, these differences increase as the emissions target gets more severe. Indeed, one of the key messages of the results is that the more severe the target and the shorter the time horizon over which it must be reached, the greater the likely economic costs.

Another important aspect of the likely costs of policy on emissions is the differential impact on different countries and regions. This impact clearly depends on the level of development of a country, its economic and industrial structure, and on its natural resource endowment and the carbon intensity of its energy system. The results indicate that, even when the degree of the severity of the target is similar, costs can diverge a great deal between countries.

Since most of the emissions over the next 15 years are likely to originate from non OECD regions, it is important that any action to tackle emissions must be internationally based. Any action by the EU alone, or even the OECD, is likely to have a very small impact on long term global carbon concentrations. Large, rapidly growing countries, such as China, India and some other countries in the Asian continent, give priority to economic development, often accompanied by massive increases in energy consumption.

There is even more of a difference in the likely costs of emission reductions among world regions than there are among European countries. A non differentiated reduction rate of 15% in OECD countries would impose marginal costs ranging from 125 $/tC in the United States, 175 $/tC in Australia-New Zealand, 200 $/tC in the European Union and as much as 350 $/tC in Japan. Thus, there is plenty of scope for an international market approach to the issue, if important practical problems can be resolved.

On the other hand, equalisation of marginal costs among the various parties results in differentiation of reduction objectives. Thus, for 200 $/tC, the reduction targets would be 25% in the United States and Australia-New Zealand, 15% in Europe, and 5% in Japan. It is important to note that projected trends up to 2010 are of crucial importance in determining the marginal cost.

The total effort in relation GDP in 2010 depends both on the marginal reduction cost in 2010 and emission dynamics in the reference projection. Thus, the corresponding cost would represent slightly less than 0.3% of GDP for the United States, due to a low marginal cost, as in the case of Europe, in this case because of the moderate increase in emissions envisaged in the reference situation. On the other hand, this cost would be one and half times higher in Japan and Canada, and up to twice as high in Australia-New Zealand. It thus appears that without the possibility of emission trading, non differentiation of objectives is likely to create substantial disparities in the charges imposed on the various parties.

Long Term Issues: the Role of Technology

While technological improvements will play an important role in the trends of energy demand and related emissions even in the short to medium term, it is clearly over a long time horizon that its impact can really be felt. The role of policy regarding technology is also different between the medium and long term. In the medium term, policy measures cannot lead to the development of new technologies but they can encourage the faster introduction and the more intensive utilisation of existing technologies. On the other hand, the nature of technological progress in the longer term can be influenced by government and industry objectives through spending on (R&D).

- The technology scenarios for the power generation sector that are examined in this study do not offer panaceas for the global CO_2 emission problem. Important though the power generating sector may be, it still accounts for about 40% of the projected world CO_2 emissions by 2030. Also, scenarios involving major technical-economic improvements in fossil fuel technologies produce weakened impacts on CO_2 emissions because at the same time as they reduce specific emissions they make these technologies economically attractive, not only vis-à-vis more polluting technologies but also more polluting ones. It is important to note that all supply-side technological improvements result one way or another in a reduction in the cost of consuming energy and hence potentially cause consumption to increase.
- Scenarios involving increased gas availability produce surprisingly weak results as far as CO_2 emissions are concerned. This is due partly to the uneven geographical distribution of the enhanced resources resulting in the gas not being available at sufficiently cheap prices where it could have made the biggest impact, i.e. in China and India, but also due to the resulting lower prices for gas and electricity.
- There is a clear need for energy saving technology breakthroughs, as these are likely to suffer less from the ambiguities and secondary effects associated with supply technologies. It is hard to see how clusters of energy technologies could by themselves make a major impact on the global CO_2 problem unaccompanied by major policy initiatives albeit market related ones. Combining technology breakthroughs with internalisation of external costs through taxation or tradable permits would magnify the impact and tend to neutralise some of the more ambiguous side effects quite apart from the fact that these policy instruments can by definition act on a much wider front.

The scenarios analysed in this study are quite revealing for the likely future of the European Union's power generation system. The EU energy system seems to be very sensitive in the period after 2020 when the massive decommissioning of old plants will necessitate the selection of new technologies for replacement generation equipment. The differential progress of different technologies is likely to influence directly a number of strategic decisions of the EU, which could lead to highly varying profiles of the future power and steam system.

The scenarios presented reveal the strategic importance of natural gas and gas related technologies for limiting CO_2 emissions and for reaching objectives on the development of cogeneration and the decentralisation of the generation market. Under all sets of energy prices examined, the gas and fuel cells scenarios result in the lowest amount of emissions by 2030. These two scenarios also have the lowest generation costs although this is partly because of the lower gas prices assumed in these scenarios. Somewhat surprisingly, the gas scenario does not have dramatic implications the degree of decentralisation by 2030. The projected generation system even in the reference case is already much more decentralised in 2030 than in 1995 and this is largely due to the penetration of gas. The additional gas generation in the gas scenario is mostly at the detriment of base load nuclear and coal plants. Of course, under the gas based fuel cell scenario, the degree of decentralisation reaches its peak.

The long term outlook of coal and nuclear is highly dependent on the technological developments in these fuels. Only under the assumptions of their corresponding scenarios do these fuels play a role significantly more important than the one assumed in the baseline scenario. Of course, EU based clean coal technological developments, could prove very beneficial to the exports of power equipment. However, the clean coal scenario indicated that technological developments in this sector could have adverse effects for the emissions within the EU. The development of nuclear energy seem to be limited in scope and gains, as restricted from load constraints and the interaction with decentralised steam generation. The incremental improvement of renewable technologies is not enough to alter the profile of the system in terms of emissions and costs.

PART I: GLOBAL AND EUROPEAN ENERGY AND EMISSIONS OUTLOOK

1 Introduction

Energy is a key issue that is present in all the sectors of modern economies. The availability of cheap, abundant and safe energy sources is indeed a requisite for sustained economic development in emerging economies. The ways in which it is consumed and used to produce welfare is the matter of several disciplinary fields. Energy technologies are important because energy is crucial in the overall economic system, not only because of the scarcity of the resources (the world energy mix is based on non renewable energy carriers), but also because of the environmental concerns. Indeed, environmental degradation is due, to a large extent, to the effects of energy production, transformation and use.

The environmental aspects of energy consumption are becoming very important, but the vulnerability of the energy system to price shocks should not be forgotten: two out of the three large global economic recessions in the last fifty years were associated with a supply shock in the energy sector.

The rediscovery of the risks of global climate change by climatologists in the middle of the seventies was reflected in the eighties by the appearance of the above question on international political agendas: statement of the Heads of State in the Hague (1989), then the Rio Conference (1992), the ratification of the United Nations Framework Convention on Climate Change (1994) and the launching of the international negotiation process within the framework of the "Conferences of the Parties". Considerable stimulus has thus been given to research on the possible consequences of climate change and the control strategies. In the case of the IPCC in particular, research has focused on three major issues: scientific aspects, effects and abatement measures, and the socio-economic aspects of climate change (IPCC, 1995). In the latter case, which involves the economic analysis of strategies to combat climate change, the two key issues in environmental economics must be addressed, namely evaluation of damage and evaluation of the cost of reducing emissions.

In general, research into the evaluation of environmental damage has attracted increasing attention in recent years, which has resulted in the development of a methodology and numerous applied studies, in particular in the energy sector (for the electricity sector, see the results of the external study, ExternE, Commission-DG XII, 1996). However, evaluation of the damage associated with climate change is hampered by intrinsic difficulties, due on the one hand to the diversity of long term local impacts and their low level of predictability, and, on the other hand, to ethical questions related to the comparability of damage in societies with extremely unequal levels of wealth (Hourcade, 1996). In the absence of a consensus on the evaluation of damage, it is therefore difficult to apply the principles of "cost benefit" analysis, advocated by the economic theory of the environment. This seeks a "pollution optimum" through equalisation of marginal pollution costs and pollution control. "Cost effectiveness" analyses have thus been favoured, aimed at minimising the cost of pollution control measures to meet a specific emission level target.

1.1 Objectives

There three main objectives in this volume: Firstly, it aims at presenting a “business as usual” (BAU or baseline) energy and emissions outlook for the European Union and for the world by major region. Such an outlook is an essential framework for a) evaluating the likely extent of the greenhouse gas problem, at least to the extent that it depends on energy, b) examining the sectors, fuels and regions that are especially important for any increase in emissions, and c) the analysis of the potential of existing and other likely policy measures.

Having arrived at this BAU energy and emissions outlook, the second objective of the volume is to examine alternative policy packages, mainly within the EU, that may be adopted in order to tackle the problem. Many individual measures and packages have been discussed in recent years within the EU and especially in the lead up to the Kyoto conference. These include regulation type measures as well as market instruments such as taxes and emission trading.

Equally important to the effectiveness of measures are their likely cost and their impact on other policy objectives such as security of energy supply, financial robustness of the public sector. While this issue has been examined from macroeconomic perspective elsewhere in more detail, some of the costs of these measures are also examined in Part II of this report.

Finally, the third objective of the volume is to examine to long term potential of technology in assisting in solving the GHG problem. To this end, the project aimed at assessing the role of technology progress breakthroughs, which are partly the result of R&D policy. Using the power generation sector as a case study, technology clusters have been defined, grouping probable technology progress cases.

Throughout the quantitative analysis presented in this volume, an additional underlying objective was to take into account the potential changes in the market structure of European energy and to examine the linkages between market structure and possible technology developments. The analysis is complemented by providing insights into very-long term systemic aspects of energy system changes also involving fuel cycles.

1.2 Models Used

All projections and quantitative assessments included in this report have been empirical and numerical at a sufficiently disaggregated nomenclature for sectors, countries, consumers, products and plants. The analytical tools that have been used include the computable general equilibrium models GEM-E3 and WARM, the econometric model HERMES, the partial equilibrium model of the European energy system PRIMES, the World energy model POLES and the technical-economic model for new technology prospects and cost benefit evaluations SAFIRE. These have been developed under the auspices of the European Commission within the JOULE-II programme.

The bulk of the analysis presented here is based on three models.

- *GEM-E3* general equilibrium macro-economic model version 2.23, covering 14 EU member states, updated in October 1997 and used for the analysis of the economic costs for the EU of reaching the pre-Kyoto targets (Fall 1998).
- *PRIMES* energy system model version 1, covering 8 EU member states, updated in June 1997 and used for the analysis of the energy system implications for the EU of reaching the pre-Kyoto targets (Fall 1998).
- The *POLES* model is a simulation model of the world energy system up to the year 2030 (for a detailed description of the model, see Criqui et al., 1996). It is a world energy system model in that economic and demographic trends in each of the large countries or regions of the world are considered exogenous, while changes in the variables which characterise energy consumption, transformation, production and prices are endogenous to the model. It is a recursive simulation model in that the dynamics are given, from the initial point then from year to year, through progressive adjustments to supply and demand variables on the one hand and price variables on the other.

Further details of the models mentioned below, with additional references, are given in Appendix II.

1.3 Uncertainties

This volume contains a very large number of seemingly precise figures on the European and world economy, energy system, emissions, and possible technological developments for the period between 1995 and 2030. Needless to say that this apparent precision is no more than an attempt at quantifying the vast uncertainty that surrounds possible future developments. Uncertainty about the future is almost generic to any part of the analysis included here. The purpose of the quantification in this volume is to help thinking about the future - it is not, and could never be, to predict the future.

1.4 Plan of the Volume

This volume consists of three parts. The remainder of this part discusses the European and world energy and emissions outlook. Chapter 2 covers the global outlook on which the European outlook of Chapter 3 partly depends. Both chapters begin with a brief exposition of the key demographic and macroeconomic assumptions from which the energy trends are derived. The key baseline trends of high energy demand and CO_2 emissions growth in non OECD countries and the continued rising significance of the power generation sector form the basis on which the policy and technology scenarios that follow.

In the second part of the volume, the medium term policy options for tackling the greenhouse gases increase is examined. The discussion is structured around the debate prior to the Kyoto meeting. The European aspects of the debate are examined in Chapter 5. A number of specific policy measures are introduced into the scenario examined in that chapter in order to evaluate the feasibility of the Kyoto target set by EU authorities. Chapter 6 looks at the issue of international difficulties of controlling CO_2 emissions and examines the comparative difficulty that different region face in cutting emissions.

The third part of the volume examines the issues involved in the longer term. The technological aspects of energy of course, dominate these issues. The analysis presented here focuses on the power generation sector. Chapter 8 examines a series of technology clusters that could play a very important role during the next thirty years. On the basis of these clusters, a number of technology scenarios are presented in Chapter 9. The results of the analysis of these scenarios are presented in Chapter 10 and Chapter 11, for the world and the EU respectively. Finally, Chapter 12 discusses some of the issues involved in formulating and EU R&D policy. Chapter 13 presents a very long-term view of energy-system changes and fuel cycles.

2 Global Outlook[2]

The global economic and energy outlook plays a critical part in the analysis and results of this volume. The energy sector is, to a large extent, already operating within a world context and any regional analysis that does not take into account international influences is likely to prove deficient. One extreme illustration of this is the oil market, which is globally integrated. It is simply not possible to examine future oil market trends in the EU without taking into account global supply and demand, which will determine future oil prices. Similar considerations apply in the case of natural gas. European gas prices and demand will depend not only on the developments in countries, like Russia, that supply gas to the EU but also on countries as far afield as China since they are likely competitors as consumers of Russian gas.

The environment is also an issue that often needs to be analysed in a context that is bigger than the region of concern. In the case of the GHG problem, the issues are global and any policies and measures undertaken by the EU must be seen in such a context. Similarly, European R&D policy must be drawn within an international framework because non EU clients of EU produced technologies often represent the bulk of the market.

This chapter then provides the international economic, energy and environmental trends within which European investment and policy decision takers will need to operate.

2 All projections in this chapter are based on the POLES model. The main authors of the chapter are Patrick Criqui of IEPE and Niko Kouvaritakis of ECOSIM, with contributions from Antonio Soria of IPTS.

2.1 Economic Outlook

For the purposes of global projections presented here eleven geographical and geopolitical regions are identified.[3] Their acronyms, as presented in Table 2-1, are used in some tables or figures of this volume.

Table 2-1: Acronyms of world regions.

North America	NOAM
Western Europe	WEUR
OECD Pacific	PACO
Eastern Europe	EEUR
Former Soviet Union	FSUN
Central & South America	CSAM
South Asia	SOAS
South-East Asia	SEAS
Continental Asia	COAS
North Africa & Middle East	NAME
Sub-Saharan Africa	SSAF

NB: "North America" does not include Mexico, and "OECD Pacific" does not include South Korea.

2.1.1 Economic and Population Trends

The population growth assumptions, presented in Table 2-2, are derived from UN projections. Due to low birth rates in industrialised countries and to the spreading of the demographic transition in developing regions of the world, they show a continuous decline in world population growth rate, from an average 1.8 % pa between 1971 and 1995, to an average of 1.2 % pa between 1995 and 2030, with 1.4 % pa from 1995 to 2010 and only 1 % pa in the last decade of the projection.

Despite the slowdown in growth rates, the world population is likely to increase by 3 billion people between 1995 and 2030, with the bulk of this increase taking place in developing regions. Thus, the balance of population among regions will be altered, although not dramatically.

[3] POLES actually models 26 countries or regions. For further details see the model reference manual : POLES 2.2. European Commission, DG XII, December 1996.

Key Features of the Demographic Projections

- The share of rich countries (defined as NOAM, WEUR and PACO) in world population will continue to decline, from 16 % today to 12 % in 2030.
- The weight of Eastern Europe and the Former Soviet Union will also decline, from 5 to 4 %.
- The total share of developing regions will increase from 76 to 83 %.
- Due to strong demographic control, China (Continental Asia) is exceptional in that its share declines from 22 to 19 % of world population.
- Sub Saharan Africa's share is expected to increase from 10 to 15 %; the population of this area will increase of about 800 million people over the 1995-2030 period.
- Although its share is almost stable, South Asia (dominated by India) will also have a strong increase in population, by 700 million, and will remain the most populated region with almost 2 billion people in 2030.

Table 2-2: Assumptions on world population growth by region, 1992-2030, million.

		1992	2000	2010	2020	2030
		M hab				
World	WRD	5424	6150	7027	7893	8713
		% of World total				
North America	NOAM	5	5	5	4	4
Western Europe	WEUR	8	7	7	6	6
OECD Pacific	PACO	3	3	2	2	2
Eastern Europe	EEUR	2	2	2	2	2
Former Soviet Union	FSUN	5	5	5	4	4
Central & South America	CSAM	8	9	9	9	9
South Asia	SOAS	22	22	22	22	23
South-East Asia	SEAS	10	10	10	10	10
Continental Asia	COAS	22	21	20	20	19
North Africa & Middle East	NAME	5	5	6	7	7
Sub-Saharan Africa	SSAF	10	11	13	14	15

Source: UN

2.1.2 Economic Projections: the GDP Dynamics

The economic projection used in this study was derived from WorldScan model projections, which have also been recently used by OECD with the LINKAGE model[4]. They are adjusted for the changes from Market Exchange Rate (MER) to Purchasing Power Parity (PPP) economic growth rates. While most economic forecasts present variations in MER-GDP, the economic variables used in the POLES database are expressed in the system of Purchasing Power Parity developed by the CEPII[5] for his CHELEM database. This is done in order to account both for the informal sector, which in most developing areas accounts for a large proportion of total economic activity, and for the actual purchasing power of households. It is thus considered that the PPP-GDP is a better driver of energy consumption and energy intensity of GDP and more suited for international comparisons than MER-GDP.

The MER-GDP and PPP-GDP projections show significantly different pictures of world economic structure and dynamics. Using PPP, the initial GDP level is higher in developing regions, while the corresponding growth rate of the economy is lower. This results both in a very different structure of world GDP and, paradoxically, in a higher aggregated growth - in spite of lower individual rates - as the share of the rapidly growing regions is more larger. This phenomenon is illustrated in Table 2-3, where the differences in structure and growth rates are shown, up to 2020 for reasons of consistency with the Worldscan projections.

Table 2-3: PPP and MER world economic structure and dynamics.

		WE-2030 - Purchasing Power Parity			Worldscan - Market Exchange Rate			PPP-MER
		1990	2020	1990-2020	1990	2020	1990-2020	1990-2020
World	WRD	**G$90 PPP**		**yagr**	**G$90 MER**		**yagr**	
		27.4	72.5	3.3%	21.6	54.6	3.1%	0.2%
		% of World GDP			**% of World GDP**			
North America	NOAM	22	16	2.2%	28	21	2.2%	
Western Europe	WEUR	23	16	2.0%	34	24	2.0%	
OECD Pacific	PACO	10	7	2.2%	15	11	2.2%	
Eastern Europe	EEUR	3	2	2.1%	2	2	3.4%	-1.3%
Former Soviet Union	FSUN	9	5	1.4%	4	4	2.4%	-1.0%
Central & South America	CSAM	9	11	4.1%	5	9	5.2%	-1.1%
South Asia	SOAS	5	7	4.6%	2	5	6.7%	-2.1%
South-East Asia	SEAS	6	10	4.8%	4	9	6.0%	-1.2%
Continental Asia	COAS	9	20	6.2%	2	9	8.7%	-2.5%
North Africa & Middle East	NAME	4	5	4.0%	3	4	4.1%	-0.1%
Sub-Saharan Africa	SSAF	2	2	3.3%	1	2	4.0%	-0.7%

---

4 In « The World in 2020, Towards a New Global Age », OECD, 1997, p. 156.

5 Centre d'Études Prospectives et d'Informations Internationales, Paris.

In both the PPP and MER projections, the resulting economic picture is one of a sustained economic growth at world level with an average word growth of more than 3 % pa between 1995 and 2030. This is not much higher than the secular trend in world output growth of 3 % pa between 1900 and 1990, according to Angus Maddison[6]. This is also not much higher than the growth experienced between 1971 and 1995, which averaged 3.1 % pa (on a PPP basis). It has to be noted that this growth rate has been reached, in the past quarter of a century, in spite of a series of crises which affected successively the different regions of the world: the oil shocks for the industrialised oil importing countries, the debt crisis in the LDCs and the decay of the Centrally Planned Economies.

As can be seen from Table 2-4, on a decade by decade basis, the 2000-2010 period might experience the highest growth rate at about 3.8 % pa on average. This is because of the assumed recovery in the former CPEs and the declining but still high growth in Asian regions. World growth rates decline to 3.3 and 2.9 % pa during the following two decades.

Key Features of the Economic Projections

- OECD countries are projected to experience moderate growth in the 2000-2010 period (about 2.5 % pa), while the growth rate will decelerate during the two following decades within the range of 1 to 2 % pa.
- Economies in transition will recover during the two first decades of the next century with growth rate around 3.5 % pa and then experience more moderate growth, above 2 % pa.
- All developing regions are expected to grow at rates higher than 3.5 % pa during the projection period.
- When compared to the 1971-1995 period, growth will accelerate in Latin America, Africa- Middle East and South Asia, while it will decelerate in South East and Continental Asia.
- Continental Asia will remain the region with the highest growth potential, specially at the beginning of the projection period, while after 2010 it is overwhelmed by other regions, specially South Asia.

---

[6] Maddison A., « The World Economy, 1820 - 1992 », OECD, Paris, 1995.

Table 2-4: World economic growth (PPP) by region, % per annum.

yagr	1971-95	1995-2030	1971-80	1980-90	1990-2000	2000-10	2010-20	2020-30
NOAM	2.5%	2.1%	2.8%	2.7%	2.0%	2.5%	2.2%	1.7%
WEUR	2.4%	2.1%	3.0%	2.4%	1.5%	2.5%	2.0%	1.7%
PACO	3.6%	2.0%	4.3%	3.9%	2.2%	2.7%	1.6%	1.2%
EEUR	0.3%	2.9%	3.6%	-0.6%	-0.4%	3.4%	3.3%	2.1%
FSUN	-0.1%	3.3%	3.1%	1.6%	-3.3%	3.9%	3.7%	2.7%
CSAM	3.3%	3.8%	5.5%	1.2%	3.5%	4.6%	4.1%	3.1%
SOAS	4.4%	4.8%	3.4%	5.6%	4.3%	4.7%	4.9%	4.7%
SEAS	6.7%	4.2%	7.6%	6.5%	5.3%	5.0%	4.2%	3.3%
COAS	7.4%	5.1%	5.6%	8.5%	8.1%	6.3%	4.3%	3.6%
NAME	3.3%	4.2%	5.1%	1.6%	3.6%	4.1%	4.5%	4.5%
SSAF	2.4%	3.7%	3.0%	2.0%	2.5%	3.1%	4.3%	4.3%
WRD	3.1%	3.3%	3.7%	3.0%	2.7%	3.8%	3.3%	2.9%

The baseline economic projection to 2030 presented above represents a reasonable balance between optimistic and pessimistic scenarios:

- it is optimistic in the sense that very few studies, particularly dealing with energy issues, incorporate hypotheses of a growth rate superior to 3 % pa on a reasonably long period;
- it is pessimistic in the sense that the supposed economic growth in OECD countries is quite weak by the end of the period (under 2 % pa), which is below historical trends;
- it is pessimistic because it supposes drastic reductions in output growth in Southeast and Continental Asia, from current « catch-up rates » (7-8 % pa) to much more modest « mature rates » (3.5 % pa);
- it is also pessimistic because the supposed economic recovery in Countries in Transition allows them to reach their 1990 per capita income level only after the year 2010.

Finally, the key characteristic of the proposed projection is probably that although regional growth rates remain moderate when taken individually, all regions are projected to grow relatively smoothly in the period to 2030. This results in an aggregated growth rate, which is above the one of the preceding period. However it has to be noted that, in the past twenty-five years, most regions of the world have undergone severe crises of one type or another. The key assumption of the Baseline projection is thus that no structural crisis will affect any of the main world regions during the projection period. This is clearly unrealistic.

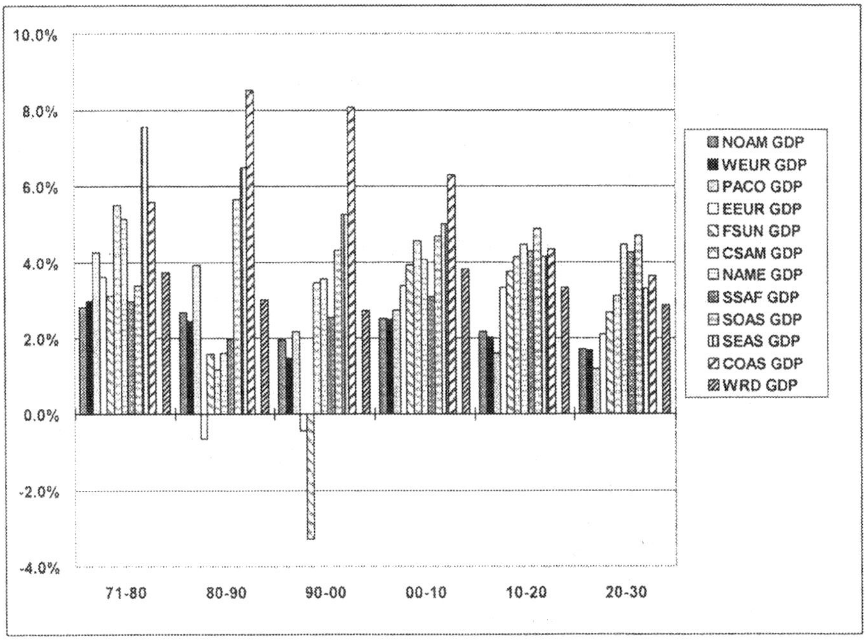

Figure 2-1: GDP growth rates by region, 1971-2030.

2.1.3 Changes in per Capita GDP Profiles

Figure 2-2 presents the trends in per capita GDP that are implied from the above projections on population and GDP. Many interesting features emerge:

- in 1970 the North America region, with a yearly per capita GDP of 15 000 $ (1990$ PPP), was clearly ahead of other OECD regions whose per capita income was close to 10 000 $. Non OECD regions were much poorer. They included an intermediate group of countries, with per capita income of 4 000 to 6 000 $, led by the Soviet Union but also including Latin America and North Africa - Middle East and, at the bottom end, regions like Asia and Sub-Saharan Africa with revenues between 750 and 1 400 $;
- by 1995, the situation was already significantly altered : the Pacific OECD had climbed to an intermediate position between North America and Europe while the former CPEs had slipped behind Latin America; South East and Continental Asia had almost caught-up with the group of intermediate income countries; two regions continued to lag, namely, Sub-Saharan Africa and South Asia;

- in 2030, the situation is projected to change substantially. While convergence among industrialised countries is likely to continue (differential growth rates in GDP and population lead to the stabilisation of the per capita income ratio at 90 % of the North American level in OECD Pacific and 70 % in Europe), it will be even more pronounced among intermediate income regions with the complete catch up of South East and Continental Asia to the group formed by Latin America and the countries in transition; South Asia's growth in per capita GDP will not be fast enough to approach the intermediate group while it may take until 2010 before the increase in per capita income of Sub-Saharan Africa resumes.

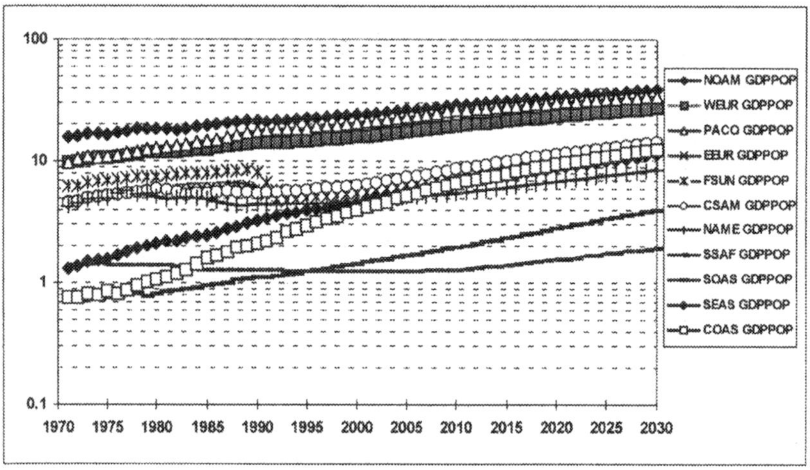

Figure 2-2: Per capita GDP in the 11 world regions (1971-2030, logarithmic scale).

2.2 Energy Outlook

2.2.1 Energy Resources and Prices

The baseline projection on natural resources (reserves, prices and distribution) is outlined in the following tables. The information contained in Table 2-5 to Table 2-10 is limited to oil and gas since the pressure on coal resources is expected to be much lower.

One of the main features of the POLES model is its ability to produce endogenous changes for international energy prices. The price mechanisms incorporated in the model do take into account the balance of supply and demand for oil, natural gas and coal: in the short run, prices depend on the demand variation and on oil supply capacities in the Gulf « swing producer » countries; in the longer run prices depend on the Reserve on Production ratios for oil and natural gas (see below).

Table 2-5: Baseline oil market outlook, production, mb/d.

OIL PRODUCTION (mbd)	1973	1980	1990	2000	2010	2020	2030	1980/ 1973	1990/ 1980	2000/ 1990	2010/ 2000	2020/ 2010	2030/ 2020
Gulf Conventional	21.0	18.8	16.3	25.6	36.9	45.9	55.3	-1.6%	-1.4%	4.6%	3.7%	2.2%	1.9%
OECD Conventional	13.3	14.5	15.2	16.9	14.8	14.5	11.7	1.2%	0.5%	1.1%	-1.3%	-0.2%	-2.1%
Other Conventional	24.3	30.2	33.7	32.6	35.6	37.4	37.3	3.1%	1.1%	-0.3%	0.9%	0.5%	0.0%
World Non Conventional	0.3	0.3	0.3	0.5	3.3	10.3	23.5	0.0%	0.0%	4.0%	20.8%	12.1%	8.6%
Total	59.0	63.9	65.5	75.7	90.6	108.1	127.9	1.1%	0.3%	1.4%	1.8%	1.8%	1.7%

Table 2-6: Baseline oil markets outlook, reserves by region, billion barrels.

OIL RESERVES (thousand mb)	1973	1980	1990	2000	2010	2020	2030	1980/ 1973	1990/ 1980	2000/ 1990	2010/ 2000	2020/ 2010	2030/ 2020
Gulf	305	338	592	585	541	460	344	1.5%	5.8%	-0.1%	-0.8%	-1.6%	-2.9%
OECD	57.8	52.5	58.6	50.5	38.1	34.8	38.6	-1.4%	1.1%	-1.5%	-2.8%	-0.9%	1.1%
Other	172.8	216.3	314.9	439.2	390.0	330.1	261.4	3.3%	3.8%	3.4%	-1.2%	-1.7%	-2.3%
World Conventional	535.8	607.0	965.9	1074.2	969.4	824.6	643.8	1.8%	4.8%	1.1%	-1.0%	-1.6%	-2.4%
Non Conventional	48.4	48.4	48.4	122.9	190.6	309.6	448.8	0.0%	0.0%	9.8%	4.5%	5.0%	3.8%
Total	584.2	655.4	1014.3	1197.1	1160.0	1134.2	1092.7	1.7%	4.5%	1.7%	-0.3%	-0.2%	-0.4%

Table 2-7: Baseline oil market outlook, oil price ($/bl) and key indicators.

OIL PRICE ($90/bl)	1973	1980	1990	2000	2010	2020	2030	1980/ 1973	1990/ 1980	2000/ 1990	2010/ 2000	2020/ 2010	2030/ 2020
World crude oil price	9.8	58.8	23.8	16.8	22.7	29.3	34.9	29.1%	-8.6%	-3.4%	3.1%	2.6%	1.8%
World RP ratio	24.5	25.9	40.6	43.3	35.1	28.7	23.4	0.8%	4.6%	0.7%	-2.1%	-2.0%	-2.0%
Average Recovery Rate (%)	16.8	16.8	16.8	18.1	21.6	25.7	30.3	0.0%	0.0%	0.7%	1.8%	1.8%	1.7%

The outlook produced by the model for the world oil dynamics is one of increasing pressure on conventional oil reserves. The increase in recovery rate, which is partly autonomous and partly price-driven in the model, allows extending the resource base, beyond the optimistic estimates of Ultimate Recoverable Resources with current technologies. But in spite of this, the increase in demand and consequently in cumulative production results in a decline of known conventional reserves in all regions of the world. This is, through the decline in the RP ratio, the prime factor for the long term increase in oil prices, which reaches 35 $/bl in 2030.

This increase endogenously weights on demand growth and it is also a strong incentive to a massive development of non conventional resources, which appears to be, in the framework of this Baseline, a requisite for the closing of the world oil balance in the long term: non conventional resources, mostly extra heavy oil and tar-sands, represent only 3.6 % of total supply in 2010, but almost 10 % in 2020 and 18 % in 2030.

Table 2-8: Baseline gas market outlook, production Gm3.

GAS PRODUCTION (Billion m3)	1975	1980	1990	2000	2010	2020	2030	1980/ 1973	1990/ 1980	2000/ 1990	2010/ 2000	2020/ 2010	2030/ 2020
OECD	728	772	774	1040	1197	1309	1368	1.2%	0.0%	3.0%	1.4%	0.9%	0.4%
of which N. America	568	582	572	742	823	913	976	0.5%	-0.2%	2.6%	1.0%	1.0%	0.7%
Former Soviet Union	269	404	742	839	1057	1293	1479	8.5%	6.3%	1.2%	2.3%	2.0%	1.4%
Middle East	37	40	94	178	304	576	847	1.5%	8.9%	6.5%	5.5%	6.6%	3.9%
Asia	24	49	116	194	264	333	401	15.8%	8.9%	5.2%	3.1%	2.4%	1.9%
Other	0	0	187	252	372	550	709	NA	NA	3.0%	4.0%	4.0%	2.6%
World	1149	1397	1914	2503	3193	4059	4803	4.0%	3.2%	2.7%	2.5%	2.4%	1.7%

Table 2-9: Baseline gas market outlook, reserves by region, trillion cubic metres.

GAS RESERVES (Trillion m3)	1975	1980	1990	2000	2010	2020	2030	1980/ 1973	1990/ 1980	2000/ 1990	2010/ 2000	2020/ 2010	2030/ 2020
OECD	15	15	18	16	16	16	17	-0.5%	1.9%	-0.9%	-0.1%	0.1%	0.5%
of which N.America	9	9	8	6	6	6	7	-0.7%	-0.7%	-2.6%	-0.8%	0.4%	1.0%
Economies in transition	23	29	50	75	71	66	58	5.2%	5.4%	4.2%	-0.5%	-0.9%	-1.3%
Latin America	2	5	8	9	10	10	9	15.5%	4.2%	1.1%	1.4%	0.2%	-0.7%
Asia	2	5	9	9	9	8	7	21.0%	5.4%	0.2%	-0.2%	-1.0%	-1.6%
Other	24	28	51	66	76	82	82	2.5%	6.4%	2.5%	1.4%	0.7%	0.0%
World	67	82	135	175	182	181	173	4.1%	5.2%	2.6%	0.4%	0.0%	-0.5%

Table 2-10: Baseline gas market outlook, RP ratios and import prices ($90/boe).

GAS PRICES	1975	1980	1990	2000	2010	2020	2030	1980/ 1973	1990/ 1980	2000/ 1990	2010/ 2000	2020/ 2010	2030/ 2020
Import prices ($90/boe)													
N. American market	20.4	30.0	10.8	15.3	21.1	27.9	32.3	8.1%	-9.7%	3.5%	3.3%	2.8%	1.5%
European market	11.3	21.4	15.2	15.5	21.6	28.9	34.2	13.6%	-3.4%	0.2%	3.4%	2.9%	1.7%
Asian market	19.2	36.5	20.6	27.0	31.9	35.5	42.8	13.6%	-5.6%	2.8%	1.7%	1.1%	1.9%
RP ratios													
N. American market	14.9	16.1	15.9	9.6	7.6	6.1	5.2	1.5%	-0.1%	-5.0%	-2.3%	-2.2%	-1.4%
European market	68.0	58.1	56.7	67.9	50.3	37.8	29.8	-3.1%	-0.3%	1.8%	-2.9%	-2.8%	-2.3%
Asian market	161.7	90.6	75.1	48.6	43.6	37.1	27.8	-10.9%	-1.9%	-4.3%	-1.1%	-1.6%	-2.8%
World	51.7	52.0	62.8	62.1	50.7	39.8	32.0	0.1%	1.9%	-0.1%	-2.0%	-2.4%	-2.2%

Three regional gas markets are identified in the model. They are interconnected in the sense that some gas exporter can sell to different markets and also that gas price dynamics include some indexation on oil price. But the gas prices are also determined independently in the different markets, as they depend on the RP ratio of the key suppliers of each region. As a result, gas prices show different profiles in the three markets. In energy equivalent terms, they are close to the oil price in Europe and North America, but higher than the oil price in the Asian region due to the strength of gas demand and the limitations of gas reserves in this region.

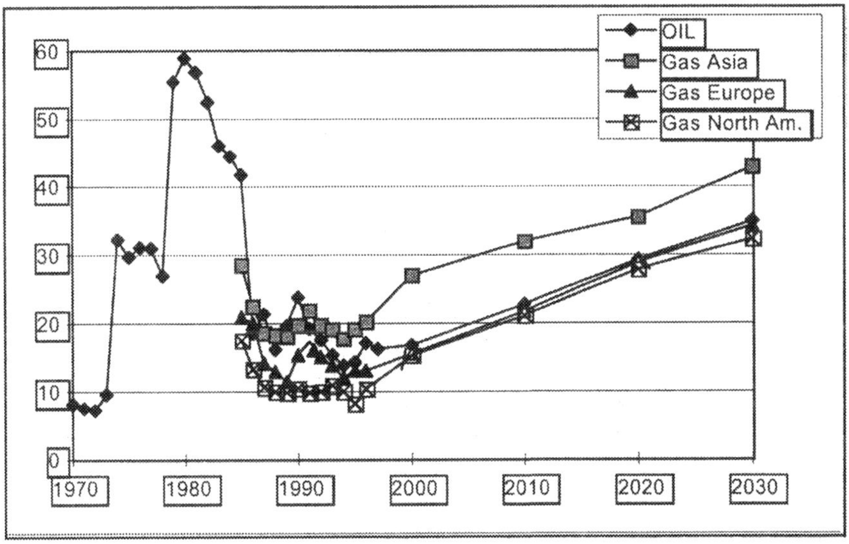

Figure 2-3: Oil and gas prices in the baseline.

2.2.2 World Primary Energy Consumption

The baseline projection up to 2030 has been produced using the main BAU technology characterisation data set discussed in later chapters (the reader is referred for details to the accompanying Annexes to this volume). With these data, the emerging world balance by major fuel is summarised in Table 2-11.

At world level, the primary energy mix trends show an increase in the share of natural gas and of coal, which, in the latter case, is mostly due to a strong development in Asia. The shares of conventional oil and of nuclear energy decline. Large hydro and new renewables maintain or slightly increase their shares in world energy supply, while that of traditional biomass strongly declines.

Table 2-11: Baseline world summary energy balance.

Mtoe						Annual % change			
	1992	2000	2010	2020	2030	1992/2000	2000/2010	2010-2020	2020/2030
Primary Production	8276	9774	12250	15075	18125	2.1%	2.3%	2.1%	1.9%
Solids	2152	2413	3423	4374	5546	1.4%	3.6%	2.5%	2.4%
Oil	3248	3770	4499	5373	6362	1.9%	1.8%	1.8%	1.7%
of which non conventional	17	25	164	514	1170	4.9%	20.8%	12.1%	8.6%
Natural gas	1670	2227	2842	3613	4275	3.7%	2.5%	2.4%	1.7%
Nuclear	457	587	648	760	861	3.2%	1.0%	1.6%	1.3%
Hydro+Geoth	191	224	279	344	412	2.0%	2.2%	2.1%	1.8%
Trad.Biomass	426	401	340	291	251	-0.8%	-1.6%	-1.6%	-1.5%
Other Renewables	133	152	220	321	418	1.6%	3.8%	3.8%	2.7%

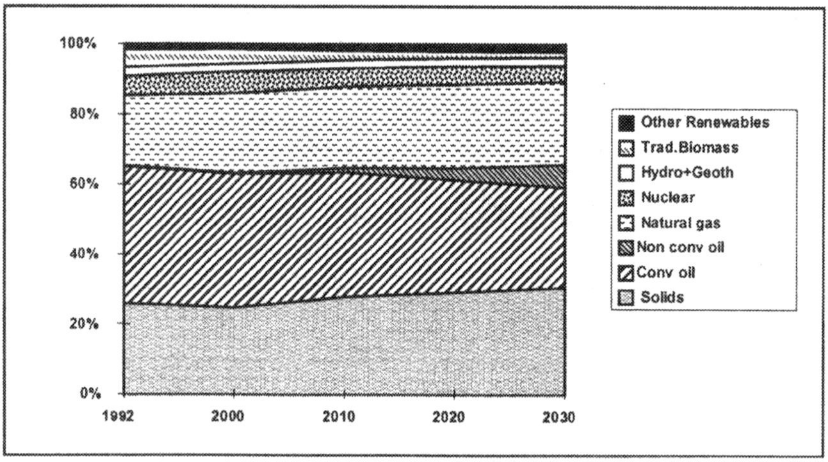

Figure 2-4: World primary energy mix.

2.2.3 Energy Balances by Main Regions

According to the baseline projection, the world energy system will undergo a major shift in the next forty years: while OECD countries still represent today half of total world energy consumption, their share will be reduced to less than one third in 2030; conversely, developing countries now represent one third of world total, their share might increase to almost 60 % in 2030.

By 2020, the Developing Asia region will by itself represent one third of total energy consumption, one fourth of total oil consumption. This shift is described in the summary energy balances provided in Table 2-12 to Table 2-16 for five regions: World, OECD, Economies in transition, Developing Asia and the Rest of developing regions. This shift is of paramount importance for many key energy issues, such as energy supply security, global capital requirements, new technology development and diffusion, and, of course, climate change.

Table 2-12: Summary energy balance, OECD countries.

OECD : SUMMARY ENERGY BALANCE - Reference Case Mtoe	1992	2010	2030	*1992/2010*	*2010/2030*	*1992/2030*
Primary Production	2997	3618	4490	*1,1%*	*1,1%*	*1,1%*
Solids	864	1060	1349	*1,1%*	*1,2%*	*1,2%*
Oil	798	756	1028	*-0,3%*	*1,5%*	*0,7%*
of which non conventional	17	17	445	*0,0%*	*17,9%*	*9,1%*
Natural gas	721	1065	1217	*2,2%*	*0,7%*	*1,4%*
Nuclear	375	476	584	*1,3%*	*1,0%*	*1,2%*
Hydro+Geoth	98	110	120	*0,7%*	*0,4%*	*0,6%*
Trad.Biomass	45	54	57	*1,0%*	*0,3%*	*0,6%*
Other Renewables	96	97	135	*0,0%*	*1,7%*	*0,9%*
Gross Inland Consumption	4161	5244	5884	*1,3%*	*0,6%*	*0,9%*
Solids	874	1063	1306	*1,1%*	*1,0%*	*1,1%*
Oil	1831	2211	2253	*1,1%*	*0,1%*	*0,5%*
Gas	841	1233	1429	*2,1%*	*0,7%*	*1,4%*
Other	615	737	896	*1,0%*	*1,0%*	*1,0%*
Electricity Generation in TWh	7688	10793	13713	*1,9%*	*1,2%*	*1,5%*
Fuel Input for Thermal PG	1003	1410	1732	*1,9%*	*1,0%*	*1,4%*
Solids	719	950	1093	*1,6%*	*0,7%*	*1,1%*
Oil	122	134	101	*0,5%*	*-1,4%*	*-0,5%*
Gas	162	326	538	*4,0%*	*2,5%*	*3,2%*
Final Energy Consumption	3061	3756	4130	*1,1%*	*0,5%*	*0,8%*
Solids	150	105	122	*-2,0%*	*0,8%*	*-0,5%*
Oil	1620	1954	2045	*1,0%*	*0,2%*	*0,6%*
Gas	709	785	743	*0,6%*	*-0,3%*	*0,1%*
Heat	22	18	19	*-1,2%*	*0,4%*	*-0,4%*
Electricity	537	817	1080	*2,4%*	*1,4%*	*1,9%*
Renewables	121	151	192	*1,2%*	*1,2%*	*1,2%*
by Sector:						
Industry	1101	1224	1241	*0,6%*	*0,1%*	*0,3%*
Transport	1011	1268	1407	*1,3%*	*0,5%*	*0,9%*
Dom. Tert. Agr.	949	1265	1482	*1,6%*	*0,8%*	*1,2%*
CO2 Emissions in Mt of C	2967	3723	4143	*1,3%*	*0,5%*	*0,9%*
Emission variation from 1990	*19*	*775*	*1194*			
Population (Million)	874	962	1013	*0,5%*	*0,3%*	*0,4%*
GDP (M$90)	15209	22918	32763	*2,3%*	*1,8%*	*2,0%*
Gross Inland Cons. / GDP (toe/M$90)	274	229	180	*-1,0%*	*-1,2%*	*-1,1%*
Elec. Gen. / cap. (kWh/cap)	8801	11225	13540	*1,4%*	*0,9%*	*1,1%*
CO2 Emissions / cap. (tC/cap)	3,4	3,9	4,1	*0,7%*	*0,3%*	*0,5%*

Table 2-13: Summary energy balance economies in transition.

Economies in transition : SUMMARY ENERGY BALANCE - Reference Case **Mtoe**	**1992**	**2010**	**2030**	*1992/2010*	*2010/2030*	*1992/2030*
Primary Production	1523	1726	2205	*0,7%*	*1,2%*	*1,0%*
Solids	414	232	296	*-3,2%*	*1,2%*	*-0,9%*
Oil	466	397	422	*-0,9%*	*0,3%*	*-0,3%*
of which non conventional						
Natural gas	597	969	1340	*2,7%*	*1,6%*	*2,2%*
Nuclear	56	81	72	*2,0%*	*-0,6%*	*0,6%*
Hydro+Geoth	24	30	35	*1,1%*	*0,9%*	*1,0%*
Trad.Biomass	16	10	7	*-2,3%*	*-1,8%*	*-2,1%*
Other Renewables	1	8	32	*12,0%*	*7,4%*	*9,6%*
Gross Inland Consumption	1434	1510	1943	*0,3%*	*1,3%*	*0,8%*
Solids	390	199	237	*-3,7%*	*0,9%*	*-1,3%*
Oil	437	321	438	*-1,7%*	*1,6%*	*0,0%*
Gas	560	862	1122	*2,4%*	*1,3%*	*1,8%*
Other	47	127	146	*5,7%*	*0,7%*	*3,0%*
Electricity Generation in TWh	2033	2560	4016	*1,3%*	*2,3%*	*1,8%*
Fuel Input for Thermal PG	411	388	533	*-0,3%*	*1,6%*	*0,7%*
Solids	155	113	145	*-1,8%*	*1,3%*	*-0,2%*
Oil	52	2	1	*-17,3%*	*-3,3%*	*-10,2%*
Gas	208	274	387	*1,5%*	*1,7%*	*1,6%*
Final Energy Consumption	1039	1167	1499	*0,6%*	*1,3%*	*1,0%*
Solids	216	84	86	*-5,1%*	*0,1%*	*-2,4%*
Oil	356	284	390	*-1,2%*	*1,6%*	*0,2%*
Gas	237	470	581	*3,9%*	*1,1%*	*2,4%*
Heat	169	154	154	*-0,5%*	*0,0%*	*-0,2%*
Electricity	131	164	258	*1,2%*	*2,3%*	*1,8%*
Renewables	30	18	40	*-2,7%*	*4,0%*	*0,8%*
by Sector:						
Industry	449	554	614	*1,2%*	*0,5%*	*0,8%*
Transport	153	140	223	*-0,5%*	*2,4%*	*1,0%*
Dom. Tert. Agr.	448	474	662	*0,3%*	*1,7%*	*1,0%*
CO2 Emissions in Mt of C	1125	1007	1302	*-0,6%*	*1,3%*	*0,4%*
Emission variation from 1990	*-164*	*-283*	*12*			
Population (Million)	412	463	509	*0,6%*	*0,5%*	*0,6%*
GDP (M$90)	2130	3428	6274	*2,7%*	*3,1%*	*2,9%*
Gross Inland Cons. / GDP (toe/M$90)	673	441	310	*-2,3%*	*-1,7%*	*-2,0%*
Elec. Gen. / cap. (kWh/cap)	4933	5530	7886	*0,6%*	*1,8%*	*1,2%*
CO2 Emissions / cap. (tC/cap)	2,7	2,2	2,6	*-1,3%*	*0,8%*	*-0,2%*

Table 2-14: Summary energy balance, developing Asia.

Developing Asia : SUMMARY ENERGY BALANCE - Reference Case						
Mtoe	**1992**	**2010**	**2030**	*1992/2010*	*2010/2030*	*1992/2030*
Primary Production	1446	2795	4335	3,7%	2,2%	2,9%
Solids	738	1882	3250	5,3%	2,8%	4,0%
Oil	309	355	296	0,8%	-0,9%	-0,1%
of which non conventional						
Natural gas	116	235	357	4,0%	2,1%	3,0%
Nuclear	21	75	138	7,3%	3,1%	5,1%
Hydro+Geoth	27	61	113	4,7%	3,1%	3,9%
Trad.Biomass	197	149	99	-1,6%	-2,0%	-1,8%
Other Renewables	38	40	82	0,2%	3,7%	2,0%
Gross Inland Consumption	1585	3579	6342	4,6%	2,9%	3,7%
Solids	767	1950	3364	5,3%	2,8%	4,0%
Oil	455	961	1677	4,2%	2,8%	3,5%
Gas	80	345	867	8,5%	4,7%	6,5%
Other	283	324	433	0,8%	1,5%	1,1%
Electricity Generation in TWh	1712	6558	14533	7,7%	4,1%	5,8%
Fuel Input for Thermal PG	353	1133	2273	6,7%	3,5%	5,0%
Solids	272	952	1922	7,2%	3,6%	5,3%
Oil	53	34	9	-2,4%	-6,5%	-4,6%
Gas	25	147	342	10,4%	4,3%	7,2%
Final Energy Consumption	1223	2495	4253	4,0%	2,7%	3,3%
Solids	437	823	1138	3,6%	1,6%	2,6%
Oil	365	845	1522	4,8%	3,0%	3,8%
Gas	46	159	372	7,1%	4,4%	5,7%
Heat	17	12	12	-2,1%	0,0%	-1,0%
Electricity	114	475	1038	8,2%	4,0%	6,0%
Renewables	228	188	182	-1,1%	-0,2%	-0,6%
by Sector:						
Industry	600	1281	2014	4,3%	2,3%	3,2%
Transport	183	443	928	5,0%	3,8%	4,4%
Dom. Tert. Agr.	440	772	1311	3,2%	2,7%	2,9%
CO2 Emissions in Mt of C	1258	3124	5571	5,2%	2,9%	4,0%
Emission variation from 1990	*189*	*2055*	*4502*			
Population (Million)	2893	3696	4461	1,4%	0,9%	1,1%
GDP (M$90)	6258	17273	38380	5,8%	4,1%	4,9%
Gross Inland Cons. / GDP (toe/M$90)	253	207	165	-1,1%	-1,1%	-1,1%
Elec. Gen. / cap. (kWh/cap)	592	1774	3258	6,3%	3,1%	4,6%
CO2 Emissions / cap. (tC/cap)	0,4	0,8	1,2	3,8%	2,0%	2,8%

Table 2-15: Summary energy balance, rest of developing countries.

Rest of developing countries : SUMMARY ENERGY BALANCE - Reference Case Mtoe	1992	2010	2030	1992/2010	2010/2030	1992/2030
Primary Production	2310	4111	7095	3,3%	2,8%	3,0%
Solids	136	249	651	3,4%	4,9%	4,2%
Oil	1674	2991	4617	3,3%	2,2%	2,7%
of which non conventional	0	147	725	38,7%	8,3%	21,8%
Natural gas	236	574	1361	5,1%	4,4%	4,7%
Nuclear	5	16	67	7,0%	7,4%	7,2%
Hydro+Geoth	42	78	143	3,5%	3,1%	3,3%
Trad.Biomass	168	127	88	-1,5%	-1,8%	-1,7%
Other Renewables	49	76	168	2,5%	4,1%	3,3%
Gross Inland Consumption	1136	1928	3960	3,0%	3,7%	3,3%
Solids	114	211	639	3,5%	5,7%	4,6%
Oil	553	1019	1999	3,5%	3,4%	3,4%
Gas	206	402	857	3,8%	3,9%	3,8%
Other	263	297	465	0,7%	2,3%	1,5%
Electricity Generation in TWh	1240	2630	7130	4,3%	5,1%	4,7%
Fuel Input for Thermal PG	188	344	1062	3,4%	5,8%	4,7%
Solids	58	105	343	3,3%	6,1%	4,8%
Oil	74	120	432	2,7%	6,6%	4,7%
Gas	54	119	288	4,5%	4,5%	4,5%
Final Energy Consumption	849	1498	2739	3,2%	3,1%	3,1%
Solids	35	64	133	3,4%	3,7%	3,6%
Oil	428	818	1436	3,7%	2,9%	3,2%
Gas	86	197	327	4,7%	2,6%	3,6%
Heat						
Electricity	88	222	594	5,3%	5,0%	5,1%
Renewables	214	203	256	-0,3%	1,2%	0,5%
by Sector:						
Industry	278	482	842	3,1%	2,8%	3,0%
Transport	245	447	838	3,4%	3,2%	3,3%
Dom. Tert. Agr.	326	568	1059	3,1%	3,2%	3,2%
CO2 Emissions in Mt of C	704	1311	2858	3,5%	4,0%	3,8%
Emission variation from 1990	*24*	*631*	*2179*			
Population (Million)	1246	1907	2730	2,4%	1,8%	2,1%
GDP (M$90)	4310	8630	18714	3,9%	3,9%	3,9%
Gross Inland Cons. / GDP (toe/M$90)	264	223	212	-0,9%	-0,3%	-0,6%
Elec. Gen. / cap. (kWh/cap)	996	1380	2612	1,8%	3,2%	2,6%
CO2 Emissions / cap. (tC/cap)	0,6	0,7	1,0	1,1%	2,1%	1,6%

Table 2-16: Summary energy balance, world.

World : SUMMARY ENERGY BALANCE - Reference Case Mtoe	1992	2010	2030	1992/2010	2010/2030	1992/2030
Primary Production	8276	12250	18125	2,2%	2,0%	2,1%
Solids	2152	3423	5546	2,6%	2,4%	2,5%
Oil	3248	4499	6362	1,8%	1,7%	1,8%
of which non conventional	17	164	1170	13,4%	10,3%	11,8%
Natural gas	1670	2842	4275	3,0%	2,1%	2,5%
Nuclear	457	648	861	2,0%	1,4%	1,7%
Hydro+Geoth	191	279	412	2,1%	2,0%	2,0%
Trad.Biomass	426	340	251	-1,2%	-1,5%	-1,4%
Other Renewables	184	220	418	1,0%	3,3%	2,2%
Gross Inland Consumption	8316	12262	18128	2,2%	2,0%	2,1%
Solids	2145	3423	5546	2,6%	2,4%	2,5%
Oil	3277	4512	6367	1,8%	1,7%	1,8%
Gas	1686	2842	4275	2,9%	2,1%	2,5%
Other	1207	1485	1940	1,2%	1,3%	1,3%
Electricity Generation in TWh	12674	22541	39393	3,3%	2,8%	3,0%
Fuel Input for Thermal PG	1955	3275	5600	2,9%	2,7%	2,8%
Solids	1205	2119	3502	3,2%	2,5%	2,8%
Oil	301	290	543	-0,2%	3,2%	1,6%
Gas	449	866	1555	3,7%	3,0%	3,3%
Final Energy Consumption	6172	8917	12621	2,1%	1,8%	1,9%
Solids	838	1075	1479	1,4%	1,6%	1,5%
Oil	2770	3900	5392	1,9%	1,6%	1,8%
Gas	1078	1610	2023	2,3%	1,1%	1,7%
Heat	208	184	185	-0,7%	0,0%	-0,3%
Electricity	871	1678	2970	3,7%	2,9%	3,3%
Renewables	594	560	669	-0,3%	0,9%	0,3%
by Sector:						
Industry	2427	3540	4710	2,1%	1,4%	1,8%
Transport	1592	2297	3396	2,1%	2,0%	2,0%
Dom. Tert. Agr.	2163	3079	4514	2,0%	1,9%	2,0%
CO2 Emissions in Mt of C	6054	9164	13874	2,3%	2,1%	2,2%
Emission variation from 1990	*67*	*3178*	*7888*			
Population (Million)	5424	7027	8713	1,4%	1,1%	1,3%
GDP (M$90)	27908	52249	96132	3,5%	3,1%	3,3%
Gross Inland Cons. / GDP (toe/M$90)	298	235	189	-1,3%	-1,1%	-1,2%
Elec. Gen. / cap. (kWh/cap)	2337	3208	4521	1,8%	1,7%	1,8%
CO2 Emissions / cap. (tC/cap)	1,1	1,3	1,6	0,9%	1,0%	0,9%

Energy demand growth will be low in the OECD region, with growth rate slightly superior to 1 % pa at the beginning of the projection period and under 1 % pa at the end. The economic recovery in the countries in transition also induces a recovery in energy demand; the corresponding increase in consumption is however moderate, this implies continued improvements in energy efficiency. Conversely, energy demand growth rates are superior to 4 % pa in Asia until 2010 and around 3 % after. In the rest of the developing regions, the demand growth remains lower than 3 % pa before 2010 but accelerates after, to more than 3.5 % pa.

In OECD regions, the supply pattern is dominated by the decline in conventional oil after 2000. This trend is however largely compensated, after 2020, by the surge of non conventional oil. In the economies in transition, natural gas continues to be a privileged source of energy and its production is more than doubled in 2030. Coal of course plays the key role in the developing Asia: its production exceeds 3 billions of toe in 2030, against less than one today. Oil and gas have a sustained development in the rest of the developing regions, which provide the major part of the supply corresponding to the increment of world demand.

2.2.3.1 Trends in Energy Intensities

In the POLES model, trends in energy intensities are dependent on the structural evolution of the economy and on autonomous technology improvements. But they are also strongly influenced by price effects. Up to 2000 and even 2010 world energy intensity decreases significantly, partly because of strong reductions in energy demand in the transition countries. The oil price increase of more than fifty percent between 2000 and 2020 puts further pressure on energy intensities and energy intensity generally declines by around 1 % pa. Intensity gains are lower by the end of the projection, largely because of more moderate oil price increases (less than ten percent in ten years).

2.2.3.2 Final Energy Demand Trends

World final energy demand which was of 6 Billion TOE in 1990 is expected to reach 9 Billion toe by 2010. The structure of the world final energy demand reflects a progressive electrification of the system, compatibly with a persistent, non declining oil share. The reason for this has to be looked for in the transport sector, which absorbed 22 % of total final demand in 1971, 26 % in 1992, and is expected to maintain its share in the future.

The transport sector accounts for more than one fourth of the final world energy and for half of the world's oil consumption. The world transport sector is indeed highly dependent on oil: 97% of the energy consumed by this sector is directly dependent on petroleum. From 1992 to 2010, world-wide annual transportation oil demand is projected to increase by 700 Mtoe (namely 14 million barrels per day), passing from around 1600 Mtoe/year to around 2300 Mtoe/year.

Changes in transportation efficiency may modify these trends. However, these changes would not be substantial if only incremental improvements on the traditional road dominated, single user explosion engine car would happen. Indeed, although technology is likely to continue to improve significantly, consumer preferences for larger and more powerful cars and trucks could completely offset the efficiency improvements.

From 1960 to 1991, the number of passenger cars and commercial vehicles almost quintupled even though global population less than doubled. The global car stock is expected to double in the forthcoming 15 years. Currently, there are great differences in car ownership ratios between countries, although these differences are likely to be reduced in the future. The rising rates of industrialisation and urbanisation will induce more transport and require increased investment for the corresponding infrastructures.

Transport demand growth will thus be much higher in the developing world than in the rest of the world: this growth will be of 2 % pa at world level, but of 5 % in Asia and of 3.5 % in the other developing regions until 2010.

The world electricity demand is expected to progress with annual growth rates superior to 3% during the next decades, although, until 2010, the average annual growth rate will reach 8% in the developing Asia and 5 % in the other developing regions. This means that, roughly speaking, global consumption will double in 2020 with respect to 1990.

To provide an idea of the dynamics of such a development, these growth rates imply that a new 250 MW power plant has to be built in the world each 3 days during this period (without considering that the existing power capacity has to be replaced). These figures put into evidence the crucial role that technology can play in the power sector in the forthcoming decades. Power generation technologies are becoming a huge market in which the technological expertise of developed countries will have to be massively transferred to emerging economies whose demand for electricity is growing a two digit annual rate.

2.2.3.3 Trends in the Electricity Systems

While the share of electricity in final energy demand will continue to increase, the power generation system will have a growing weight as a major converter of primary energy.

In the Baseline projection, the renewable sources of electricity will still be dominated by large hydro plants. Their future development will be sufficient to maintain their share in total primary supply, but not in electricity generation, which grows more quickly. As for nuclear energy, the outlook is one of relatively slow growth, with an increase in nuclear electricity production, but however insufficient to maintain the share of nuclear energy in primary supply and in electricity generation.

The electricity sector will thus continue to use large quantities of fossil fuels. In the OECD region, while oil will continue to decline in power generation, coal will provide part of the increment in fossil fuel inputs, but natural gas will play a rapidly increasing role. In developing Asia, the role of coal in power generation will remain predominant: in 2010, about 2 Gtoe of coal could be burnt in thermal power plants in Asia. The situation might be more balanced in the other developing regions, where coal, oil and natural gas would contribute each to 30 to 40 % of fossil fuel inputs of power plants.

Further details on the baseline projections of the electricity generation sector are provided in Part III of the volume.

2.3 CO_2 Emissions Outlook

2.3.1 Trends in Carbon Emissions

While world CO_2 emissions may incur a 50 % increase in 2010 and more than a doubling in 2030, the pattern of emissions by region will be profoundly altered by this date, as illustrated in Table 2-17 OECD countries represent today about half of total energy-related CO_2 emissions and the developing regions, altogether, less than one third. In 2030, the picture will be reversed, with OECD's share reduced to 30 % and that of the developing regions amounting to 60 %, as the Economies in transition will only have a small increase of emissions in absolute value and a strong decrease in relative share.

Table 2-17: World CO_2 emissions by main region.

	1990	1992	2010	2030
World (MtC)	5986	6054	9164	13874
OECD	49%	49%	41%	30%
Trans Eco	22%	19%	11%	9%
Dvg Asia	18%	21%	34%	40%
Dvg Oth	11%	12%	14%	21%

The new balance in regional CO_2 emissions will of course result from differentiated demographic and economic dynamics. But it also comes out from a movement of convergence in per capita CO_2 emissions among regions. Today, per capita emissions are eight times higher in OECD regions than in Asia. Although they might continue to increase in the industrialised world, the relative increase will be much higher in the developing one. As a result, the per capita emission ratio may be reduced to a factor of less than four in 2030, as illustrated in Figure 2-5. The commitments accepted by the Annex I countries after the Kyoto Conference, will, if they are taken seriously, increase this movement of convergence, which in the Baseline projection remain however moderate.

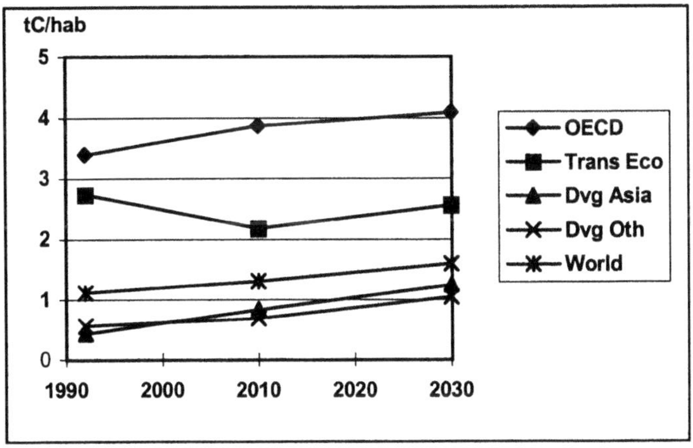

Figure 2-5: Per capita CO_2 emissions, by main regions.

As a consequence of the on going dynamics in energy consumption and CO_2 emissions, the Baseline projection shows that, if a « flat rate » stabilisation or slight reduction from 1990 level may appear an ambitious but realistic target for the Annex I countries (the OECD and the Economies in transition), such type of target can obviously not be applied to developing countries, the emissions of which should double or triple, in 2010 and the again double in the following twenty years.

Of course the increase in emissions is particularly strong in Asia and this phenomenon is due to strong demographic and economic dynamics - if however the on going crisis is soon superseded - but also to a highly coal intensive energy resource base for that region.

Table 2-18 gives the absolute increases in CO_2 emissions between 1990 and 2010 and between 2010 and 2030. It clearly shows that the bulk of emission increases will come from developing regions and that curbing the emissions in the Annex I countries will in no way be sufficient to aim at any level of world stabilisation.

It can thus be considered that the question of when and how to associate non Annex I countries to the effort of CO_2 emission control will soon come out high on the international negotiation agenda.

Table 2-18: Emissions increases from 1990.

(MtC)	1990	2010-1990	2030-2010
World	5986	3178	4710
OECD	2949	774	419
Trans Eco	1290	-283	295
Dvg Asia	1069	2055	2448
Dvg Oth	679	632	1548

2.4 Comparisons of Projections of the World Energy System to 2030: Similarities and Differences

Empirical studies aimed at forecasting developments in the world energy system in the medium and long term are undoubtedly less frequent today than at the time of the oil crisis. Nevertheless, a number of specialised organisations have continued to work in this field, in particular the International Energy Agency (IEA, 1996), the U.S. Department of Energy (DOE-EIA, 1997) and the World Energy Council (WEC-IIASA, 1995). The first two organisations seek to clarify the energy environment in the medium term, using projections, which, though sometimes contrasting, largely reflect a "consensual vision" of current developments in the field. The WEC has a different and probably more ambitious goal, insofar as its last study - carried out with the IIASA - was intended to draw attention to the expanding range of possibilities in the longer term (2020 and 2050).

Comparison of the various reference projections or central scenarios is still not without problems, since the accounting conventions used are not identical; in particular where traditional renewable energies are concerned. Nevertheless, this comparison reveals a fairly significant convergence of results concerning world consumption of all types of energy, at any rate up to the year 2020. All projections point towards 12-13 Gtoe in 2010 and 15-16 Gtoe in 2020.

Table 2-19: Results of four reference cases for the world energy system.

IEA-OECD		"Capacity Constraints"		
1996	1993	2000	2010	1993-2010
Population				1.4%
GDP				3.2%
TPES	8.4	9.7	12.3	2.2%
Coal	2.3	2.6	3.2	2.0%
Oil	3.3	3.8	4.7	2.1%
Nat Gas	1.7	2.1	2.8	2.9%
Nuclear	0.6	0.6	0.7	1.0%
Hydr + Ren	0.6	0.7	0.9	2.5%
CO2 (MtC)	5.9	6.7	8.6	2.3%
Oil Price ($/bl)		17	25	

EIA-DOE		"Reference Case"		
1997	1995	2010	2015	1995-2015
Population				
GDP				3.1%
TPES	9.1	12.8	14.0	2.2%
Coal	2.3	3.1	3.4	1.9%
Oil	3.5	4.9	5.3	2.1%
Nat Gas	1.9	3.2	3.6	3.2%
Nuclear	0.6	0.6	0.6	-0.1%
Hydr + Ren	0.7	1.1	1.2	2.2%
CO2 (MtC)	6.2	8.8	9.7	2.2%
Oil Price ($/bl)		20.4	21.0	

WEC-IIASA		CME "A1"		
1995	1990	2020	2050	1990-2020
Population	5.3	7.9	10.1	1.3%
GDP	20.9	46.9	101.5	2.7%
TPES	9.0	15.4	24.8	1.8%
Coal	2.2	3.7	3.8	1.7%
Oil	3.1	4.7	7.9	1.4%
Nat Gas	1.7	3.6	4.7	2.5%
Nuclear	0.5	0.9	2.9	2.0%
Hydr + Ren	1.6	2.5	5.5	1.5%
CO2 (MtC)	6.0	9.5	12.9 [11,7]	1.5%
Oil Price ($/bl)				

POLES		Baseline March 1998		
1998	1992	2010	2030	1992-2030
Population	5.4	7.0	8.7	1.3%
GDP	27.9	52.2	96.1	3.3%
TPES	8.6	12.6	18.7	2.1%
Coal	2.1	3.4	5.5	2.5%
Oil	3.3	4.5	6.4	1.8%
Nat Gas	1.7	2.8	4.3	2.5%
Nuclear	0.5	0.6	0.9	1.7%
Hydr + Ren	1.1	1.3	1.7	1.2%
CO2 (MtC)	6.0	9.2	13.6	2.2%
Oil Price ($/bl)		22.7	34.9	

Units: population in billions, GDP trillions of $, energy in billions of toe (Gtoe); POLES units have been adjusted as concerns hydro for consistency with other forecasts.

However, this apparent convergence of global results masks potential inconsistencies, insofar as assumptions concerning prices - when they are made explicit - are far from identical. Thus, the DOE projection is based on the assumption of virtually stable oil prices up to the year 2015, whereas in the IEA projection and the reference scenario the price of oil will have already increased by 40% by 2010. It will be seen that in the latter projection prices increase sharply after 2010 owing to the decline in growth of conventional production and the need for rapid exploitation of non conventional resources.

The results according to energy source are also less in agreement. However, by 2010 - in other words the date now used as a reference in international negotiations on the greenhouse effect - the world energy supply should probably reckon with slightly more than 3 Gtoe of coal, like natural gas, and between 4.5 and 5 Gtoe of oil. On the basis of the carbon coefficients of each energy source, this would represent worldwide energy-related CO_2 emissions of about 8.7 GtC, compared with 6 GtC in 1990. There is obviously a significant discrepancy between the stabilisation objectives announced in the international negotiations and projections "without policies" which give an increase in emissions of approximately 45% over 20 years, from the reference date of 1990.

Table 2-20: Factors to explain the rise in CO_2 emissions in 7 energy scenarios (by adding mean annual growth rates).

[t.c.a.m.]	[CO2] =	[CO2/CEP] +	[CEP/PIB] +	[PIB/POP] +	[POP]
AIE - CC - 2010	2.3%	0.1%	-1.0%	1.8%	1.4%
AIE - ES - 2010	1.7%	0.0%	-1.5%	1.8%	1.4%
DOE-Ref-2015	2.2%	0.0%	-0.9%	3.1%	
IPCC-IS92a-2020	1.7%	-0.2%	-1.0%	1.5%	1.4%
CME - A1 - 2020	1.5%	-0.3%	-0.9%	1.4%	1.3%
CME - C1 - 2020	0.2%	-0.6%	-1.4%	0.9%	1.3%
POLES - Base - 2030 (PPP)	2.2%	0.1%	-1.2%	2.0%	1.3%

NB1: PEC = Primary energy consumption.

NB2: The use of buying power parities to account for GDP in the projection leads to a certain increase of world growth, because it increases the weight of the rapid growth developing regions.

Table 2-20 shows that only in very differing scenarios it is possible to escape this constant rise in energy CO_2 emissions. This is the case for example in the WEC-IIASA "ecological" scenarios -C- in which increases in CO_2 emissions can be contained thanks to a combination of three factors:

- appreciably lower economic growth (2%/year instead of 3%);
- a marked acceleration in energy intensity improvements (1.4%/year instead of 1%);
- and strong "decarbonising" of supply (increased use of nuclear power or renewable energy, depending on the case).

The feasibility and even the desirability of this type of scenario could obviously be called into question. In any case, such scenarios would be highly unlikely to correspond to what would be a "spontaneous" development of the energy systems. On the contrary, they most likely already imply extremely voluntarist policies. In this respect, they are scenarios which illustrate a possible solution to the problem but which do not clearly define the methods and policies involved.

2.5 Conclusions

The baseline produced with the POLES model provides a consistent and detailed description of the world energy situation in 2010 and 2030. Among the main features of this outlook, the following issues have to be emphasised:

- the weight of the developing regions will increase rapidly in the world economy and energy system; in 2010, they will represent half of world GDP and some years later half of world total energy consumption;
- in the next forty years, most part of world energy consumption increases will take place in the developing world;
- this will have an impact on international energy markets, as oil demand will largely grow along the development of transport and car equipment in that regions;
- while oil resources are large and might be extended by technological progress, they are however limited; a transition to non conventional liquid fuels will have to take place after 2010;
- this may imply oil prices higher than current level in order to close the world balance, while sustaining supply and limiting demand; natural gas prices may also increase, as gas demand will be strong in every region of the world;
- world CO_2 emissions will be on an upward trend; while « flat rate » reductions provide ambitious but reasonable targets for industrialised countries and the economies in transition, they are fully inadequate for CO_2 policies in the developing world;
- the expected increase in developing country emissions is very large and stems directly from the strong dynamics in energy demand, particularly for transport and electricity; as concerns this latter case, coal use in power generation in Asia appears as one of the main sources of incremental emissions in the next decades.

The world of energy will probably be very different after 2010 of what it is today. This is already the case in this baseline projection, the main characteristics of which are a higher level of demand, international prices and CO_2 emissions.

3 European Outlook[7]

3.1 The Baseline Scenario: the European Economy to 2030

The baseline scenario[8] simulates a dynamic path of the EU economy up to 2030. It is derived from exogenous assumptions about the evolution of technology progress associated to production factors, the change of the world context (prices and demand) and a continuation of current pattern of public finance policy.

3.1.1 Introduction

The purpose of the present study is to provide a preliminary long run scenario up to 2030 for each EU member states for use as the baseline in several studies of energy and environment policy analysis for the European Union member states. The present projection aims to consistently delve into considerable detail at the level of sectoral disaggregation provided by the GEM-E3 model.

Deriving projections for the longer term is very difficult given the considerable uncertainties involved. Two are the main difficulties of such an undertaking.

- The first is projecting aggregate regional GDP and population growth. Several studies are available concerning the medium and long term (usually up to

---

[7] The principal authors of this chapter are Pantelis Capros, Leonidas Mantzos and Panagiotis Georgakopoulos of NTUA.

[8] The baseline scenario is the one previously prepared for the European Commission and reported in "European Macro-economic Projections for Baseline Scenario", March 1997, Report of NTUA to the EC.

2020) covering the whole planet such as with the LBS-EGEM model (which however does not include Europe) or the world projections with the World Scan model, as these where prepared for the OECD, where Europe is just one region. This latter study was the basis for the present projection.

- The second, and perhaps even more elusive, difficulty reflects the sectoral changes. As the past has shown these changes can often be very dramatic and move in different directions in different countries. No study is available at the level of disaggregation needed for the energy models, so this part of the projection after 2000 is original.

The present scenario draws from the macro-economic and sectoral projections available for the short term (up to 2000) and then uses the aggregate world assumptions derived with the World Scan model, extending them to 2030. From a sectoral perspective, there has been attempted to build a separate "story" describing the evolution of each EU country. The projections were made in two steps:

- assuming gradual conditional convergence of the EU economies in terms of per capita income, the GDP of each country was derived;
- understanding the present situation of each country and identifying already existing trends, the driving forces of growth for each economy were identified and served to derive sectoral growth.

3.1.2 Short Run Projections: 1995-2000

Three were the main sources consulted for the preparation of the projections for the period 1996-2000.

The *DGII projections* for each member state, from which GDP, private consumption, consumers' price indexes, GDP deflator, interest rate and exchange rate were taken.

Sectoral projections were taken from the *DRI study "Europe in 1999 - Economic analysis and Forecasts"*. These projections where only available for 6 EU member states and covered only some of the industrial sectors. Information on services was derived from the *HERMES projections* (again for six countries), which were also used to cross evaluate the DRI assumptions. Both studies only gave one average figure for the whole 1996-2000 period, so assumptions had to be made about the changes in each particular year. The assumption of a linear trend was mostly followed. For the other EU countries no sectoral forecast where available, so current trends were assumed to continue, in such a way however, that the EU total for each sector matches the DRI forecasts.

Table 3-1: GDP projection up to 2000.

Short run GDP projections (annual growth rates)						
	1995	1996	1997	1998	1999	2000
Austria	2.81%	1.00%	1.60%	2.50%	2.90%	3.10%
Belgium	0.90%	1.40%	2.19%	2.66%	2.90%	3.04%
Germany	1.90%	1.41%	2.16%	2.80%	3.02%	3.21%
Denmark	2.76%	2.08%	3.06%	2.96%	2.86%	3.11%
Finland	4.21%	2.13%	3.67%	3.11%	3.39%	3.56%
France	2.14%	1.09%	2.09%	2.66%	2.85%	3.27%
Greece	2.00%	2.37%	2.51%	2.78%	3.12%	3.51%
Ireland	9.84%	7.75%	5.81%	5.22%	4.79%	4.60%
Italy	2.96%	0.84%	1.44%	2.57%	2.74%	3.03%
Netherlands	2.08%	2.45%	2.77%	2.95%	3.06%	3.07%
Portugal	2.18%	2.51%	2.81%	3.13%	3.48%	3.68%
Spain	2.85%	2.10%	2.67%	3.17%	3.66%	3.71%
Sweden	3.01%	1.72%	2.11%	2.49%	2.78%	2.96%
UK	2.38%	2.31%	3.00%	3.00%	2.94%	2.99%
EU Average	**2.4%**	**1.6%**	**2.3%**	**2.8%**	**3.0%**	**3.2%**

3.1.3 Long Run Projections: 2001-2030

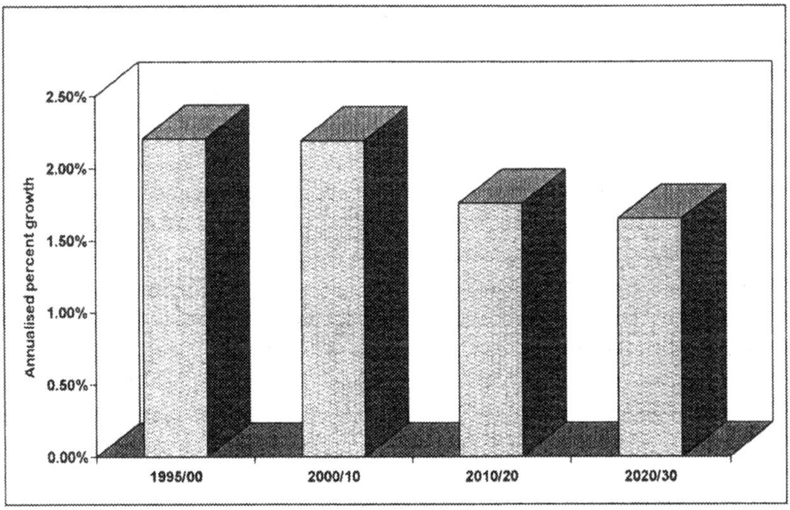

Figure 3-1: GDP per capita in the EU.

The assumption about average EU growth was based on the "High growth" variant of the latest *World Projections*, which performed with the *World Scan* model for the OECD. In the baseline scenario average EU growth is assumed to increase by 2.4% in the decade 2001-2010, slowing to 1.8% in the following decade 2011-2020 and stabilising to 1.7% thereafter. The EU population is assumed to increase slightly until the first decade of the next millennium also due to immigrants. After 2010 the rate of growth falls going to 0% after 2020.

Table 3-2: Country difference from EU average in per capita GDP.

	2000	2030
Austria	9.31%	3.60%
Belgium	6.23%	3.00%
Germany	4.30%	3.70%
Denmark	35.35%	25.00%
Finland	26.19%	20.00%
France	11.85%	3.70%
Greece	-67.23%	-55.00%
Ireland	-0.22%	7.00%
Italy	11.94%	9.50%
Netherlands	3.29%	3.60%
Portugal	-57.96%	-40.00%
Spain	-28.17%	-16.00%
Sweden	28.54%	23.00%
UK	-1.71%	1.10%

The GDP projection was then elaborated by member state following an assumption of gradual convergence of per capita incomes within the EU. Even in 2030 this convergence is however, far from complete. Based on the projections of changes of intra EU differences in per capita income, the general EU GDP growth and demographic assumptions, the implied GDP growth per country was obtained. The assumption of convergence implies higher growth rates for the cohesion countries (Greece, Ireland, Portugal, Spain), growth above EU average for the UK and lower growth levels for the rich Scandinavian countries (Denmark, Sweden and Finland). This trend is already present to some extent as can be seen from the high growth rate experienced by Ireland, Spain and Portugal in the period 1985-95. In these comparisons across EU member states market exchange rates have been used (relative to the ECU), rather than purchasing power parity indicators.
By keeping then constant the average EU growth as given in Figure 3-1, and applying the convergence process given in Table 3-2, GDP growth rates by

country are computed in a consistent manner. These are shown in Table 3-3 and Figure 3-2 that follow.

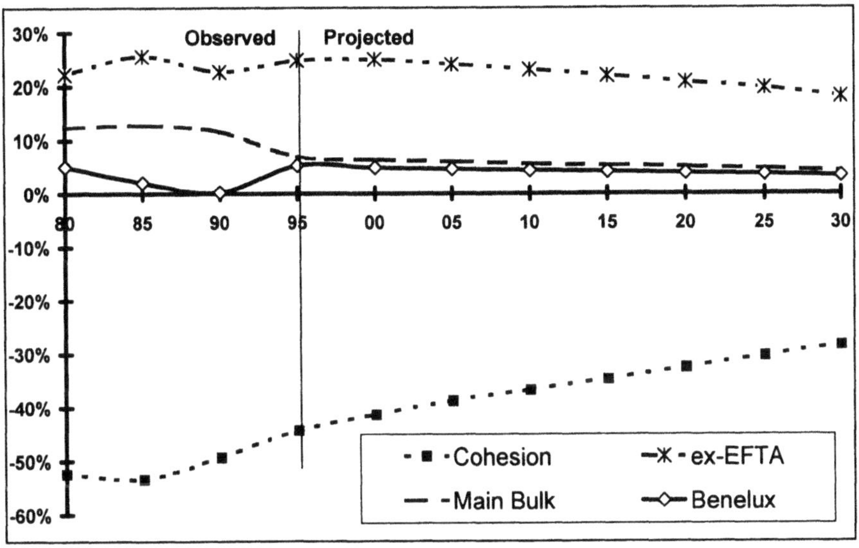

Figure 3-2: Evolution of growth rates in the EU, 1980-2030.

Table 3-3: Annualised percent change in GDP.

	Observed			Forecast						
	1980-85	1985-90	1990-95	1995-2000	2000-05	2005-10	2010-15	2015-20	2020-25	2025-30
Austria	1.31%	3.22%	2.28%	2.22%	2.19%	2.15%	1.74%	1.59%	1.51%	1.49%
Belgium	1.02%	2.85%	1.51%	2.44%	2.29%	2.23%	1.82%	1.67%	1.59%	1.57%
Germany	1.23%	3.23%	1.84%	2.52%	2.50%	2.26%	1.80%	1.65%	1.62%	1.62%
Denmark	2.19%	1.82%	1.68%	2.81%	2.19%	2.10%	1.61%	1.44%	1.38%	1.16%
Finland	2.85%	3.39%	-0.69%	3.17%	2.40%	2.32%	1.84%	1.62%	1.56%	1.47%
France	1.42%	2.99%	0.94%	2.39%	2.30%	2.21%	1.85%	1.67%	1.58%	1.58%
Greece	1.49%	1.68%	1.40%	2.86%	3.40%	3.34%	2.99%	2.96%	2.95%	2.60%
Ireland	3.33%	5.57%	6.30%	5.63%	3.78%	2.68%	2.18%	1.97%	1.87%	1.81%
Italy	1.42%	3.00%	1.33%	2.10%	2.19%	2.06%	1.80%	1.63%	1.58%	1.53%
Netherlands	1.28%	3.06%	1.82%	2.86%	2.64%	2.55%	2.12%	1.86%	1.78%	1.75%
Portugal	0.82%	8.04%	1.72%	3.12%	3.76%	3.61%	3.09%	2.85%	2.76%	2.71%
Spain	1.46%	4.45%	1.51%	3.06%	2.91%	2.84%	2.39%	2.20%	2.16%	2.15%
Sweden	2.24%	2.80%	0.92%	2.41%	2.38%	2.24%	1.70%	1.53%	1.50%	1.50%
UK	3.47%	3.25%	1.27%	2.85%	2.59%	2.52%	2.00%	1.68%	1.63%	1.61%
EU Average	**1.73%**	**3.22%**	**1.43%**	**2.57%**	**2.48%**	**2.35%**	**1.93%**	**1.73%**	**1.68%**	**1.66%**

Fiscal and Monetary Policy

It is assumed that monetary unification around 2000-2005 will tend to eliminate fluctuations in interest and exchange rates and lead to a gradual convergence of prices. Additionally a combination of monetary policy and the high intra EU level of competition are assumed to keep price increases and inflation below the 1995 rate.

This process further implies that growth of private consumption will be lower than average GDP, leaving more room for financing investments, in the short run. Gradually a progressive shift is projected leading after 2020 to equality of growth rates.

Tight fiscal policies are assumed to prevail in the next decade aiming to reduce public deficits. After that time a shift of the role of the government towards regulation instead of provision will result in increase of the public sector at a level lower than average GDP.

Long Run Sectoral Projection[9]

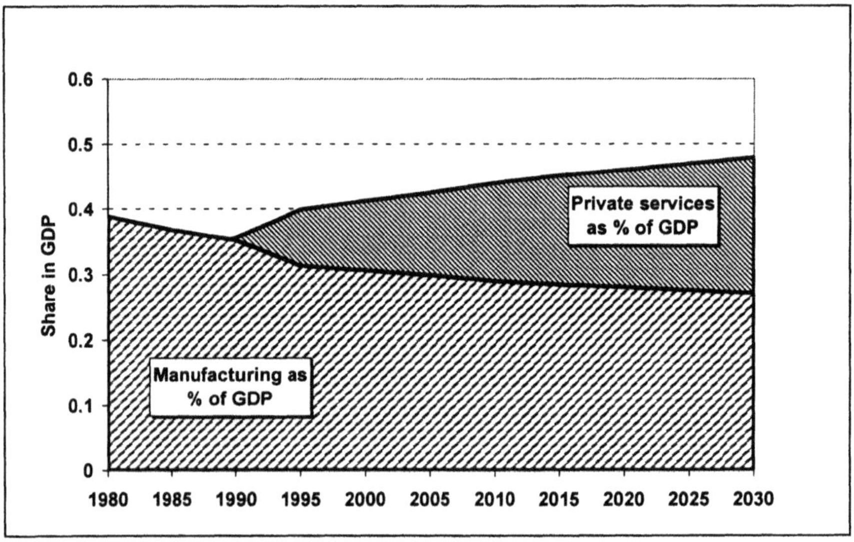

Figure 3-3: Evolution of services and manufacturing in Germany.

From a sectoral point of view, the baseline assumes a continuation of current trends, without entailing extreme situations. Specialisation of countries occurs but in most cases not dramatically, services increase their dominance but do not

[9] For detailed macroeconomic and sectoral projections see Appendix II.

encompass the whole economy. Industrial activity is rather stabilised following a period of re-structuring. New industrial activities with high value added and lower material base are projected to emerge in most countries. Considerable differences are assumed in the evolution of the economies in the different countries. Some of the main assumptions of the scenario are as follows.

- In Germany and France industrial growth is assumed to occur mainly through engineering, chemicals and the non ferrous sectors (at a lesser degree). Building materials, paper and pulp and non metallic minerals will also increase but less than GDP. Production will stagnate in construction, agriculture, textiles and the iron and steel sectors. The share of services will continue to be higher than GDP.

- Similar trends are also projected in the UK. Activity in the manufacturing sector is maintained close to the 1990 levels, mainly through chemicals, equipment and paper and pulp. In Italy a general slowdown of industrial activity is projected, except in building materials, and equipment goods. Activity in textiles and food processing is preserved. Growth is balanced benefiting almost all sectors. Increase in services is also projected not to be very high compared to the EU average.

- In Belgium manufacturing in general increases more than GDP up to 2005 and keeps some of its potential thereafter. Growth comes also from specialised activities in non ferrous and non metallic minerals, but mainly from chemicals, engineering and construction. Services increase, but less spectacularly than in most other countries.

- Austria and the Netherlands also keep industrial activity, through engineering, metals and chemicals (in the Netherlands) or through construction and a balanced growth of most other sectors (in Austria).

- In the Scandinavian countries it is assumed that the paper and pulp sector will have a considerable increase (especially in Sweden and Finland), while engineering also increases above GDP. Building materials and construction on the other hand, will stagnate, or increase marginally (Denmark). The growth in the public sector is assumed to be very low in all these countries.

- Cohesion countries experience high growth rates. In the first years up to 2005, the main sources of industrial growth are construction, building materials also because of the cohesion funds. After that time growth is also diverted to food, textiles and trade. Ireland is an exception, since most of the increase in industrial output is assumed to come from engineering, special activity in metals and chemicals, following an inflow of foreign investments. Similar trends are also projected to occur in Portugal. The increase in the share of market services in all these countries is assumed to be above 4% until 2010 and falls gradually thereafter.

The main assumptions of the projections from a sectoral viewpoint are summarised in the following table.

Table 3-4: Main sectoral assumptions of baseline scenario.

Sector	***Assumptions***
Iron and Steel	Following the decreases observed between 1990-95, production continues to fall undergoing extensive restructuring, to achieve stability after 2020.
Non ferrous	Following considerable restructuring between 1990-95, the sector rebounds achieving stabilisation and growth. This effect is present in Germany, the Netherlands, Spain and the UK.
Chemicals	Production of basic chemicals is projected to be maintained in some countries and especially in France, Germany, and the Netherlands. The higher value added brands of the sector (such as pharmaceuticals) are assumed to flourish benefiting from an increase both in domestic demand and in exports. Growth rates similar to GDP are experienced by Germany, France, Ireland, Italy, Netherlands, and Belgium. In the southern cohesion countries, on the other hand, production is assumed to stagnate or increase only marginally.
Paper and pulp	The sector is assumed to face a significant increase in demand especially from the rest of the world. The EU is assumed to continue to be a net exporter. Growth in this sector is assumed to be one of the main driving forces of economic growth in the Scandinavian countries: about 3.5% per annum for Finland and Sweden between 2000 and 2020.
Non metallic minerals	Activity in traditional building materials is projected to increase especially in the South, benefiting from an increasing demand from the construction sector, especially in the first decade of the next millennium. In the North activity will be mostly diverted to specialised products with high value added (such as high quality ceramics etc.).
Engineering	After a decade of relative stagnation and restructuring, the sector is assumed to strongly rebound being one of the main driving engines of economic growth for the most important European economies: Growth above GDP is projected for Germany, France, UK, Ireland and Italy especially up to 2020. Significant increases are also projected for Belgium and the Netherlands.
Construction	The outlook for construction appears better in the countries of the south where it will be significantly higher than GDP, also because of the cohesion funds. In Portugal and Greece for example the increase in construction in the decade 1995-2005 is of the order of 4% and only slightly lower in Spain.

Table 3-4: Continued.

Textiles	After almost a decade of considerable restructuring that lasts up to 2000, the sector is assumed to rebound in the southern countries: Greece, Italy, Portugal and Spain all face a 1.5-2% growth per annum between 2000 and 2030. In the other countries production is assumed to stabilise in level lower than 1990.
Food and Agriculture	Some growth is expected but less than GDP. These sectors are generally assumed to undergo significant restructuring. Higher increases are projected in the south (especially in Greece and Portugal where after 2005 these sectors are expected to start growing significantly again) and to some extent in France.
Energy	The growth in the energy sector is expected to be of the order of 1-2% for most countries, growing at about half the GDP rate.
Services	The role of services is assumed to increase significantly in all countries. Their growth rate is everywhere higher than average GDP. Their share in the economy increases over time but at a decreasing speed, leading to some stabilisation after 2020. Still increase in activity in private service remains more than 2% per annum until 2030. The increase in the service sectors (especially tourism) is more pronounced in the South.

3.1.4 Energy and Emissions Outlook

The baseline scenario is similar to the pre-Kyoto scenario constructed in March 1997 for DG-XVII by means of the MIDAS model. The macro-economic and world energy price assumptions are exactly the same.

3.1.4.1 Overview of Results

The following graphics illustrate the results for the baseline. They show that in the baseline there are substantial energy efficiency gains accompanied with a noticeable penetration of natural gas in power generation, renewables and distributed heat in the end-use sectors. This of course leads to a moderate growth of CO_2 emissions in the baseline (8% more in 2010 compared to 1990, for EU-15).

While GDP grows in EU-8 by 2.42% per year, primary energy demand is found to increase at an annual rate of 0.88% and CO_2 emissions grow even slower, by 0.49% per year. The decoupling of energy from economic growth is due first to the restructuring of economic growth, as this is more based on high value added commodities (and thus less energy and material intensive), and secondly to continued technological progress in the energy use and conversion. The combined effect of these mechanisms results in a steady decrease of the overall energy intensity indicator of -1.5% per year in EU-8 (1995-2010).

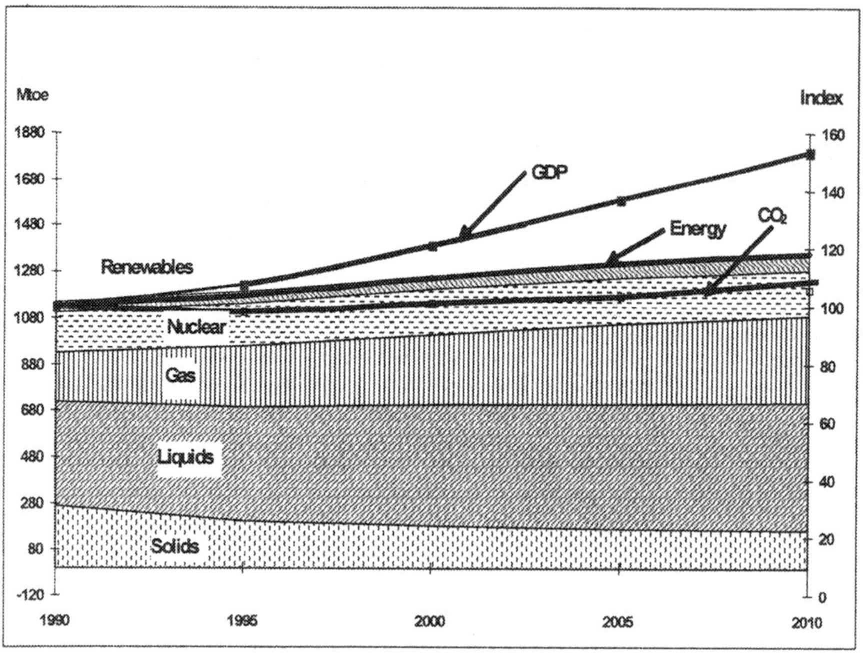

Figure 3-4: Baseline scenario primary energy, GDP and CO_2.

3.1.4.2 Primary Energy Demand

The fuel mix in primary energy changes smoothly, mainly allowing for substitution of solid fuels by natural gas. Nuclear and liquids preserve there shares, while the penetration of renewables is noticeable but limited (5.5% of primary energy demand in 2010, being 3.9% in 1990).

The changes in fuel mix imply a moderate decrease of the carbon intensity (-0.4% per year), contributing to the decoupling of CO_2 from energy demand growth. This effect is however significantly smaller in size than the energy intensity effect.

3.1.4.3 Final Energy Demand

In final energy demand, transports (1.69% per year) and tertiary (1.30% per year) continue to grow significantly, while stagnation of energy demand is projected for industry (0.70% per year) and households (1.42% per year). The projection shows a penetration of distributed heat (mainly produced by cogeneration), a decline of solid fuels and a continued, even though moderated, penetration of gas and electricity. The part of liquids is maintained, because of the their use in transports.

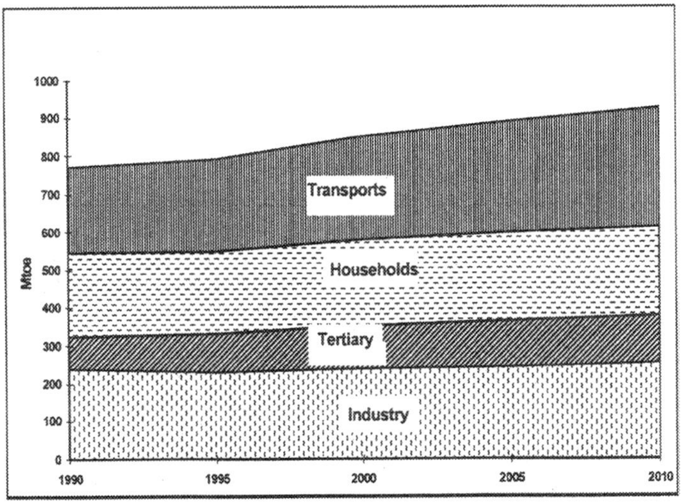

Figure 3-5: Baseline final energy demand.

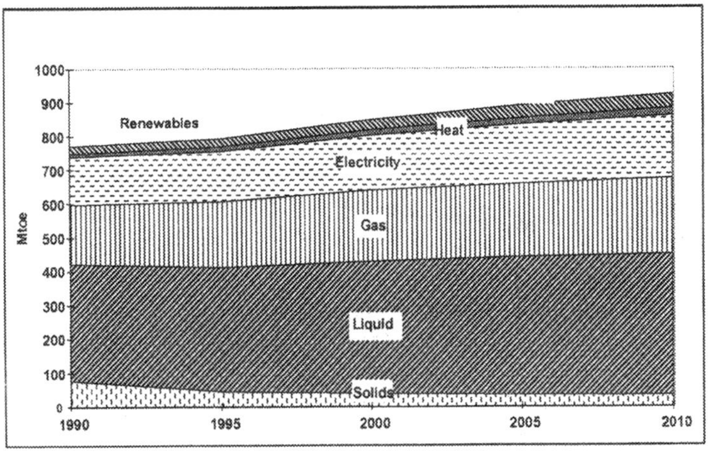

Figure 3-6: Fuel mix in baseline final energy demand.

3.1.4.4 Power Generation

Power generation is projected to increase in EU-8 by 1.3% per year. The sectoral evolution involves mainly a relatively high growth of independent generation (it represents 12.5% of total power capacity in 2010, to be compared to 5.4% in 1990), and an important shift in favour of natural gas used in combined cycle power plants (more than 2/3 of all new power capacities). The progress in gas turbine technology allows also a significant increase of cogeneration (12.5 GW more capacity in 2010).

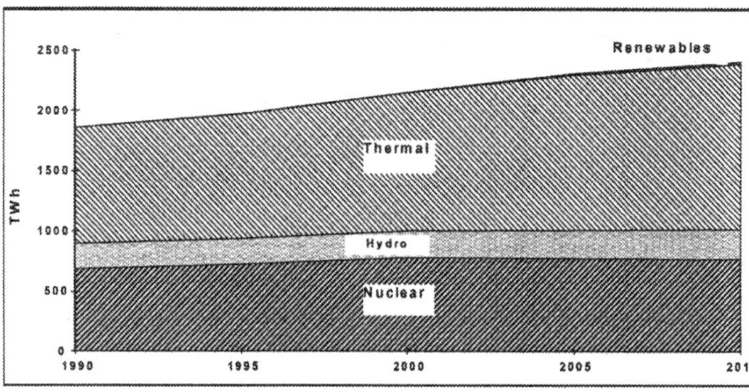

Figure 3-7: Power generation by source.

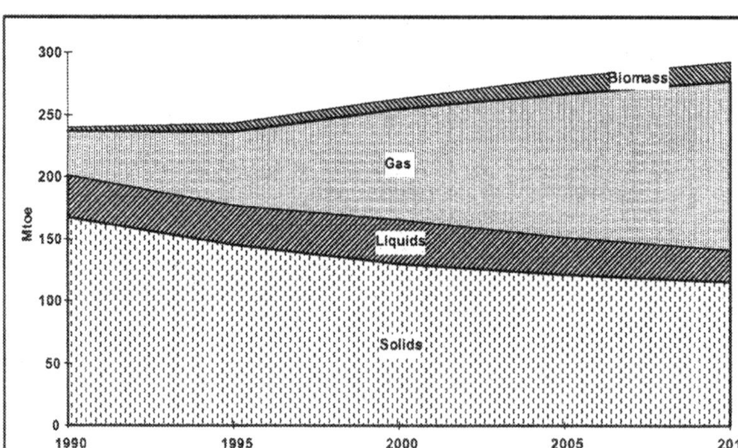

Figure 3-8: Fuel mix in thermal power generation.

3.1.4.5 Co₂ Emissions

The evolution of CO_2 emissions by sector reveals in the baseline that transports and tertiary are mainly responsible for the increase, which leads in total for EU-8 at 5.7% more emissions in 2010 compared to 1990. Emissions from independent CHP also increase but this of course is compensated by the efficiency gains in the end use. Central power generation is almost stable regarding emissions.

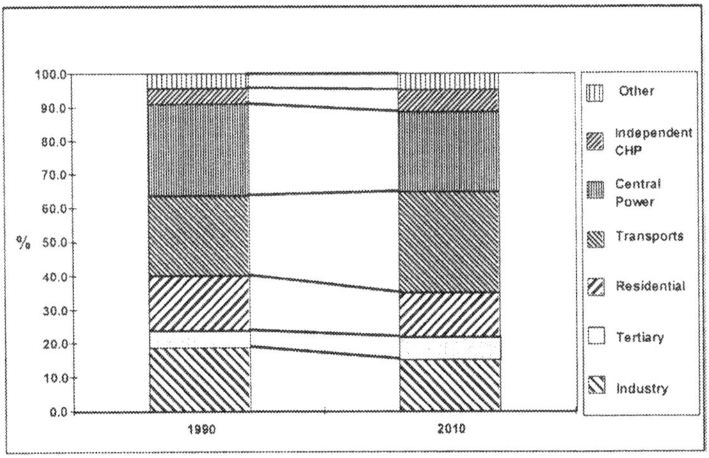

Figure 3-9: Structure of CO_2 emissions.

Among final consumption sectors it is the tertiary and transportation sectors that show the fastest growth in emissions. As shown in Table 3-5, the residential and industrial sectors will emit less CO_2 in 2010 than they did in 1990.

Table 3-5: Emission index by sector.

CO_2 Emission Index	1990-2010
Industry	86.7
Tertiary	134.2
Residential	86.8
Transports	132.8
Central Power	92.1
Independent CHP	148.4
Other	119.0

PART II: MEETING THE KYOTO TARGETS: THE ROLE OF POLICY

4 Introduction

This part of the volume deals with technology and policy options to tackle the problem of rising emissions in the period to 2010. It was seen in the first part of this volume that, under baseline assumptions, CO_2 emission was likely to grow substantially in the period to 2010. Even in Europe emissions were projected to increase by almost 8% above their level of 1990. Yet over the past couple of years there has been a great deal of debate on how to reduce emissions even in the medium term. This debate reached a turning point in the Kyoto Conference on climate change where all developed regions of the world undertook firm commitments and set targets for reductions in CO_2 emissions by 2010. The European Union agreed to decrease the CO_2 emissions in 2010 by 8% compared to the level of emissions in 1990. This would involve a reduction of nearly 16% when compared to the level of emission projected for that year under baseline assumptions.

The task of meeting the Kyoto targets should not be underestimated. It is taken for granted, of course, that the reduction in emissions is not to occur through reduced economic growth. While in most developed economies there is a tendency for energy demand to grow somewhat less than economic growth, there is a very strong tendency for any sizeable economic expansion to be associated with an increase in energy demand of about 1% less than GDP. Furthermore, it is important from the results of the first part of this chapter, that energy patterns are almost certain to continue to be dominated by CO_2 emitting fossil fuels in the period to 2010. Even if renewable fuel technologies were more competitive than they are currently, it would take much longer than 15 years for their massive introduction. This issue will be discussed at greater length in Part III.

Hence, reducing CO_2 emissions for the EU as a whole over the next 15 years would amount to no less than breaking the historical link between economic growth and energy use. This link has been very strong in most countries throughout the post war period and it has survived even the upward energy price shocks of the 1970s.

One additional factor that makes the Kyoto targets difficult to reach is the limited scope of contribution by new technologies. New technologies can either break the link between energy and growth, through equipment that is much more energy efficient, or make energy much less polluting, through energy from renewable sources or through much more efficient generation of final energy. However, even if these new technologies were fully available at present, there is simply not enough time to replace the bulk of energy using capital stock and equipment by 2010 without causing substantial economic damage.

The difficulties mentioned above are not insurmountable in the longer run. Annex I includes a discussion of most aspects of the energy system that gives rise to emissions and a large number of measures is analysed in the next chapter. Even by 2010, there is a great deal that could be done to limit emissions and, perhaps, meet the Kyoto targets, as the two scenarios presented in the next chapter indicate. Technology plays a key role in these scenarios. However, most of its contribution will be in the form of integrating faster existing energy efficient technologies or accelerating the introduction of known advanced technologies. As it will be seen in the next chapter, much of the reduction in emissions by 2010 can only be reached through these means of improving energy efficiency.

Even if reaching the Kyoto targets is feasible, it is very important to examine the likely economic costs of this task. This is the purpose of the last part of the next chapter and the whole of the second chapter in the part of the volume. It is shown that, both for the EU and for the rest of the world, there are large differences between the average and marginal costs of meeting any target. Furthermore these differences increase as the emissions target gets more severe. Indeed, one of the key messages of the results is that the more severe the target and the shorter the time horizon over which it must be reached, the greater the likely economic costs.

Another important aspect of the likely costs of policy on emissions is the differential impact on different countries and regions. This impact clearly depends on the level of development of a country, its economic and industrial structure, and on its natural resource endowment and the carbon intensity of its energy system. The results of the next chapter for 8 EU member states indicate that, even when the degree of the severity of the target is similar costs can diverge a great deal between countries.

The international dimensions of emissions target for 2010 are very important and are discussed in the second part of this part of the volume. The environmental threat is perceived with different degrees of concern by governments, decision-makers and the general public in different world regions. Amongst the developed nations, the European Union has taken a leading role in recognising the problem as an important issue to be solved, and in proposing co-ordinated efforts to achieve a cost-efficient solution. As was seem in Chapter 2, most of the emissions over the next 15 years are likely to originate from non OECD regions.

Consequently, any uncoordinated action by the EU, or even the OECD, is likely to have a very small impact on long term carbon concentrations. Large, rapidly growing countries, such China, India and some other in the Asian continent give priority to economic development, often accompanied by massive increase in energy consumption. Joint implementation may provide the opportunity to reduce the CO_2 emissions in developing countries, possibly the most important emitters in the future.

The results of the second chapter, included in this part, indicate that there is even more of a difference in the likely costs of emission reductions among world regions than there are among European countries. Thus, there is plenty of scope for an international market approach to the issue, if important practical problems can be resolved. The scope for joint implementation over the period to 2010 is likely to be limited.

The work presented here was concluded before the Kyoto summit. Indeed, the major motivation of the analysis presented in the next two chapters was to assist the debate within the EU by providing quantitative estimates of the impact and cost of a number of policies and measures.

5 European Policies for Meeting the Kyoto Targets[10]

5.1 Introduction

The objective of this chapter is to evaluate, through the use of PRIMES, the economic potential of a number of policies and measures (PAM) for limiting the EU's level of CO_2 by 2010. Consistency analysis and quantified re-evaluation were the aims of the model-based analysis.

The policies and measures analysed here are based on the objective set by the Council of Ministers in March of 1997 which set a target for CO_2 emission reduction by 15% in 2010 compared to 1990 emissions. The decision included also targets per member-state defined following negotiations. The individual targets are summarised in the following table.

[10] The principal authors of this chapter are Pantelis Capros and Leonidas Mantzos of NTUA.

Table 5-1: Emission reduction targets by country.

2010 Emissions	Reduction from 1990
AU	-25%
BE	-10%
DK	-25%
FI	0%
FR	0%
GE	-25%
GR	31%
IR	15%
IT	-7%
LX	-30%
NL	-10%
PO	43%
SP	17%
SW	5%
UK	-10%
EU	-10%

The decision of allocation concerns a 10% reduction for the whole of the EU. An additional 5% is considered to be achieved through common policies and measures, not yet identified.[11]

Also, the decision concerned the basket of greenhouse gases, including methane, N_2O and CFC. It is estimated that it is cost effective to reduce emissions of the non CO2 emissions, up to the level of -3% expressed in terms of CO_2 equivalent. This would leave -12% as a target for CO2 emissions alone.

The Council had been supported by an ad hoc expert group that considered a set of policies and measures (PAM) that are necessary to be engaged for reaching the CO_2 target. The consideration of the PAM package allowed for a definition of targets per sector and by country that have been further considered in the negotiations for the burden sharing among member states.

The proposed PAM package covered the following.

1. Energy efficiency in heavy industry.
2. Equipment (domestic appliances, motors, air compressors, heat pumps, etc.).
3. Cogeneration, district heating.
4. Renewables.
5. Transports (cars, modal).

[11] Following the Kyoto negotiations, the EU reduction target has been set at 8%.

6. Buildings, thermal integrity.
7. Taxation (minimum excise taxes on energy).
8. Removal of subsidies.
9. No other measures for power generation.

The approach has been highly regulatory: through a bottom-up approach, technical and economic potential has been determined by sector and a target has been set. There has been no consideration of the policy instruments that would be necessary to achieve these targets.

To illustrate how challenging is the emission reduction target, a set of recent energy scenarios for the EU, which have been constructed for DG XVII, are in Figure 5-1.

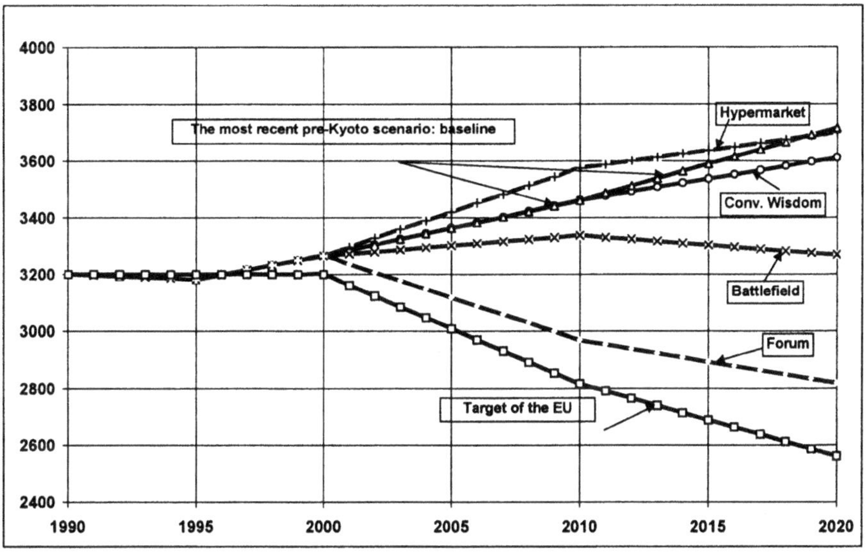

Figure 5-1: CO_2 emissions projections for EU-15.

The figure shows that the Kyoto target (arbitrarily extrapolated beyond 2010) is below the Forum scenario, which has been specially designed to prioritise environmental concerns. The Battlefield scenario shows a stabilisation of emissions, but this is due to economic recession. Hypermarket involves the highest growth of CO_2 emissions, while the recent pre-Kyoto scenario (March 1997, made with MIDAS) projects a trajectory quite close to that of the older

Conventional Wisdom scenario. The gap between the scenarios and the targeted trajectory increases over time, beyond 2010.

5.2 Scenario Assumptions

5.2.1 The KATREN Scenario

A new scenario was constructed, named "Kyoto Advanced Technologies and Renewables" (KATREN). This scenario considered the set of Policies and Measures defined by the EU ad hoc climate group and assumed the imposition of targets by sector as regulation constraints. The scenario does not assume any additional technological options compared to the baseline, as the baseline scenario that serves as reference already includes policies and efficiency gains.

The PAM sectoral targets take the form of sectoral efficiency gains (for example in % of energy use) or obligations for inputs in the output of power and/or steam. These regulatory targets are imposed at the level of a sector, represented in the model as an economic agent that optimises his behaviour (for example minimises his own energy costs). When facing the regulation constraints, regarding efficiency or fuel obligations, the agent can act through a variety of options, as for example he can alter his fuel mix, invest in advanced technology equipment, invest in energy savings, etc. After adjusting his energy structure, the agent addresses a modified energy demand to the rest of the energy system and/or changes his price settings. These alterations act throughout the energy system and provoke secondary effects that end up in the equilibrium of the energy markets. Therefore, the potential of energy efficiency or the of use of carbon-free resources, are affected by the adjustment of markets and may effectively be very different from those estimated in engineering studies, as those supporting the ad hoc expert group on climate change. That kind of consistency is ensured by using a model such as PRIMES. However, because of computational complexity, the market equilibrium is evaluated by the model separately at the country level, ignoring trade effects. Also, as the dynamics of the model follow myopic anticipation, the analysis ignores eventual effects that would be due to intertemporal effects (rational expectations).

As the quantification of the potential of the PAM package was the aim of the model-based analysis, the study had to re-estimate the bottom-up engineering information. This information is introduced in the classification scheme of the model, in the form for example of technical-economic characteristics of vintages of technologies for the future, or non linear cost curves for direct energy savings, etc. The model results depend heavily of course on these technical-economic data, as these define the potential at a micro-scale.

For this purpose, the engineering information in PRIMES has been extensively revised by requiring to a variety of sources, other than the background material used by the ad hoc expert group. These data, further used to attribute numerical values to the regulation constraints (in KATREN), are based on engineering studies (IKARUS, Utrecht's papers, MEGAWATT, Joule-Thermie sectoral projects, SAVE data bases for energy efficiency, and TERES-II for renewables) and consultation of experts.

Only implicit policy measures are considered in KATREN and are interpreted as regulation, agreements, removal of barriers etc. The scenario is formally defined as:

- a set of efficiency constraints by end-use category on top of efficiency gains achieved in the baseline scenario, quantified on the basis of bottom-up engineering studies, considering a set technology progress options, from which a target value is initially selected on the basis of a pay-back period indicator. As mentioned, these act globally on a sector and leave certain degree of freedom for the sectoral adjustment;
- non fossil fuel obligation regulation constraints (on output) promoting the use of renewables, that have been quantified according to TERES II "internalisation of externalities" case (revised to exclude some apparently extreme forecasts). Again, these obligation constraints act globally on the power sector, however separately for centralised power generation and for independent generation) allowing for a flexible adjustment of the sector, including the choice of renewables to use;
- similar obligation constraints, separate constraints for utilities and independent producers, imposing a certain share of cogeneration (CHP) in the total production of steam;
- regulation that imposes high performance cars in transports;
- in the KATREN scenario there is no changes of costs, subsidies and taxes.

To comply with the regulation constraints and obligations, the agents in the KATREN scenario will draw from a technical potential formulated in the model by the existence of explicit advanced technologies, renewables, energy saving measures, etc. The technical-economic characteristics of this technological potential are exactly the same in KATREN as it is in the baseline scenario. In the latter, that potential has been poorly exploited because of relative costs. It is expected that the regulation will oblige the agents to draw more from that technological potential, but the way to do it is left to the market optimisation as simulated by the model.

The only exception for this technological potential is car transports for which KATREN can select an efficient type of car, while this was more limited in the baseline scenario.

In more detail, the KATREN scenario is defined as follows.

Heavy Industry Processes

- Agreement on energy efficiency gains defined by type of process (aluminium, cement, blast furnace, electric arc, etc.). Possibility to act on direct savings accelerated adoption of advanced technologies or fuel mix.

Equipment, Appliances, Cross Cutting Technologies

- Regulation that imposes acceleration of the adoption of more efficient appliances, most of which are electric (in all sectors). This regulation can be interpreted as a result of policies, such as labelling, DSM programmes higher preference on more efficient appliances, etc.

Thermal Integrity of Buildings

- Regulation on new buildings only for households. The dynamics depend on the low capital turnover of buildings.
- Retrofitting allowed only for tertiary buildings.

Car Transports

- Regulation on car efficiency (only 5lt/100km cars will be available after 2002 for new purchases). There is no other choice for consumers when purchasing a new car after 2002.

Cogeneration - CHP

- Obligation to produce more steam from cogeneration plants as a percentage of total steam production. This of course limits the use of boilers.

Renewables

- Higher minimum output share (from TERES-II "internalisation of externalities - revised") in power generation and steam generation per country, without specifying the type of renewable source.

- The scenario does not assume any additional availability of waste energy and biomass, and there is no special program for biofuels.

Power Generation

- No further regulation, other than renewables and CHP. The sector bears indirect effects, only.
- No additional nuclear capacity is allowed (from baseline). The non fossil fuel obligation cannot be satisfied through new nuclear energy capacity. The horizon of 2010 is of course too close from now to allow for a reorientation of nuclear policy in Europe.
- By considering security of supply policies, the KATREN scenario imposes that total natural gas requirements of a country should not exceed the level of natural gas (primary energy) as projected the baseline scenario. As the KATREN scenario involves high effort in electricity conservation, and as the power generation sector is not regulated regarding the fossil fuels, it is unlikely that the natural gas upper limit is binding.

All technologies and their costs are in KATREN as in the other scenarios and the baseline.

5.2.2 The CO_2Tax Scenario

PRIMES has been used also to construct a set of Carbon-Tax-Equivalent scenarios (starting from the baseline) to compute the shadow-marginal cost of the emission reduction target by country.

In each scenario of this type, several runs are made to determine the value of a carbon tax that is necessary to operate throughout the energy system so as to reach a predetermined level of CO_2 emission reduction. Given that PRIMES is a market equilibrium model, the required level of carbon tax is exactly the measurement of the system's marginal abatement cost associated to a reduction level. Therefore, the so-called CO_2Tax scenario is not a taxation scenario, in policy terms, but simply serves to the evaluation of the marginal abatement cost.

It must be also emphasised that the CO_2Tax scenario leads by construction to an optimal allocation of the CO_2 reduction effort. On the contrary, the KATREN scenario, as being a regulation scenario, is sub-optimal. This optimality rule theoretically applies to allocation for both the countries and the sectors. However, in the present CO_2Tax runs, we have limited the optimality only among the sectors of each country, considering the allocation by country as given following the decision of the Council on 3.3.1997. Of course, the country allocation needs not to be optimal. Therefore, the CO_2Tax scenario overestimates the optimal marginal abatement cost, compared to the case that the CO_2 effort would be allowed to be flexible among the countries, as well. The inefficiency of the allocation (regarding the countries) has been estimated to be around 36% in marginal abatement cost terms.

It must be also noticed, that the CO_2Tax scenario affects directly the power generation sector, for which there has been no regulation in the KATREN scenario, other than renewables and CHP. Also it is assumed in the CO_2Tax that there is no additional technology option from the baseline, in terms of technological potential, as it is also the case of the KATREN scenario. The only new potential in the CO_2Tax scenario, is the new efficiency norm for cars (as in KATREN) the imposition of which has been necessary to ensure comparability of results across scenarios.

In the power generation sector some constraints are preserved, as in the KATERN scenario. This concern: the constraint on the expansion of nuclear capacity beyond that of the baseline; and the upper limit on natural gas use (again beyond the level of the baseline). In the CO_2Tax scenario, the natural gas constraint may be binding, as the power generation sector is directly affected by the carbon tax.[12] However, it is expected that there will be electricity conservation in the CO_2Tax scenario as well.

Several runs for a range of levels of CO_2 emission reduction targets allowed the construction of a marginal abatement cost curve.

5.3 KATREN Scenario Results

5.3.1 Overview

The analysis with PRIMES for the KATREN scenario covered 8 EU countries, namely Belgium, Netherlands, France, Germany, Italy, Spain, UK and Sweden, that represent about 90% of the EU-15 primary energy demand. EU-8 represents 89% of the EU-15 emissions of CO_2. The rest of the EU have generally higher potential of renewables but do not use nuclear energy (except Finland).

---

[12] It should be recalled from the results presented in the first part of this volume that in the baseline scenario there is already substantial penetration of natural gas in power generation.

To obtain in EU-15 -10% of CO_2 emissions in 2010 compared to 1990, the EU-8 must reduce by about -11.3%. In our baseline scenario, EU-8 increases CO_2 emissions by 5.7% (from 1990), compared to 8% for EU-15.

As mentioned above, the KATREN is composed of a set of arbitrarily set regulations and obligations per sector or energy use. These have been set following a bottom-up engineering approach and have been selected on economic grounds, compared to the corresponding technical potential. By construction, the set of regulation is different from an optimal allocation of the CO_2 emission reduction effort.

The KATREN scenario, as constructed with PRIMES, reaches -10.6% of CO_2 emissions in 2010 compared to the 1990 level. This does not mean that it is impossible to obtain more emission reduction. It would be sufficient to impose a more severe regulation. As the marginal efficiency of the regulation in reducing emissions increases with the severity of the regulation, it would be needed proportionally higher regulation to reach a more ambitious reduction target.

Compared to the baseline, the KATREN scenario abates in 2010 about 450 Mt of CO_2 for EU-8 (or 120 Mt of carbon, which is about 150 Mtoe of primary energy or -11.4% of primary energy demand of EU-8). The emissions are reduced by -15.4% in 2010 from the baseline scenario.

5.3.2 Primary Energy Demand

As GDP growth remains unchanged from the baseline, the KATREN scenario achieves the reduction of emissions mainly by improving dramatically the energy intensity. In all demand sectors, as in cogeneration and renewables, the KATREN scenario shows substantial gains in energy efficiency. This concerns all energy forms, including electricity the use of which is highly rationalised. To a lesser degree, KATREN reduces emissions also through fuel substitutions. As additional nuclear capacities are restricted, the improvement in the fossil fuel intensity of primary energy demand is due to more penetration of renewables. This improvement is significant, when compared to baseline, but limited in magnitude. The KATREN scenario involves substitutions within fossil fuels as well, as the shares of less carbon intensive fossils increase. The possibilities for further substitution in favour of natural gas, from their baseline level, are limited and fully exploited in KATREN.

In primary energy demand terms, the demand for all energy forms decreases (as total primary energy needs decrease as well), except renewables that increase, representing in 2010 7.8% of total energy instead of 5.4% in the baseline. Natural gas requirements decrease in volume, keeping an almost constant share of 28%. Electricity conservation in final demand (-11% in 2010 from baseline) induces a decrease in the use of all energy forms for power generation, including nuclear energy (-12%), except renewables that increase substantially.

5.3.3 Power and Heat Generation

The increased contribution of renewables is, of course, a result of the non fossil fuel obligation that has been imposed as a regulation in KATREN. However, due to the shrinking of the electricity sector, as a consequence of demand-side savings, the development of renewables is significantly lower than projected both in the work of the ad hoc group on climate change and TERES II (they were both based on an engineering type of approach).

A similar mechanism is observed for the development of cogenerated heat and power. Although there has been a regulation favourable to its deployment in KATREN, the savings of steam and heat requirements in the end-use sectors restrict the potential of CHP. Another reason for this is that investors face an unfavourable environment for exploiting economies of scale (this relates to both the size and the load factors for potential applications of CHP).

Despite these side effects of energy efficiency gains in final demand, investment in renewables and cogeneration are impressive. Additional investment (compared to baseline) in renewables for power generation amounts to 34 GW in EU-8, representing an increase of 72% from baseline. Similarly, the increase of investment in cogeneration corresponds to a doubling of capacities from baseline (25 GW additional capacities).

There is also a structural shift in favour of independent generation that is more important in KATREN than in the baseline scenario. Electricity generation from IPP increases by 20% while centralised generation decreases by -13% in 2010. This is supported by massive investment in gas turbine technologies (mainly equipped with steam injection) in industrial and tertiary applications. The adjustments in the power sector (both centralised and independent) are accompanied by an acceleration of investment and hence premature scrapping of older plants (compared to the baseline). This of course induces higher electricity production costs.

5.3.4 Final Energy Demand

Final energy demand in industry decreases in 2010 by 8.3% from baseline, while industrial activity remains unchanged. This is achieved by direct energy savings, mainly in heat uses, by an acceleration of scrapping in favour of advanced technology equipment (in processes, motor drives, etc.), but mainly by substantially increasing the generation of steam from cogeneration plants (its share is tripled, reaching 12.5% of industrial energy demand). The penetration of cogeneration is detrimental to all fossil fuels, but mainly to natural gas: it is preferable to use more gas in industrial cogeneration than in direct combustion. This mechanism evidently increases the share of self-generation of electricity in industry (by 33%). Despite the direct electricity savings (13% of savings compared to 6.5% in the baseline), the share of electricity in industrial energy use increases slightly, as a consequence of the opportunity cost of self-generating electricity.

The changes in the tertiary sector are different from the above, despite the similarity related to the substantial energy efficiency gains obtained in this sector. In this sector, it seems to be preferable to give priority to electricity demand savings than to rely on distributed heat. This is related of course to relative costs, as distribution costs of steam are high in this sector, but also it is due to the limited opportunity of deploying cogeneration, as the load in this sector is highly seasonal. As a consequence, the sector prefers to save energy directly, including in heat uses and specific electricity. In addition, there is substitution in favour of natural gas. Final energy demand in the tertiary sector decreases by -11.5% from baseline, leading to a compensation of the high-energy growth of the sector as projected in the baseline.

Households act mainly through their preference for advanced technology appliances, resulting in significant savings in the specific electricity uses. Savings in heat uses are rather limited, confirming the difficulties faced by households in seeking energy efficiency gains.

High uncertainty prevails in the analysis for the transportation sector and road transport in particular. The bottom-up engineering studies specified that it is technically possible to commercialise a new type of car (with internal combustion engine) that would be controlled by a computer to reach a specific average consumption of 5 lt per 100 km. Such a car, if its production were generalised, would induce a moderate increase of the car purchase price (15%) and would achieve as much as 40% savings in the energy bill for the consumer. At normal discount rates, it is easy to calculate that the annualised cost of transportation for the consumer would be lower than that of the conventional car. Therefore, the consumers should directly choose the new car when replacing older cars. However, one should also consider the welfare costs for the consumer, as the car being controlled by the computer does not permit fast driving etc. Given that such costs are not accounted for in the model, the simulations would entail massive adoption of the new car even in the baseline. One would not need a regulation to impose that car if the analysis is confined to engineering calculations. If we believe that this would not be the case in reality, then a very high discount rate should be used in the transport sub-model of PRIMES. Given that the auto-oil programme of the EC was considering a regulation imposing that new car type in new purchases, we decided to include such a regulation only in the KATREN scenario (and in the CO_2Tax scenario) and not in the baseline. This is equivalent to endowing the policy scenarios with an additional cost-effective technological potential, leading to negative costs in complying with environmental constraints. We will further discuss this issue in evaluating the global cost of the CO_2 policy.

In the presence of such regulation (new cars only available for new purchases after 2002), the model projects a substantial decrease of final energy demand in transports (-17.5% in 2010 from baseline). The average car efficiency improves by 20% in 2010. Small changes occur in the other transport means and there are little effects from inter-modal transport combinations (this related also to the fact that PRIMES has ran separately by country), as these are not regulated.

5.3.5 CO_2 Emissions

As mentioned above, the reduction of CO_2 emissions in KATREN is mainly a result of the combined effect of energy efficiency gains, mainly in final demand and through cogeneration. Substitutions in favour of renewables and some limited substitutions in favour of natural gas also contribute to the reduction in emissions. The results on CO_2 reductions in the KATREN scenario are summarised in Table 5-2 and Table 5-3.

Table 5-2: Key results of the KATREN scenario.

KATREN - Annual rates of Change	1995-2010
Overall Energy Intensity	-2.30
Fossil Fuel Intensity of GIC	-0.16
Carbon Intensity of Fossil Fuels	-0.53
CO_2 Emissions per Unit of GDP	-2.97

Table 5-2 shows, in annualised growth terms, a decomposition of CO_2 reduction in KATREN (as a ratio to GDP growth) in three components: energy intensity, fossil fuel intensity of primary energy demand and carbon intensity of fossil fuels.

Table 5-3 summarises the findings of KATREN regarding CO_2 emission reduction. All sectors reduce emissions, except independent power and steam production. This of course allows for higher carbon efficiency gain in final energy demand. The transportation sector contributes substantially, but this is related to the rather uncertain assumptions about the regulation in that sector. Industry also performs impressively, while tertiary and households seem to have lower possibilities. The decrease of emissions from centralised utilities must be mainly attributed to a substitution effect, as emissions and activity in independent production increases, and, to a lesser degree, to the increasing share of renewables.

Table 5-3: Sectoral emissions, comparisons of KATREN scenario with baseline.

In 2010 for EU-8					
CO_2 Emissions in Mtn	**Baseline**	**Shares**	**KATREN**	**Shares**	**% Diff. From Baseline**
Industry	466.6	15.5%	372.8	14.7%	-20.1%
Tertiary	194.8	6.5%	177.0	7.0%	-9.1%
Residential	396.9	13.2%	365.3	14.4%	-8.0%
Transports	899.4	29.9%	738.4	29.0%	-17.9%
Central Power	716.2	23.8%	545.5	21.5%	-23.8%
Independent CHP	192.8	6.4%	221.1	8.7%	14.7%
Other	139.1	4.6%	121.9	4.8%	-12.4%
TOTAL Emissions	**3005.8**		**2542**		**-15.4%**

Table 5-4 shows a decomposition of CO_2 emission reduction by type of PAM (policies and measures) classified according to the agenda of the ad hoc group on climate change of the Council. This decomposition needs a complex calculation on the numerical results of the model and involves certain assumptions. The table is comparable to a similar table with which the ad hoc-group's document concludes. The differences must be attributed to two reasons: first to the differences in the numerical estimates of the bottom up engineering information; secondly to the indirect effects induced by the market equilibrium at the level of the whole energy system, as the ad hoc group did not check energy system consistency, contrary to PRIMES analysis. For the latter effects, the contributions of renewables and CHP are illustrative: they differ substantially from the figures of the ad hoc group, the difference being attributed to the consequences of market and system effects, as explained in previous sections.

The decomposition table shows a high contribution of transports. This again emphasises the importance of conducting additional in depth studies for the transports sector that are necessary to reduce uncertainty about the potential in this sector. It is evident that if this potential is smaller than that of the above table, then there will be important difficulties in reaching the CO_2 emission reduction target, as the car regulation is cost-effective in engineering cost terms, as mentioned previously.

Table 5-4: Contribution of policy and measures, KATREN scenario.

Contribution of Policies and Measures		KATREN	
2010		Mtn of CO_2	Shares in %
Process and Heat	Heavy Industry	68.5	14.8
Appliances	Tertiary	28.5	6.2
	Residential	47.2	10.2
Cars	Transports	161.1	34.8
Buildings	Domestic Sector	37.6	8.1
Steam and Power	CHP	49.1	10.6
Renewables	Intermittant	38.9	8.4
	Biomass, waste	43.0	9.3
Other effects	Power generation	4.5	1.0
Other Indirect	(not distributed)	-15.0	-3.2
TOTAL		**463.4**	**100.0**
Target in % reduction from 1990	-11.30%		
Diff. Of Baseline form target in Mtn of CO_2	488.8		
Additional effort required in Mtn	25.4		

The contribution of CHP and renewables is also impressive, ranking second after transports. The rest of emission reduction is attributed to direct energy efficiency gains in the end-use of energy, equally distributed between industrial processes and domestic appliances.

The KATREN scenario shows an economically and technically feasible package of standards and regulation that serves to approach the emission reduction target. Of course, additional sensitivity analysis is required to refine the regulation assumptions and their allocation to sectors and energy uses. The scenario also revealed important side effects, regarding CHP and renewables in particular, that cannot be neglected when considering the energy markets and their interactions.

5.4 The CO_2Tax Scenario

5.4.1 Overview

As mentioned above, the purpose of building a CO_2Tax scenario was to estimate the marginal abatement cost of CO_2 emission reduction and then construct a marginal abatement cost curve for a range of emission reduction targets. Mathematically the imposition of the carbon tax is exactly equivalent to the imposition of a global (to the whole energy system) emission reduction constraint, the shadow price of which is exactly equal to the level of carbon taxation that is necessary to meet the emission reduction. In the following we use the term "carbon tax" to denote the shadow price of this global emission constraint.

The CO_2Tax scenario is optimal, in the sense of the allocation of the emission reduction effort by sector and energy use. To this respect, it is useful to compare with the allocation in the regulation scenario (KATREN), as this would reveal inefficiencies of the regulation. The two scenarios are in fact comparable as they include exactly the same technological potential in all sectors. For this reason, the car regulation in transports has been also included in the CO_2Tax scenario. The CO_2Tax scenario involves carbon taxation to the inputs of power generation as well, while the KATREN scenario did not include any regulation for power generation except the obligations for renewables and cogenerated steam. This difference will have important consequences for the results.

The CO_2Tax scenario presented below has computed a level of carbon tax that is necessary to achieve -12.4% of emissions in EU-8 in 2010 compared to the level of 1990. This reduction in EU-8 is approximately equivalent to a reduction close to -15% (from 1990) in the whole EU-15, and corresponds for EU-8 to -17.2% fewer emissions from baseline. Obviously the CO_2Tax scenario corresponds to a higher effort in emission reduction, than KATREN, as the latter achieved for EU-8 -10.6% reduction in 2010 from 1990.

To understand the results of the CO_2Tax scenario, one has to recall that in this scenario it is possible to act more in the power generation sector, than KATREN. Thus, certain expensive actions on the demand-side, which in the KATREN scenario were preferable to cheaper actions in the power generation, can be avoided. Actions in this sector (such as for example inter fuel substitutions) are only restricted by a global upper bound on the use of natural gas (that should not exceed the level of the baseline scenario) and the impossibility of investing in new nuclear capacities, beyond the level of the baseline scenario. There are also certain restrictions that characterise the inertia of the sector, as for example fuel contracts that hold for the use of indigenous resources, such as domestic coal in Germany. Despite these restrictions, the results show that the adaptation flexibility of the power sector is high and that neither the global natural gas constraint nor the nuclear investment restriction is binding. This is related to the fact that in the policy scenarios there is an important conservation of electricity in demand (that shrinks the volume of the sector, hence the needs for fuels and new investment), while the planning horizon is limited to medium term (2010) that is small compared to the lag of nuclear decisions.

Consequently, the CO_2Tax scenario achieves lower energy efficiency gains in final demand than KATREN, but involves higher restructuring in power generation regarding in particular the fuel mix of centralised utilities.

5.4.2 Primary Energy Demand

In primary energy demand terms (gross inland consumption), the CO_2Tax scenario projects lower energy requirements, by -9.6% in 2010 from baseline, compared to -11.4% in KATREN.

The fuel mix in primary energy demand differs as well. The CO_2Tax scenario projects a dramatic decline of the use of solid fuels (-36% in 2010 compared to baseline, while this reduction is -19% in KATREN). It projects also the use of natural gas and nuclear to be higher than that of KATREN, even though both are used at levels lower than those of the baseline in 2010 (-4.7% for natural gas and -1.7% for nuclear). The primary demand for liquid fuels remains in CO_2Tax at a level close to that of KATREN, but the use of renewables increases significantly in volume (7% more in CO_2Tax compared to KATREN, in 2010) and as a share of total gross inland consumption (8.1% in CO_2Tax).

5.4.3 Power and Heat Generation

The changes in primary energy demand are also induced by the results for the power generation sector. As in CO_2Tax scenario there is less electricity saving than in KATREN, power generation decreases in 2010 only by -3.7% from baseline (-11.1 % in KATREN). Generation from nuclear energy is maintained at almost the level of the baseline, but renewables contribute less than in the KATREN scenario (intermittent sources in particular).

Substantial changes, compared to KATREN and the baseline, occur in the fuel mix of centralised thermal power generation. Natural gas represents almost 50% of thermal generation in 2010 (35% in KATREN), while the share of solid fuels declines up to 32% (45% in KATREN). Biomass and waste increase also in CO_2Tax, but at a lesser degree.

To achieve this restructuring, the CO_2Tax scenario projects higher investment in centralised power generation than the baseline (and of course KATREN). Given that electricity production decreases, higher investment indicates that important premature replacement of plants takes place. Additional investment mainly concerns gas combined cycle and clean coal technologies.

The CO_2Tax scenario involves also lower development, compared to KATREN, of the independent power and steam production activities. The shifts in favour of independent production, which are observed in KATREN, are much less important in the carbon tax scenario. In this scenario, independent power generation sticks in 2010 at the level of the baseline scenario, while in KATREN it has increased by 20%. Steam production from CHP is also stable from baseline, and much lower than in KATREN. Boilers preserve their share, as in the baseline, but they change their fuel mix. However, independent renewable plants double from KATREN, but they remain small in magnitude. This evolution in CO_2Tax scenario indicates that the choice of policy, that is regulation or taxation, may have important consequences on the structure of the power generation industry. Regulation induced high restructuring of the industry, in favour of decentralised production, while carbon taxation (or in other terms the global optimality condition) induced no further decentralisation (from baseline) but important restructuring regarding the fuel mix in centralised power generation.

These results can be explained in economic optimality terms. The CO_2Tax scenario induces higher operation costs for power generation, but the centralised generators have the advantage of exploiting economies of scale and the possibility to rely more on carbon free resources. In these circumstances, the increase in the fuel bill partly compensates the gains from the production of steam by decentralised generators, an effect that undermines their competitiveness. In such a case, that is within the CO_2Tax context, it seems that there is more opportunity to invest in direct savings in heat uses (as it is shown in the results for industry) than to save through the CHP, as decentralised power generation is less competitive (because of costs and economies of scale).

Despite these changes, the CO_2Tax scenario is, regarding power and steam generation (aggregating centralised and decentralised), more efficient for CO_2 emissions than KATREN. Therefore, when CO_2 is considered, the side effects on CHP promotion induced in the presence of optimality conditions of CO_2Tax scenario should not be considered negatively, in the perspective of the whole energy system.

5.4.4 Final Energy Demand

The CO_2Tax scenario indicates that, to reach the global CO_2 target, the industrial sectors rely more on direct energy efficiency gains in the end-use. Compared to KATREN, the industrial sectors invest more on advanced technologies and saving measures and less on distributed heat and cogeneration. The sectors prefer to use fossil fuels directly in combustion, rather than through cogenerated steam. Regarding fuel mix, the industrial sectors rely more on natural gas and electricity and less on solid fuels, compared to baseline. To this respect this was also the case in the regulation scenario.

Similar trends are observed for the tertiary sector, regarding cogeneration, that was however already low in KATREN as well. In the CO_2Tax scenario, the global optimality allocates an emission reduction effort to tertiary sector, significantly lower than in KATREN. Energy demand is reduced only by -2.8% in 2010 (from baseline), compared to -11.5% in KATREN. This allows for significantly higher electricity demand in this sector, while demand for gas changes less.

Energy demand by households also decreases less in CO_2Tax than in KATREN (2.8% instead of -8.4%). Again this allows for higher electricity demand.

The changes in final energy demand should of course be considered in relation with the changes in power generation. The additional CO_2 emission reduction gains permitted in power generation under the CO_2Tax scenario, allowed for a relaxation of constraints in the final energy demand sectors, especially in tertiary and households where the compliance costs are higher. This further allowed for higher electricity demand, that provided opportunities for economies of scale in the centralised power sector, a factor that if combined with the higher fuel bill of generators, lead to cancelling out the further deployment (as it was the case in KATREN) of an independent power and steam generation sector.

5.4.5 CO_2 Emissions

The allocation of emission reduction effort by sector, under the optimality condition of CO_2Tax scenario, is further illustrated in Table 5-5 that decomposes emissions by sector. The table shows lower contribution of industry, tertiary and households and higher contribution of the power generation sector, if compared to the KATREN sectoral allocation of emission reduction. In the CO_2Tax scenario, the independent power generation sector also reduces emissions, as structural substitutions in the power sector are less important than in KATREN.

Table 5-5: Sectoral emissions, comparisons of CO_2 scenario with baseline.

In 2010 for EU-8					
CO_2 Emissions in Mtn	**Baseline**	**Shares**	**CO_2 Tax**	**Shares**	**% Diff. From Baseline**
Industry	466.6	15.5%	407.0	16.0%	-12.8%
Tertiary	194.8	6.5%	184.0	7.2%	-5.5%
Residential	396.9	13.2%	379.6	14.9%	-4.4%
Transports	899.4	29.9%	726.3	28.6%	-19.2%
Central Power	716.2	23.8%	509.3	20.0%	-28.9%
Independent CHP	192.8	6.4%	160.3	6.3%	-16.9%
Other	139.1	4.6%	122.4	4.8%	-12.0%
TOTAL Emissions	**3005.8**		**2488.9**		**-17.2%**

The emission reduction can be further decomposed in components that separate energy efficiency gains from substitutions. This is illustrated Table 5-6 that shows that energy efficiency gains are again mainly responsible for emission reduction. Compared to KATREN, this table also shows a more significant contribution of substitution in favour of carbon free energy forms and a further substitution within fossil fuels, in favour of less carbon intensive fuels.

Table 5-6: Key results of CO_2 scenario.

KATREN - Annual rates of Change	1995-2010
Overall Energy Intensity	-2.16
Fossil Fuel Intensity of GIC	-0.32
Carbon Intensity of Fossil Fuels	-0.65
CO_2 Emissions per Unit of GDP	-3.11

5.5 Comparison of Scenarios

The tables and graphics that follow further illustrate the comparison of the KATREN to the CO_2Tax scenario.

Table 5-7 shows that the restructuring in favour of independent generator and CHP does not takes place in CO_2Tax, contrary to KATREN, while renewable energy is developed more in the CO_2Tax context.

Table 5-7: Scenario comparisons.

Indicators for 2010 EU-8	Baseline	KATREN	% Diff. From Base	CO_2 Tax	% Diff. From Base
Market Shares					
electricity	20.1%	20.6%		21.6%	
natural gas	24.5%	24.3%		26.0%	
Electricity Production					
from central utilities	92.8%	90.4%		93.0%	
total generation in TWh	2417.67	2149.75	-11.1%	2329.06	-3.7%
Steam Production					
from CHP	25.5%	66.3%		26.3%	
total production in TWh	577.34	546.06	-5.4%	490.23	-15.1%
Renewables					
in primary energy in Mtoe	73.9	93.4	26.4%	99.6	34.8%
as a % of gross inland cons.	5.4%	7.8%		8.1%	

Table 5-8 further illustrates the findings regarding the structure of the power generation sector. It shows also that the CO_2Tax scenario is more efficient than KATREN in the power industry. The same table summarises the changes in energy intensity ratios by sector. It shows that tertiary and households improve less in CO_2Tax than in KATREN.

Table 5-8: 2010 Indicators, scenario comparisons.

Indicators for 2010 EU-8	Baseline	KATREN	CO_2 Tax
Central Utilities			
average use of equipment	53.0%	49.8%	49.5%
investment over Capital Ratio	3.6%	0.7%	6.4%
Independent Producers			
average use of equipment	36.3%	39.8%	33.5%
investment over capital ratio	3.3%	7.6%	-4.3%
Indicators in End-Use			
productivity ratio	10.6%	9.0%	9.4%
investment over capital ratio	7.7%	8.0%	7.8%
Energy Intensity Ratios (in annual growth rates for 1995-2010)			
industry	-1.3%	-1.9%	-2.0%
tertiary	-1.5%	-2.3%	-1.7%
households	-1.8%	-2.4%	-2.0%
passenger transports	-0.7%	-1.9%	-1.9%
goods transports	-0.8%	-1.2%	-1.2%
Overall Intensity Ratio	-1.5%	-2.3%	-2.2%

Table 5-9 shows a decomposition of the change in CO_2 emissions from 1990 to 2010 in the context of the three scenarios. The table reveals that the CO_2Tax case acts more than KATREN on reducing carbon intensity of the fossil fuel aggregate and the intensity of total primary energy demand in fossil fuels. They both act mainly on energy efficiency, providing impressive improvement of the overall energy intensity.

Table 5-9: Key results across scenarios.

1990-2010, EU-8	Baseline	KATREN	CO_2 TAX
Economic Growth	42.8%	42.8%	42.8%
Global Energy Intensity	-26.0%	-38.1%	-36.0%
Fossil Fuel Intensity of GIC	-0.8%	-3.5%	-5.8%
Carbon Intensity of Fossil Fuels	-10.4%	-12.4%	-14.2%
Total Emission Change	**5.6%**	**-11.2%**	**-13.3%**
% changes in logarithmic terms			

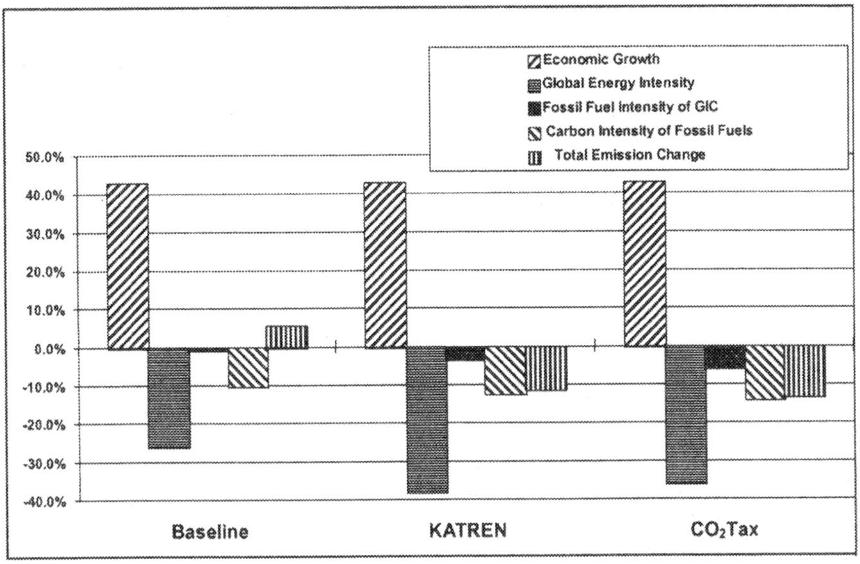

Figure 5-2: Changes between 1990 and 2010, scenario comparisons.

Table 5-10: Contributions of policies and measures.

Contribution of Policies and Measures 2010, EU-8		KATREN		CO_2 TAX	
		Mtn of CO_2	%	Mtn of CO_2	%
Process and Heat	Heavy Industry	68.5	14.8	14.8	17.1
Appliances	Tertiary	28.5	6.2	6.2	0.8
	Residential	47.2	10.2	10.2	4.9
Cars	Transports	161.1	34.8	34.8	33.2
Buildings	Domestic Sector	37.6	8.1	8.1	1.5
Steam and Power	CHP	49.1	10.6	10.6	-0.8
Renewables	Intermittant	38.9	8.4	8.4	6.4
	Biomass, waste	43.0	9.3	9.3	11.6
Other effects	Power generation	4.5	1.0	1.0	27.0
Other Indirect	(not distributed)	-15.0	-3.2	-3.2	-1.7
TOTAL		**463.4**	**100.0**	**516.7**	**100.0**
Target in % reduction from 1990		-11.30%			
Diff. Of Baseline form target in Mtn of CO_2		488.8			
Additional effort required in Mtn		25.4		-27.9	

The decomposition of emission reduction by type of policies and measures (PAM package) differs across the scenarios. Of course, in the CO_2Tax scenario there is no PAM package, as in the KATREN scenario. However, Table 5-10 decomposes emission reduction by PAM type in order to further illustrates the comparison of scenarios.

The table shows how important becomes the contribution of the power generation sector within the context of the CO_2Tax scenario. The importance of transports, and the resulting uncertainty prevails in both scenarios. The role of renewables is confirmed in both scenarios. Expensive effort in the end-use sectors, in buildings in particular, are avoided in CO_2Tax. In this scenario, the decline of CHP contribution is noticeable. Table 5-11 shows the same decomposition separately for each member-state (EU-8)

Table 5-11: Sectoral decomposition by country.

PRIMES, EU-8, 1990-2010	Belgium		Germany		France		Italy	
	KATREN	CO_2 TAX	KATREN	CO_2 TAX	KATREN	CO_2 TAX	KATREN	CO_2 TAX
Heavy industry	21.7%	23.0%	14.1%	17.0%	15.3%	12.5%	10.5%	25.9%
Appliances	11.6%	7.4%	21.1%	8.8%	7.3%	3.8%	12.6%	3.1%
Buildings	7.7%	1.8%	8.2%	1.9%	10.9%	0.8%	6.9%	0.3%
Transports	27.1%	23.0%	29.9%	27.0%	43.7%	54.3%	29.7%	26.4%
Renewables	14.0%	8.1%	12.4%	9.0%	8.1%	11.9%	32.6%	23.3%
Steam and Power	6.8%	-1.4%	14.3%	-0.9%	12.5%	-1.8%	8.6%	-0.5%
Power Generation and others	11.1%	38.2%	0.0%	37.4%	2.1%	18.4%	-0.8%	21.5%
TOTAL of CO_2 abated in Mtn	**20.7**	**28.3**	**148.6**	**169**	**60.6**	**49.5**	**75.5**	**93.1**
as % of Baseline	16.5%	22.5%	16.5%	18.8%	15.0%	12.3%	16.2%	20.0%
as % of target at -10% for the EU	74.0%	101.2%	92.0%	104.7%	155.0%	126.6%	78.1%	96.3%
Remaining effort in Mtn CO_2	**7.3**	**-0.3**	**12.9**	**-7.5**	**-21.5**	**-10.4**	**21.2**	**3.6**

PRIMES, EU-8, 1990-2010	Netherlands		Spain		Sweden		UK	
	KATREN	CO_2 TAX	KATREN	CO_2 TAX	KATREN	CO_2 TAX	KATREN	CO_2 TAX
Heavy industry	20.5%	21.5%	18.0%	7.8%	20.3%	13.8%	13.1%	9.7%
Appliances	22.0%	8.7%	13.2%	1.9%	0.0%	6.9%	22.6%	2.5%
Buildings	7.3%	5.4%	2.7%	0.2%	9.8%	3.8%	10.6%	0.2%
Transports	21.2%	17.7%	36.7%	40.2%	45.1%	38.8%	46.7%	53.1%
Renewables	18.9%	35.4%	30.7%	39.5%	25.6%	35.0%	11.4%	10.0%
Steam and Power	8.1%	-0.8%	5.4%	0.0%	3.0%	-0.6%	10.2%	-0.6%
Power Generation and others	1.9%	12.1%	-6.7%	10.4%	-3.8%	2.5%	-14.5%	25.2%
TOTAL of CO_2 abated in Mtn	**25.9**	**49.7**	**47.9**	**46.3**	**13.3**	**16**	**70.9**	**64.8**
as % of Baseline	13.9%	26.7%	17.0%	16.4%	19.1%	22.9%	12.3%	11.2%
as % of target at -10% for the EU	52.9%	101.4%	114.3%	110.5%	90.0%	108.3%	122.6%	112.0%
Remaining effort in Mtn CO_2	**23.1**	**-0.7**	**-6.0**	**-4.4**	**1.5**	**-1.2**	**-13.1**	**-7.0**

5.6 Extrapolation to EU-15

Table 5-12 summarises this extrapolation for the KATREN scenario only.

Table 5-12: Extrapolation of the KATREN scenario for EU-15.

In 2010 for EU-15	Baseline	KATREN	% Diff.
CO_2 Emissions in Mtn	3490	2944	-0.15645
CO_2 Emissions as % change from 1990	0.095	-0.076	
Energy Intensity gain compared to baseline	100	88.8	
Contribution of Renewables as % of GIC	0.065	0.094	
Contribution of Policies and Measures			
heavy industry		14.8%	
appliances		14.9%	
buildings		75.0%	
transports		35.0%	
renewables		19.6%	
steam and power		9.9%	
power generation and others		-1.7%	
		100.0%	
Total of CO_2 abated in Mtn		546.0	
Necessary to meet -15% target (-12% for CO_2)		674.0	

The extrapolation from EU-8 to EU-15 is difficult as the member states not covered in the present analysis with PRIMES have various specific features, related for example to higher renewables, the absence of nuclear, etc. However, these member states represent only 10-15 % of the whole EU.

The extrapolation has used certain ratios and results from MIDAS model-based scenarios, but could not avoid some arbitrary assumptions.

5.7 Evaluation of Costs

5.7.1 Definitions and Assumptions

The objective of this section is to propose an evaluation of the energy system costs implied by the CO_2 emission reduction target for 2010 and for EU-8. Costs may be conceived either in average cost terms, or in marginal cost terms.

The average cost involves an estimation of the annualised total energy system cost, which results from satisfying the primary energy requirements of the consumers. If such a cost is evaluated for the baseline scenario and for a policy scenario, then the difference would indicate the additional (positive or negative) total energy system cost of the policy involved in the policy scenario. By dividing this additional total cost by the amount of CO_2 emissions abated in the policy scenario (on the same year for which the annualised cost is evaluated), one can compute a proxy to the average system cost (it is a proxy as the calculation is not exact in intertemporal terms).

The marginal cost, associated to the policy scenario, is defined in the usual manner: it is equal to the average cost of the last unit of CO_2 abated, corresponding to the cost of the more expensive measure taken in the policy scenario, evaluated at the margin. It has been mentioned, that given the market equilibrium formulation of PRIMES, the marginal cost is computed as the shadow price of a global emission constraint the allocation of which as an effort attributed to each sector is an endogenous result of the model, found through market equilibrium. This shadow price of the global constraint acts as if a pure carbon tax is imposed to the use of all fossil fuels (in the form of excise tax proportionate to their emissions of carbon dioxide). It must be noted that the computation of the marginal cost of the policy does not involve the determination of any total or partial energy cost.

For the calculation of the average cost of the policy, methodological difficulties arise. These are mainly due to the overall structure of PRIMES, as this is a market equilibrium model (simulating the simultaneous balance of demand and supply in all markets) and not a global optimisation model (as for example EFOM used in the Crash studies, or MARKAL used in the ETSAP program of IEA). The global optimisation models use by construction a single global objective function, seeking for its minimisation, defined at the level of the whole energy system and including costs of primary energy and all other operation and capacity costs, excluding of course fuel purchasing costs by the end-users. On the contrary PRIMES evaluates total energy costs at the level of each economic agent (industry, tertiary, households, transports, centralised power generation, etc.), where the model accounts for operation costs, final energy fuels purchase costs and capacity costs. To evaluate a global energy system cost, PRIMES has to aggregate over the energy sector.

This aggregation is problematic, given the heterogeneity of definitions of cost elements per sector. For example, the transport sub-model considers the total cost of all vehicles, which of course are not used for energy purposes but for transportation. Similar is the case in appliances and buildings in the domestic sector. On the contrary, the model does not need to consider the value of the whole capital stock used in industry for production purposes, but limits the evaluation to capital necessary to use and convert energy only. So the capital cost differs by sector, in PRIMES, as it is difficult to separate energy-related capital to capital providing non energy services.

An additional problem, in the aggregation of sectoral total costs, relates to market equilibrium and the consideration of final energy forms in the sectoral costs. If we consider that all costs faced by the energy supply sectors are reflected in the selling prices of fuels, then the aggregation has to be made only for the end-use sectors, to avoid double counting. This presupposes however, that there are no temporary changes in the rates of profit of supply sectors and no change in the sectoral structure of their tariffs, as for example would be the case if cross-subsidisation policy is followed. Although in PRIMES change in profit rates and change in tariff structure are possible, in the present scenarios we blocked this mechanism. Thus, the supply sectors use constant mark-ups and therefore we can aggregate energy costs only over the final consumers to approximate the concept of the total energy system cost.

An additional problem relates to the computation of annualizing. As the policy scenarios generally involve higher capital cost (higher investment) that serve to reduce operation costs (due to lower fuel costs because of higher energy efficiency), the way to compute the annualised cost is very important. This involves discounting.

The PRIMES model uses different discount rates by sector, so as to correctly simulate the behaviour of the economic agents. If one wants to use a uniform discount, say 8%, this would also alter the baseline scenario. A set of model computations of costs is presented below, including value of capital stock (at replacement prices), annuity for capital, investment requirements, operating costs, fuel costs and tax payments. These elements are computed for industry, tertiary, households, transports and power generation (centralised, as the decentralised generation is included in the demand sectors). To estimate total costs at 8% discount, we take the cost elements from the results of the model and we re-aggregate them discounting with 8%. Of course this is done ex-post without affecting the results of the model.

Generally the model's discount rates are higher than 8% (it is 8% only in centralised power generation), ranging from 12% in decentralised power decisions to 25 % in small household consumers. We consider that these discount rates correspond to reality, as the aim is to mimic the real behaviour of consumers or suppliers. The use of a uniform discount rate of say 8% in the results only (without any effect on energy results) can be understood as a way to appreciate the policy results from the perspective of the society.

5.8 Average Cost Computations

The following table summarises the cost evaluation.

Table 5-13: Cost evaluation of scenarios.

Averaged net cost per tn of carbon avoided in ECU'90/tnC	**KATREN**	**CO_2TAX**
Industry	161	197
Tertiary	79	70
Households	268	384
Transports	-56	-283
Total without transports	173	221
TOTAL	87	39
Average net cost at 8% uniform discount rate		
Total without transports	96	70
TOTAL	48	12
Marginal cost in ECU'90/tnC		297

5.8.1 Model's Discount Rate

In this case we aggregate the cost elements per sector according to the values of the discount rates as used in the model in simulating their behaviour.

The average cost is conceived at the whole energy system, aggregating the costs for the final consumers. Because of the particular nature of the regulation in car transports (see previous sections), where the total differential cost of regulation is negative, we provide cost estimates both for the total and for the consumers other than transports.

The highest cost case is when considering the model's discount rates, the (sub-optimal) PAM package (KATREN scenario) that excludes power generation and the allocation by country as defined in 3.3.1997, which also is sub-optimal in cost-engineering terms. In this case, by excludes the transport sector the average cost would be 190 ECU'90/tn Carbon (or 52 ECU'90/tn of CO_2[13], requiring 50 billion ECU of 1997, representing 0.67% of 2010 GDP). If we include the transport sector, the average cost is about 90 ECU'90/tn Carbon leading to a cost of 22 billion ECU of today (0.3% of 2010 GDP).

[13] We extrapolate this cost to all EU and the whole energy sector.

If the effort by sector is optimally allocated, including power generation, as it is the case of the CO_2Tax scenario, the average cost decreases substantially. It is estimated to be (at model's discount rates) as low as 39 ECU'90/tn Carbon (or 10 ECU'90/tn CO_2, representing 9.2 billion ECU, or 0.12% of 2010 GDP), when including transports. If we exclude transports, average costs are important: 221 ECU'90/tn Carbon (or 58 ECU'90/tn CO_2, representing 55 billion ECU, or 0.7% of 2010 GDP).

The average costs faced by the final consumers, other than transports are significant. Depending on the optimality of effort allocation by sector, industry bears a cost ranging between 161 and 197 ECU'90/tn Carbon, tertiary between 70 and 79 ECU'90/tn Carbon, and households between 268 and 384 ECU'90/tn Carbon. In this calculations all taxes are excluded.

5.8.2 At 8% Discount Rate

At 8% discount rate, applied ex-post to the results of the model (without altering the scenario results), average costs decrease substantially.

The PAM package (KATREN scenario), including transports, represents an average cost of 48 ECU'90/tn Carbon (instead of 90) and 96 ECU'90/tn Carbon if transports are excluded.

The optimal allocation by sector, as in CO_2Tax scenario, leads to an average cost of 12 ECU'90/tn Carbon (instead of 38) or 70 ECU'90/tn Carbon when transports are excluded.

5.9 Marginal Cost Computations

The results show the following:

- the marginal abatement costs are significantly higher than the average costs. In case of optimal allocation by sector, which is for the CO_2Tax scenario, marginal costs are 297 ECU'90/tn Carbon (-11.3% of CO_2 in EU-8), compared to average cost of 38 ECU'90/tn Carbon;
- the distribution of marginal costs per country differs substantially. This is of course related to the pre-defined allocation of the emission reduction effort, as it follows the decision of 3.3.97 which is a result of negotiation involving consideration of several factors on top of cost-engineering. In case of optimal allocation of effort by country, as in a different CO_2Tax scenario not presented in this paper, the marginal abatement cost is reduced to 190 ECU'90/tn of Carbon (36% more efficient);

- the marginal abatement cost curve (this is estimated through some model runs and extrapolations) is steeply increasing, especially beyond -8 to -9% of CO_2 emissions in 2010 (compared to 1990). The average cost curve (at 8% ex-post discount rate) is increasing smoothly, as in the attached figure.

Table 5-14: Marginal costs by country.

Marginal Costs in case of allocation follwing the decision of 3.3.97	**ECU'90 per tn of Carbon avoided**
Belgium	413
France	149
Germany	317
Italy	303
Netherlands	526
Spain	180
Sweden	291
United Kingdom	213
Total EU-8	297

	Baseline	**KATREN**	**% diff.**	**CO_2 Tax**	**% diff.**
Annual Investment Requirements *(demand & power, in billion ECU'90)*	686	762	11.1	733	6.9
Annuity of Capital *(demand only, in billion ECU'90)*	1719	1785	3.8	1739	1.2
Annual Net of Taxes Operation Costs *(demand only, in billion ECU'90)*	591	548	-7.2	576	-2.5
Annual Net of Taxes Operation Costs *(power sector, in billion ECU'90)*	48	41	-15.9	47	-2.9
Total Tax Revenue, in billion ECU'90	198	164	-17.2	327	65.5
as % of fuel bill	34.3	32.6		56.8	

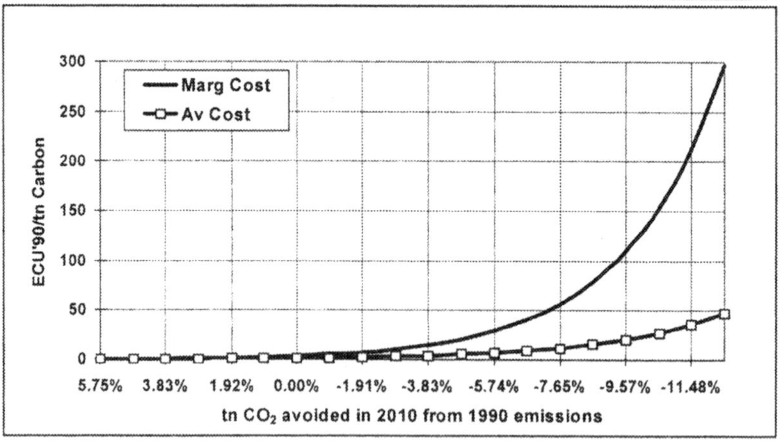

Figure 5-3: CO_2 abatement cost curves, EU-8.

5.10 Other Economic Effects

The following table shows two noticeable effects of the policy: investment requirements increase substantially; in the KATREN scenario the energy tax basis declines. These effects may have significant macroeconomic consequences.

Table 5-15: Financial implications.

	Baseline	KATREN	% Diff.	CO2 TAX	% Diff.
Annual Investement Requiremens					
(demand and power, in billion ECU'90)	686	762	11.1	733	6.9
Annuity of Capital					
(demand only, in billion ECU'90)	1719	1785	3.8	1739	1.2
Annual Net of Taxes Operation Costs					
(demand only, in billion ECU'90)	591	548	-7.2	576	-2.5
Annual Net of Taxes Operation Costs					
power sector, in billion ECU'90)	48	41		47	-2.9
Total tax revenues, in billion ECU'90	198	164	-15.9	327	65.5
as % of fuel bill	34.3	32.6	-17.2	56.8	

6 International Dimensions[14]

6.1 Introduction

This chapter uses the POLES model in order to attempt an examination of the comparative costs of emission reductions in countries and regions outside the EU. For comparison purposes, the POLES results for the EU, which is examined in greater detail in the previous chapter, are also included. As in the previous chapter, the methodology selected was to determine the marginal and total costs incurred in achieving the reduction targets by introducing a "shadow carbon-tax" in the various countries and regions identified in the model.

The model is used to construct marginal cost curves for emission reductions and then to clarify the positions of the various Annex II parties in the negotiations. An analysis then follows of the various possible mechanisms for reaching the reduction targets.

One of the conclusions is that the currently proposed "flat rate" emissions reduction framework leads to significant discrepancies in marginal and total costs, and that more differentiated solutions would probably improve the global cost-effectiveness of the mechanism. Among these solutions, those aimed at equalisation of effort (total sectoral reduction cost/GDP) in the various countries would come closer to the marginal cost equalisation optimum. They would thus

[14] This chapter was initially prepared as a contribution to the EC pre-Kyoto process. Since then, emission targets have changed. There has also been additional information since then, which has been incorporared into POLES snd which has affected somewhat the model projections. However, the approach used in this chapter concerning the principles of the cost reduction curves and their analytical implications remains valid. The principal writers of this chapter are Patrick Criqui of IEPE and Niko Kouvaritakis of ECOSIM.

offer "second best" possibilities, perhaps more readily acceptable to the various parties.

The analysis proposed here fall within the realm of the cost-effectiveness analysis of emission control strategies. A global but sectoral approach has been adopted, since the analysis is based on a model of the world energy system up to the year 2030, the POLES model. In the first part, we present the main characteristics of the model used followed by an analysis of the limits as well as the interests of a sectoral approach to the problem. In the second part, we present the results of an assessment of sectoral emission reduction costs. To this end, the "reference projection", corresponding to a scenario "without new policies", is analysed. The concept of a "shadow carbon-tax" is then used to calculate the sectoral marginal costs of reduction, differentiated by region. Finally, in light of these costs, various plans incorporating quantitative objectives and different distributions of the emission reduction effort are analysed and an attempt is made to determine the consequences for current negotiations on greenhouse warming.

The POLES model analysis of the cost of international strategies to reduce emissions is carried out in two stages:

- first, by defining a reference or baseline scenario representing changes which might be expected in the energy situation without these strategies;
- next, by examining the results of a series of alternatives, corresponding to different shadow carbon-taxes representative of the various levels of effort to reduce emissions.

In Chapter 2 the characteristics of the baseline scenario and the key results were presented. It was also seen in that chapter that the baseline results, at least for CO_2 emissions, are reasonably consistent with those of other world energy forecasts. The remainder of this chapter describes in detail the results of the alternative scenarios, which incorporate a shadow carbon-tax. The chapter concludes with an attempt to use these analyses to draw up guidelines for the formulation of international objectives and strategies to combat the greenhouse effect.

6.1.1 Limits and Interest of a Sectoral Approach to the Evaluation of the Reduction Strategies

Structuring the Economic Debates within the Framework of the IPCC

Most of the studies undertaken within the framework of Group III of the IPCC (socio-economics) have used a "cost-effectiveness" approach, involving the analysis and comparison of the results of various emission control cost studies (IPCC, 1995). Debates have focused on two questions, that of the extent of low- or zero-cost reduction potentials ("no-regret" measures) and that of the macro-economic costs or benefits of emission control policies (double-dividend). Among the general conclusions which can be drawn from these studies, the most significant is perhaps that the considerable discrepancies in the results are due less to the diversity of the models than to fundamentally different assumptions on the potential effectiveness of public measures (Hourcade, 1996):

- a critical factor for the "no-regret measures" results is the assumption that public policies can reduce, without apparent cost, the significant and persistent discrepancy between the observed and theoretical energy efficiency of the actual energy using equipment;
- "double-dividend" results depend on the assumption that governments can redeploy tax revenues obtained from carbon taxes or permit trading systems for the benefit of the entire economy and - particularly in Europe - for employment.[15]

Any appraisal of the cost of strategies to reduce or limit greenhouse gas emissions obviously depends on the choice of the assumptions adopted on these two questions. The cost may be high in the absence of both "no-regret" and "double-dividend" measures. In the opposite case, however, the cost may be low or even negative.

One of the limitations of sectoral approaches is undoubtedly that they do not help address these two questions. This is true of the approach adopted with the POLES model. Since it is a sectoral model, it cannot be used to assess the impact of changes in energy variables on the rest of the economy. It also is a partial equilibrium model, in which internal adjustments are made primarily through national and international energy prices. The model is constructed in such a way that it cannot be used to identify no-cost potentials, so that the policies for mobilising such potentials can be introduced only via exogenous hypotheses, which are, to a large extent, arbitrary.

The fact that the questions of no-regret potential and double dividend cannot be tackled is certainly occasioned by the intrinsic constraints of the model, but it also stems from a double choice:

- the choice not to obscure the results by taking into account hypotheses concerning measures which should already have been implemented for their own sake, but which in reality have not;

[15] Malinvaud, 1993.

- the choice to pursue an analytical type of procedure, in other words, one which, breaks down and clearly identifies the different categories of costs and benefits related to measures envisaged.

Relevance of an Approach Based on Sectoral Costs

It thus appears that the limits of the POLES model in no way detract from its usefulness in evaluating GHG emission reduction or limitation policies. From a sectoral point of view, it is clear that substantial reductions in emissions compared with the reference can only be obtained through an improved effort in terms of energy efficiency - triggered in the model by an increase in end-user prices - and greater development of low-emission technologies. In most cases, real costs will be linked to these technological substitutions.

The model can be used to simulate - sector by sector and country by country - an increase in the price of energy via taxes and then to calculate the corresponding technological substitutions and emission reductions. This increase can be defined as the "shadow carbon-tax" making it possible to obtain the simulated reduction. The shadow tax corresponds to the marginal reduction cost (MRC) for the energy sector. This concept forms the basis of the systematic study of emission reduction costs presented below.

The interest of the POLES model is that it is sufficiently detailed to be used to compare the marginal reduction costs and associated potentials in a coherent way - in terms of tonnes of carbon - for the various energy sectors or services, in the different countries or regions identified. It is thus possible to highlight some of the structural data which are determinant in calculating the costs of reduction strategies. Similarly, the cumulative total of rising marginal costs can be used to calculate the total cost of a reduction programme for a given country, which then makes it possible - still using a sectoral approach - to look at the question of the charges imposed on the various parties in the negotiations concerning the Convention on Climate.

From an Assessment of Sectoral Costs to an Assessment of Macro-Economic Costs or Benefits

Of course, sectoral costs represent only part of the costs and benefits associated with emission reduction strategies. However, there is nothing to prevent a sequential analytical procedure from being used with other tools to study the costs and benefits arising from the energy developments obtained, either for other sectors or for the economy as a whole. In this connection, it would be possible and justifiable here to reintroduce the concept of no-regret measures by taking into account in particular the positive effects of CO_2 emission reductions in terms of the decrease in other external costs: local and urban pollution, cost of congestion and harmful effects on health in the transport sector.

Similarly, on the assumption that the reductions would in fact be obtained through the use of economic instruments such as carbon-taxes or permit trading systems - provided that permits were sold or open to bids - it would be a good idea to take into account the potentially positive impacts on the economy of recycling the revenues thus obtained through efforts to combat the greenhouse effect. Evaluation of macro-economic costs or a possible double-dividend is therefore essential, but this exercise can be carried out separately once the structural changes made necessary in the energy systems, as well as the policies and instruments and associated costs, have been clearly identified with the help of sectoral models.

6.1.2 Effect of a Shadow Carbon Tax and Analysis of Reduction Potential According to Region

By introducing a shadow carbon tax into the POLES model, it is possible to calculate changes in emissions in relation to the reference case presented earlier.

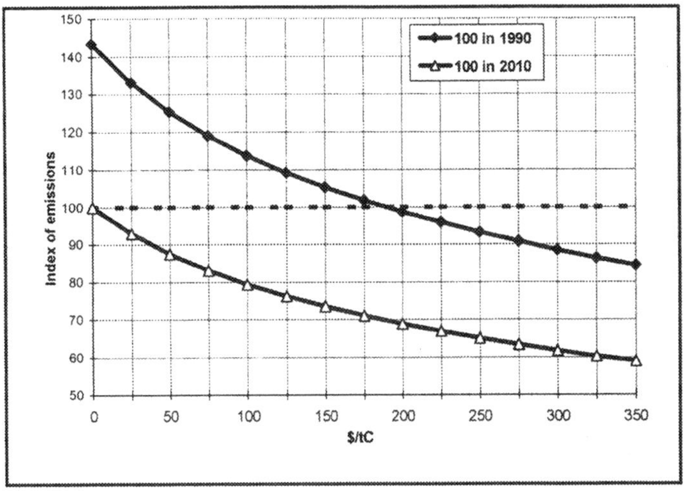

Source: POLES model - July 97.

Figure 6-1: Index of world emissions of CO_2 in 2010 for various carbon tax levels.

This is illustrated, at world level, in Figure 6-1 on the basis of an emission index of 145 in 2010, the emission level decreases with the introduction of a shadow tax of from 25 $/tC up to 350 $/tC. If it is read vertically, this figure also represents the curve for sectoral marginal costs of emission reduction. It can thus be seen that stabilisation of emissions at the 1990 level would correspond to a shadow tax of 180 \$/tC, or, in other words, that the equivalent objective of a 30% reduction in emissions in relation to the 2010 reference would be achieved at a marginal reduction cost (MRC) of 180 \$/tC.

The amount of the shadow carbon tax for a given reduction target reflects in fact all the changes imposed on the energy sector, as simulated by the model: cost of improved energy efficiency, of greater development of non-fossil energy, and of stronger penetration of fossil sources with low carbon intensity. From one region of the world to the next, the MRC will therefore depend essentially on the reference situation in each of these fields. In a seemingly paradoxical way, the more unfavourable this initial situation, the easier it will be to reduce emissions, while the more virtuous a country is with respect to CO_2 emissions, the greater will be the cost for that country of making further reductions. This situation could be described as the "Norwegian syndrome".

Marginal Reduction Costs Differentiated according to World Regions

In the model, the principal variables affecting the differences in MRC in 2010 are:

- the carbon intensity of the energy supply, and, in particular, the relative importance of coal in the electricity sector;
- the price structure and rates charged to end-users (in countries with low energy taxation, a carbon tax will have a significant direct impact on consumption);
- resources and development potential of renewable energy.

Figure 6-2 shows the considerable differences between the major regions of the world:

- for the OECD regions, reduction costs with respect to the 2010 reference would be higher than the world average in Europe and Pacific Asia, but lower in North America;
- the diversity of costs would be even more marked for developing regions, with reduction costs much lower than the average in China, Southern Asia and sub-Saharan Africa (all regions with high coal potential), but higher in south-east Asia, North Africa and the Middle East, and Latin America.

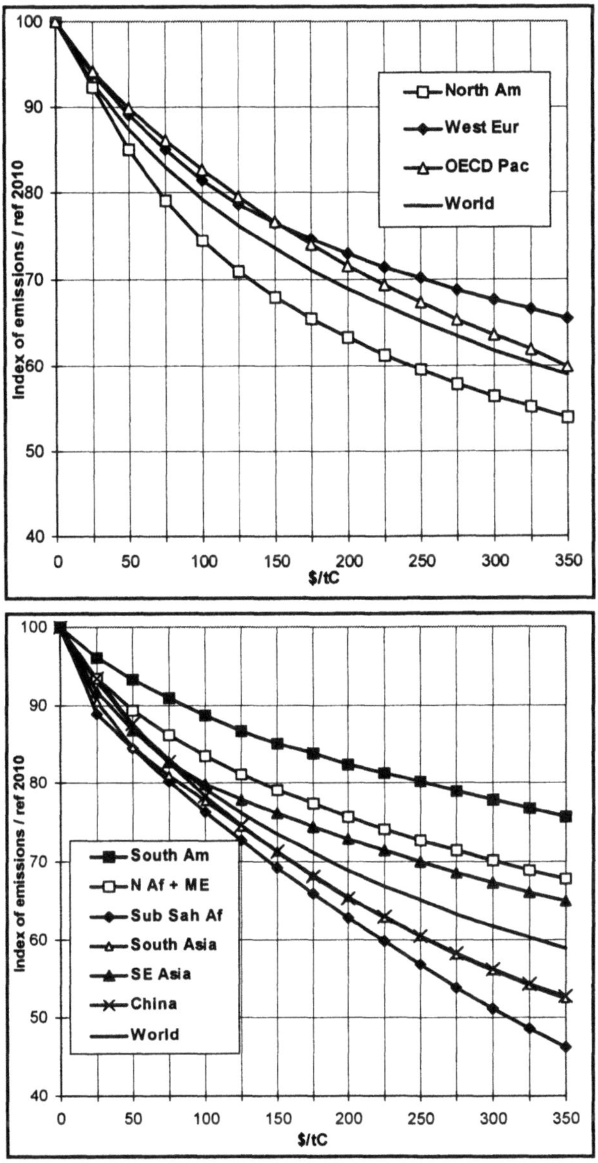

Source: POLES model - July 97.

Figure 6-2: Marginal reduction cost curves in the major world regions (in relation to reference year 2010).

Thus, the analysis whereby reduction costs would be systematically higher in the industrialised countries than in developing countries is not confirmed in every case by a sectoral analysis taking into account the structural characteristics of the energy systems. This conclusion is important in that proposed permit trading systems and joint implementation schemes are based on an *a priori* vision and, as will be seen, will therefore have to be considered on a case by case basis. In fact, low cost reduction potentials exist in every country, opening up the way for "emission trading". However, it is not certain that massive reductions would be less costly in the South than in the North.

Marginal Reduction Costs and the Structure of International Negotiations

The results of the exercise on the introduction of a shadow carbon tax can also be examined for One State or an organised group of States. These results can thus help throw light on some of the issues at stake in the international negotiations, in particular the costs associated with quantitative objectives in the form of a flat-rate reduction compared with a basic year. Thus, the marginal cost of achieving the objective proposed by the European Union before the Kyoto COP-3 can be analysed for the major Annex I partners: a 15% cut in emissions by 2010 compared with the 1990 level.

Figure 6-3 shows that a non differentiated reduction rate of 15% would unquestionably impose extremely different marginal costs on the various parties: 125 $/tC in the United States, 175 $/tC in Australia-New Zealand, 200 $/tC in the European Union and as much as 350 $/tC in Japan. On the other hand, equalisation of marginal costs among the various parties should result in differentiation of reduction objectives. Thus, for 200 $/tC, the reduction targets would be 25% in the United States and Australia-New Zealand, 15% in Europe, and 5% in Japan.

Finally Figure 6-3 demonstrates the crucial importance of projected trends up to 2010 in determining the marginal cost. Thus, Australia-New Zealand has the most gradual marginal cost curve. On the other hand, the very steep rise in anticipated emissions by 2010 means that the marginal cost for this region for a 15% reduction is closer to the European level than to the American level. But here it is only a question of equalisation of marginal costs; when the total effort for each country is taken into account (total cost expressed as a percentage of GDP for example) new configurations can appear, and these must be identified for the purpose of analysing the terms of the negotiations.

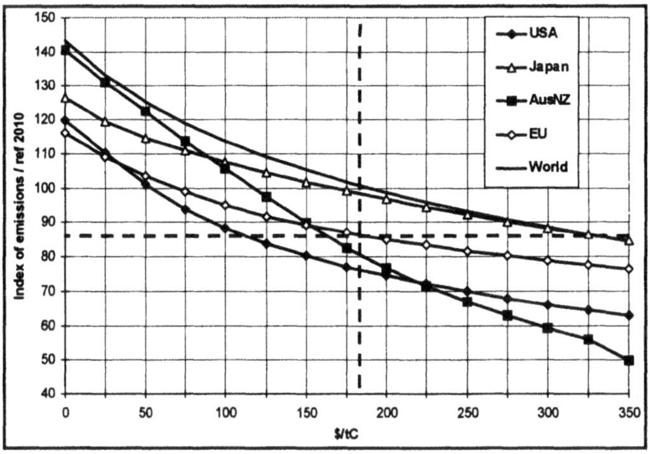

Source: POLES model - July 97.

Figure 6-3: Marginal reduction cost curves for the major partners of annex I (in relation to reference year 2010).

Equalisation of Marginal Costs or Equalisation of the Total Reduction Effort ?

While the marginal cost associated to a reduction program corresponds to the cost of the last ton saved, the total cost of the same program is equal to the sum of the different and increasing costs of each ton saved. The final stage in characterising the costs of emission reduction strategies thus involves calculating this sum of the increasing marginal costs in order to deduce from this figure the total cost of a reduction programme. Total cost configurations as a function of marginal cost levels (Figure 6-4) seem at first glance to be counter-intuitive. Some countries with low marginal reduction costs show a high total cost, and vice versa. This result can be explained as follows:

- for a given marginal cost level, much greater investment in reduction measures will be made in a country with a low marginal cost;
- the total cost of the programme will then be high, the volume effect of reductions overriding the effect of unit cost.

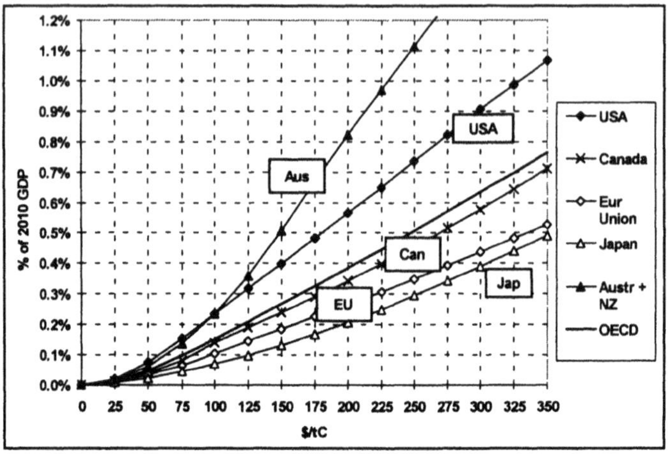

Source: POLES model - July 97.

Figure 6-4: Total cost of reduction as a function of the marginal cost level.[16]

Figure 6-4 can be used to analyse the position of each country with respect to one of the objectives proposed prior to the Kyoto conference, namely to reduce emissions by a flat rate of 15% compared with 1990 levels:

- for the United States and the European Union, the total costs might be comparable (less than 0.3% of estimated GDP in 2010), although the corresponding marginal costs differ considerably, with the EU level almost twice that of the United States;
- in Canada and Japan, on the other hand, the same objective could be achieved only with higher marginal costs and a total cost of about 0.45% of GDP;
- finally, Australia appears destined to incur by far the highest total cost, almost 0.6% of GDP, even though its marginal cost is lower than that of the Community.

The simulation of reduction programmes through the introduction of a shadow carbon-tax thus provides coherent information on the relative costs for the various countries. These results can then be re-examined with reference to the various plans proposed for the international negotiation, in order to asses the respective limits and advantages, in terms both of efficiency and acceptability by the various parties.

---

[16] On the figure, the country markers are placed at the marginal cost and total cost levels corresponding to a flat-rate reduction objective of 15% compared with 1990 levels.

Hypothesis 1: a "Flat-Rate" Reduction in Emissions of 15% Compared with the 1990 Level in OECD Countries

Here, it is assumed that all the OECD countries - the EU being regarded as a "bubble" - must reduce their emissions to 15% below 1990 levels, with no possibility of emission trading outside their territory. The imposed emission level for the OECD countries in 2010 would then be 2.5 GtC, i.e. 0.45 GtC below the 1990 level, but 1.05 GtC below the reference projection. As seen earlier, the differences in marginal costs obtained in the model for such a reduction per country are then significant, ranging from 125 $/tC for the United States to 200 $/tC in the European Union and as high as 350 $/tC in Japan. The figure is even higher in the non Union region of Europe, which includes Turkey, a country where there is high potential for increased emissions.

The total effort in relation GDP in 2010 depends both on the marginal reduction cost in 2010 and emission dynamics in the reference projection. Thus, the corresponding cost would represent slightly less than 0.3% of GDP for the United States, due to a low marginal cost, as in the case of Europe, in this case because of the moderate increase in emissions envisaged in the reference situation. On the other hand, this cost would be one and half times higher in Japan and Canada and up to twice as high in Australia - New Zealand. It thus appears that without the possibility of emission trading, non differentiation of objectives is likely to create substantial disparities in the charges imposed on the various parties.

Hypothesis 2: the Same Reduction, but at World Level, through a Joint Implementation System or a Generalised Emission Permit Trading System

Another hypothesis, which remains theoretical, involves assuming that the same reduction in 2010 (-1.05 GtC, or -12%, in relation to the world reference situation) could be imposed on OECD countries but that they could obtain this reduction at a lower cost through joint implementation or the purchase of emission reduction certificates in other regions of the world. If such an emission trading system could be generalised to all the countries and all the sectors, marginal costs could be equalised at the world level. This equalised cost would be then much lower than the cost calculated in Hypothesis 1, since a world reduction of 12% in relation to the reference could, in theory, be obtained at a marginal cost slightly below 50 $/tC. Thus, in their territory, OECD countries would carry out those emission reductions costing no more than 50 $/tC - for 460 MtC, i.e. slightly less than half of the total objective - and would finance or purchase outside their territory measures or rights corresponding to the other half of the total reduction objective.

But it seems unlikely that joint implementation measures or emission trading systems could be completely generalised. They would probably be limited to specific operations and sectors, an obvious candidate being the electricity sector, which is characterised by a limited number of operators and large technical units.

An emission trading system for the electricity sector alone would obviously involve higher marginal costs, since the trading base would be reduced. A more precise evaluation would therefore be called for. The figure of 50 $/tC would then provide only a reference for the base level of the marginal reduction cost, or the price of emission reduction certificates, corresponding to the objective concerned.

Hypothesis 3: Equalisation of Marginal Costs within the OECD Countries, through a Tax (Hyp. 3.1) or Permit Trading Systems (Hyp. 3.2)

This hypothesis is probably more operational than the previous one since it could be implemented in two quite radically different - but possibly realistic - ways: through the introduction of a common carbon-tax in the different countries or permit trading between countries that have agreed to restrictive quantitative reductions. The results of the model show that emissions could be reduced to 15% below 1990 levels at a marginal cost of about 170 $/tC for the OECD countries as a whole. The total cost of this reduction would represent approximately 0.3% of the GDP of the OECD in 2010.

When it comes to the burden sharing issue between the OECD countries, it is crucial to know whether this should be achieved through equalisation of marginal costs by means of a common tax or through permit trading. In the first case (Hyp. 3.1), the countries with the highest reduction potential at low or moderate cost would contribute more, so that paradoxically their total cost in relation to GDP would be higher than that of the other countries. Thus, total potential at a cost of less than 170 $/tC would represent a reduction of 23% compared with 1990 in the United States, for a total cost of almost 0.45% of GDP. For the EU, the reduction would be 13% for a cost of 0.22% of GDP and for Japan only 1% for 0.16% of GDP. Although the rejection of a carbon tax by the United States is probably based on other reasons, there are clearly limits to an international approach based on a single tax, but this does preclude "local" implementation of such a system.

In the second case (Hyp. 3.2), it is assumed that each country is required to contribute to a 15% reduction compared with its situation in 1990, but can buy or sell the reductions corresponding to the difference between the national marginal cost and the equalised marginal cost. The situation will then differ in each country, depending on whether their marginal cost for a 15% reduction is lower or higher than 170 US dollars per ton of carbon ($/tC):

- the United States and Australia would be in a position to offer permits/have the possibility of selling permits, for approximately 75 and 10 MtC respectively;
- on the other hand, Canada, the EU and Japan would require permits, for approximately 10, 20 and 50 MtC respectively.

If the Eastern European countries, which now form part of Annex I, were included in a joint implementation or permit trading system, the trading base would be widened and the equalised marginal cost probably reduced, thereby modifying the position of the OECD countries. However, inclusion of these countries poses some important problems, both in terms of modelling and in reality. These problems concern the question of the total amount of the permits attributable to countries where emissions have fallen by 20 to 30% in recent years, due simply to the economic transition they have undergone. The risk is then one of fictitious permit trading or "hot air trading".

Hypothesis 4: Differentiation of Quantified Objectives to Equalise Effort

As an alternative to the solutions of a single tax or permit trading systems, which, although possible, would still entail significant implementation problems, there is the possibility of differentiation of agreements to take into account, in a global manner, the present and future characteristics of each country, in line with the thinking adopted by the European Council. Such a solution has neither the simplicity of the non differentiated flat-rate approach, nor the efficiency of a system of marginal cost equalisation through a tax or trading system. Nevertheless, it would be possible to define objectives which are acceptable to the various parties and which do not entail costs that are far too high compared with those of the solution involving minimisation of total costs. It would be a question of seeking a "second best" solution, which would satisfy the different parties.

Various equity criteria can be used to formulate a system of differentiated objectives (see in particular the proposals of Colombier, Dessus and Laponche, 1997). As a first approach, the simple solution of equalising the ratio of the total reduction cost expressed in terms of estimated GDP in 2010 has been used here. As mentioned earlier, equalisation of marginal costs within the OECD countries would make it possible to achieve the objective of a flat-rate reduction of 15 % for a total cost of approximately 0.3% of the 2010 GDP for the region. It is then possible to calculate the emission reduction level on the basis of each OECD country devoting 0.3% of its 2010 GDP to the reduction effort. (This figure could be compared to the objective - declared but seldom attained - of industrialised countries to contribute 0.7% of their GDP in the form of aid to developing countries).

The results of the model show that emission objectives corresponding to this kind of reduction effort can, in certain cases, reduce the disparities between the solution of equalised marginal costs and the flat-rate solution:

- for the United States, the reduction would be 15%, in other words the same as in the flat-rate solution;
- for the EU, it would be slightly higher, at about 17%;

- in the Pacific OECD region, it would be lower, with a 10% reduction in Japan and only stabilisation in Australia - New Zealand.

The total reduction in the OECD countries would be of the order of 410 MtC compared with 1990.

For an identical cost, the solution of equalising marginal costs results in a total reduction of 440 MtC. In other words, each individual country would have to agree to devote 0.33% of GDP to the emission reduction effort in order to obtain the same quantitative result. In this "second best" solution, loss of efficiency appears fairly limited - being about 10% in relation to the solution of marginal cost equalisation - and, all things considered, acceptable. It could therefore open the way either to differentiation of quantitative objectives or, through duality, to identification of a common effort level acceptable to the parties, with results which would not be too far removed from the theoretically most efficient solution.

6.2 Conclusion

Any method to determine quantitative objectives or to decide how the efforts to combat global climate change should be divided, has to take into account intrinsic difficulties stemming from the fact that such efforts have to be made over time and must therefore be assessed with respect to projected variables. The "flat rate with no trading" approach (Hyp. 1) may involve objectives which are not acceptable to some of the parties.

To a certain extent, the approach which combines a flat rate and permit trading (Hyp. 3.2) overcomes this difficulty, because it is also based on measurable initial emission levels as well as on market mechanisms which, through the trading system, are meant to guarantee the efficiency of the measures and thereby reduce the total cost. Nevertheless, it is not certain that the flat rates would be acceptable to all the parties, depending on their expected future emission rates. Obviously, the problem of the reference can never completely be eliminated.

As for the solution of joint implementation or emission trading - which represent respectively the "industrial" and "commercial" versions of the same basic principle - it is by no means certain that there would be sufficient operations and transactions to ensure a real convergence of marginal costs (Hyp. 2). It should be noted that the joint implementation operations currently listed in the "pilot phase" would correspond to total reductions of 2 MtC, whereas it has been shown that the reductions mentioned for the OECD countries in 2010 are of the order of 1000 MtC with respect to the reference situation.

When considering approaches aimed at differentiating objectives to allow for changes in the characteristics of the energy systems, the problem of agreeing on a reference must be settled. Even if it is possible to settle for emission standards for different types of energy services, the potential growth in such services in the various countries during the period concerned will remain a moot question. The European experience in this field definitely shows that agreement can be reached within a group of countries, but it is not certain that this precise agreement, though deemed acceptable, can be considered an equity criterion in terms of sharing the burden.

Equalisation of relative costs (total cost/GDP, Hyp. 4) is not a solution, which is free of problems, in particular because it does not guarantee systematic convergence of emission levels. However, it does have a certain number of advantages:

- first, this solution can be formulated in terms of either financial effort (allocation of a certain percentage of GDP to emission reduction) or the corresponding reduction objectives, provided that in this case there is a minimum of agreement on the evaluation of anticipated costs (the issue of a reference level, of course, arises again);
- second, according to the results of the model, this solution produces results which are fairly close to the solution which is in theory the least expensive (Hyp. 3) and it does not preclude the possibility of permit trading;
- last, the industrialised nations have already reached an agreement in principle concerning another issue of world concern: they have agreed to allocate a certain percentage of their income to aid for developing countries. It is thus possible to conceive of a solution of "0.3% of GDP to help combat climate change", comparable to the "0.7% in development aid".

PART III: CONTROLLING EMISSIONS IN THE LONGER RUN: THE ROLE OF TECHNOLOGY

7 Introduction to PART III

The basic issue to be analysed in this part of the volume is the role of technological progress in stemming the growth of emissions in the longer term. While technological improvements will play an important role in the trends of energy demand and related emissions even in the short to medium term, it is clearly over a long time horizon that its impact can really be felt. This is because the process of technological innovation towards specific targets, such as reduced carbon intensity of the energy system, can normally only occur after a prolonged and sustained effort and it involves many phases, like research and development (R&D), initial demonstration, commercialisation and wider diffusion. R&D can be very time consuming and its results are, to some extent, unpredictable. Even after a successful R&D phase, there is a long time interval before the new technological advance is fully commercialised and accepted by the market. The longevity of much of the energy stock results in long delays between the introduction of a new technology in the market and it having a material impact on energy use patterns.

The role of policy regarding technology is also different between the medium and long term. In the medium term, policy measures cannot lead to the development of new technologies but they can encourage the faster introduction and the more intensive utilisation of existing technologies. On the other hand, government and industry objectives through spending on (R&D) can influence the nature of technological progress in the longer term. In other words, if governments want to control emissions they can encourage suitable technological developments and increase the chances of appropriate technological breakthroughs through increased R&D funding of suitable technologies.

Because of data limitations and the limited scope of this study, the following chapters examine the long term role of technology only in the case of the power generation sector. One of the key conclusions to emerge from the first Part of this volume was that, at least over the next 30 years, the use of electricity would grow much faster than other fuels in final consumption. Hence, reducing emissions from the power generation sector could, on its own, play a significant role in tackling the problem of greenhouse gases. The sector also presents itself as a convenient case study because of the well defined and documented technologies that are currently in use and are likely to evolve over the next 30 years.

This part of the volume is structured as follows.

The next chapter examines the emerging clusters of electricity generation technologies. The discussion includes not only the present tech.-economic status of these technologies, but also their prospects in the medium to long term (2000-2030).

The following chapter presents the baseline trends in electricity demand and describes the five technology scenarios to be analysed. The next two chapters then analyse the impact of these scenarios on the global and European power generation system. The results are presented on a technology-by-technology basis and include the analysis of the environmental costs and benefits associated with each scenario. Chapter discusses some of the issues that emerge for European R&D policy.

8 Power Generation Technology Clusters: Present Status and its Potential[17]

This chapter presents a general discussion on the present state and likely future developments in a number of groups of related power generation technologies. These groups, or clusters, of technologies then form the basis of the technology scenarios presented and analysed in later chapters.

Apart from fundamental changes in technology, the other defining features of the scenarios presented later is the degree of centralisation of the ESI.

8.1 Nuclear Industry: a Paradigm in Crisis

Nuclear technology has experienced an important diffusion in the 1970s and 1980s, as a result of an important technology push, originally carried out by governments of the major industrialised countries. However, the importance of nuclear energy for the worldwide production of electricity (17% in 1995) cannot conceal the difficulties that its diffusion has undergone in the whole world.

According to the projections made at the end of the 1970s, the worldwide nuclear capacity should have reached 1,300 GW in 1990 and 3,600 in the year 2000. It was 325 GW in 1990 and will reach 366 GW in 2001, at best. It is important to assess the changes in the field of the selection of electric production techniques that would encourage a future boost in the diffusion of nuclear technology. It has been argued, that nuclear energy could provide a solution to the increase of world-wide energy needs, particularly in emerging, rapidly-developing countries. Also it

---

[17] The principal authors of this chapter are D. Finon of IEPE (Nuclear section), M.Witeley of ESD (Clean coal section), S. Mima of IEPE (Fuel cell section) and Ph.Menanteau of IEPE (Wind power section).

its greater diffusion could help meet the limitations of greenhouse effect gas emissions agreements revive its development.

The development of nuclear technology has followed a "technology push" model. This model, less and less used, presents three main characteristics:

- it is linked to complex technologies;
- its projects are defined according to technological objectives, regardless of economic and commercial feasibility;
- due to nationalistic and military reasons only a very small number of large, national, industrial firms can appropriate the technology.

The selection of competitive nuclear technologies was realised quickly, with a characteristic “lock-in” phenomenon adaptable to American techniques for light water reactors, in spite of their rather poor performance. BWR and PWR techniques had benefited from a prior military impulse (submarine reactors, production of fissile material). They also benefited from US government support in the geopolitical arena (control over the risk of proliferation permitted by American supply of enriched uranium) (Bupp and Derian, 1978: Cowan, 1990).

The introduction of hybrid nuclear technology through steam turbines in the 1960s and 1970s led to a change in the electricity production paradigm. This required the exploitation of economies of scale, leading to ever larger size nuclear plants with substantial capital requirements. The capital intensity and large scale optimal plant of nuclear power generation have become weaknesses in recent years.

As a, that started at the end of the 1960s, the commercial diffusion of nuclear power slowed down after the mid-1970s in industrialised democratic countries, except for France and, to a lesser extent, Japan. This was the result of a crisis in social acceptance of nuclear power due to new societal values and priorities (e.g., safety, health, life quality, participation, etc.). Nuclear development has been a major focus of new social conflicts because it embodies different critical aspects of political control in industrialised societies, and also due the specific nature of its risks, from which particular fears can also arise.

Apart from reduced social acceptance, the nuclear industry also had to cope with the significant slackening of the growth in electricity demand, from a 7% in the 1960s to a 1-2% more recently. This has resulted in the appearance of significant over-capacities. The overall result was the cancellation of many orders for nuclear plants, especially in the United States (119 cancellations from 1975 to 1978).

In summary, the weaknesses faced by the nuclear industry are:

- technological complexity
- safety concerns

- institutional weaknesses
- large scale of the optimal plant

8.1.1 Incremental Innovation in Nuclear Development

As a result of the above tends a number of advanced developments in nuclear technology have been abandoned. Efforts for super-generators and for a new management of wastes have been also been abandoned in most industrialised countries.

A process of incremental innovation, i.e., improving the adaptation of 1st generation reactors, has substituted the "technology push." Most current effort by the industry aim at preserving or restoring social confidence through the introduction in the market of reactors that meet stronger safety norms. This is attempted through:

- reinforcing isolation systems
- a better prevention of serious accidents produced by fusion of the nucleus
- reducing accident probability
- simplifying exploitation

The aim is to reduce costs and investment risks, by simplifying the design of the reactors, and by reducing the cost of components and instrumentation. This may lead back to a "learning" process through standardisation, which would result in a reduction of the time and cost of installation. New developments are based on the greater use of quality steel, simpler materials, which are less sensitive to operating errors, and information technology.

Greater modularity and divisibility of nuclear equipment would be more compatible with present constraints of financing and requirements for a faster pay off period (see the AP 600 Westinghouse and the AB 600 General Electric developments, partly financed by the US Department of Energy).

"Evolutionary" LWR

The nine present manufacturers of nuclear equipment are increasingly able to offer, often through partnerships, better-designed PWR or BWR technologies (see Table 8-1). Some of these improved technologies are already reasonably developed. General Electric, in alliance with Toshiba and Hitachi, is the most advanced since the construction of two large MW ABWR reactors in Japan (first start-up in 1997). The first N4 (PWR) reactors by Framatome also belong to this category. Meanwhile, the creation of the European Pressurised Water Reactor

(EPR) prototype, studied by Framatome and Siemens since 1990, in the framework of the REP 2000 project, could be built in France from 2000 to 2005.

Passive Safety LWR Reactors

These reactors are only at the design stage. They are studied especially in countries where their completion has been blocked, in order to commercialise reactors of a completely different design and undermine objections. "Passive safety" is ensured through major thermal inertia of the primary circuit and of the weakest and specific fuel potential. Thus, the evacuation of the heat in case of an emergency stop is easier.

Table 8-1: Stage of the development of the advanced LWR techniques and passive safety techniques.

	Concept in development	Agreed concept	Project prototype	Prototype in construction	Advanced development stage
Advanced reactors (evolutionary)	European Pressurised Reactor: EPR ABB-Atom: BWR 90	ABB-CE: System 80 + (1300 MW PWR)	Westinghouse -Mitsubishi: APWR (2X1420 MW in Japan) Commande: 1998	General Electric ABWR 2 prototypes of 1350 MW (Japan) Start up: 1997-1998	Framatome: N4 (4 reactors) 1996-1998
Passive safety reactors (revolutionary)	General Electric: SBWR Westingh-Mitsub: SPWR Siemens: SWR 1000 European Passive Plant	Westinghouse:	AP600		

8.1.2 Renewed Technology Push? The Heritage of Advanced Reactor Programmes

A number of research programmes have been carried out in France and Japan concerning liquid metal fast breeder reactors (LMFBR). The importance of accumulated responsibilities and the size of the investment undertaken is likely to maintain these programmes alive for a decade, after which they are likely to disappear because of the lack of "demand pull".

The technique studied at the CERN (Geneva), which is backed by the European Commission and the Spanish industry, is at a preliminary stage. The sub-critical character of the nucleus prevents fusion risk, as the fission reaction is interrupted by particle flux suspension. Also, its particular neutronics allows the reduction of the actinide production and the consideration of long-term waste transmutation. Moreover, the thorium-U233 cycle presents higher resistance to proliferation, but without radical rupture compared to traditional reactors.

The implementation and diffusion of such a nuclear reactor, a new paradigm indeed, would require the mastering of a vast number of techniques: powerful linear accelerators adapted to continuous long-term functioning, lead cooling technology (metallurgy, pumps, valves), controlling Th-U233 combustible elements (manufacture, exposure to irradiation), improved techniques for the management of wastes, etc. Consequently, new specific skills would have to be developed for nuclear R&D agencies, electromechanical enterprises, and companies within the nuclear sector.

In industrialised countries, there are potential "demand pull" factors that could provoke the renewed diffusion of nuclear technology. These include the need for replacing installations that become obsolete and the possibility of higher gas prices that would reduce the comparative advantage of gas-combined cycle units. In developing countries, which face potential limits to their resources, the huge expected growth in their need for power generation capacity would be the most likely factor for a renewed demand for nuclear plants. The nuclear industry worldwide could, of course, be a beneficiary of the concern for global warming.

8.1.3 Concerns with the Social Acceptability of Nuclear Power

The issue of nuclear wastes is likely to remain controversial in the foreseeable future. There are no clear principles on the necessary scientific evidence for the acceptability of a definitive storage installation, and the related duration and inter-generation issues (Berkhout, 1991). With no alternatives available, temporary surface storage is currently considered to be a necessary evil. A more definitive solution is yet to be found. The present lack of diffusion of nuclear technology limits research on the end-of-cycle waste problems.

The industrial organisation of the electric sector, which was based on regulated monopoly, had favoured nuclear technology diffusion mostly for two reasons. First, it allowed companies to postpone the costs and risks of storage and, by ensuring a stable rate of return, this model encouraged capital investments with a longer payback time. Second, it allowed the socialisation of the risks and the costs associated with any accidents, the management of wastes and eventual decommissioning. In other words, any unforeseen additional costs could be wholly transferred to the price paid by the consumer, or financed by the taxpayer.

Competition regulations, in a private - property rights framework, seem to be incompatible with the expansion of nuclear power. Firstly, competitive firms tend to privilege modular investments and they prefer shorter payback periods (of the order of 5 to 6 years, instead of the 11 or 12 years admitted in the framework of regulated monopoly). Secondly, the combination of competitors/private operators forces the disclosure of real costs. Therefore, it is no longer possible to rely on a complete socialisation of unforeseen costs and of those risks associated to nuclear development, even when states can somehow guarantee a financial-risk cap in the event of nuclear accident. Finally, private investors no longer perceive the associated economic risk carried by rival operators -not by the client- as identifiable and assessable.

8.1.4 Clean Coal Technologies

A significant amount of world wide electricity production comes from coal combustion. In Europe, more than a quarter of electricity comes from solid fuels. The majority of coal is burnt in pulverised fuel (PF) boilers heating water to drive steam turbines. However, in today's terms, these PF turbines are relatively inefficient with high levels of emissions. These disadvantages have driven research into efficiency improvements (better turbines, higher boiler temperatures), alternative methods of combusting coal (gasification, fluidised bed combustion), co-firing with other fuels, and even the conversion of coal into alternative types of fuel (e.g. liquefaction, pyrolysis). This report analyses the technological changes that have occurred within the combustion process itself to improve the economic, efficiency, and environmental performance of electricity from coal combustion.

Clean coal technologies are not a new concept, as an understanding of the theory behind these applications has been in the public domain for decades. However, the process of transferring the theory into practical commercial applications has proved to be difficult and still proves to be the main stumbling block behind their lack of widespread implementation. There are many reasons behind this problem, the main ones of which are covered in this report.

There are many factors forcing change in the European and worldwide markets. These are occurring at a rapid pace, predominantly by technological change, but also from other economic and political factors. The dynamics of the changes are provided in more detail during the rest of this section. There are many factors influencing leaps in technological progress, each of which lead to improvements in various aspects of clean coal combustion. Some of the forces behind technological change within coal-based power generation sector are listed below.

Economics

In Europe, growth in electricity demand has been relatively low over the last two decades. Therefore, the factor determining the construction of new plant has tended to be the need for replacement of retired power stations. There are two major elements that determine the economics of clean coal plant. The first, fuel costs, is uncontrollable, while the second, capital costs, and can be directly affected through research and development. If clean coal is to become the first choice in new plant construction, it has to become economic against the alternative technologies such as traditional coal-fired pulverised coal-fired steam turbines and gas-fired combined cycle power plants.

Fuel Costs

Deep mined coal in Europe is expensive in comparison to other fossil fuels, so the various indigenous coal industries have only survived due to government support, be it in the form of a direct subsidy or through price guarantees. However, cheap imports are available, so the cost of fuel is not necessarily a factor against clean coal technology. It should be noted that, the relative stability of the fob price of coal does not eliminate volatility in the imported price of coal, due to the large share of shipping costs in the final price.

Competition

Competition between various generating technologies is intense in Europe. The future for coal-fired generation is clean coal. Therefore, as Europe has broad experience in coal burning technologies, intense research is being carried out to make clean coal competitive against alternative technologies, especially with respect to the export market to Asia and the rest of the world. Private ownership of the utilities has significantly changed the culture of these organisations. Over the last few decades, investment policy has become much more financially based. This is currently a factor against clean coal. On the other hand, this gives the component manufacturers the opportunity to invest in clean coal plant themselves, which frees them from the constraint of having a single/few electricity supplier(s) within the region. However, despite no clean coal plant having yet been built in Europe as a result of privatisation, there is scope for this to happen. Bearing in mind the liberalisation of electricity markets, the change towards third party access is more associated with choice to the consumer. This allows the concept of product differentiation to be introduced. It is possible that clean coal electricity can be marketed as a premium source of electricity, which will encourage the development of this new market. Further research would be necessary in this area to investigate the possibility of clean coal electricity as a niche market and to ascertain the type of niche that this technology can hold.

Environment

Coal is a relatively dirty fuel compared to other fossil fuels. This is in terms of carbon dioxide emissions (global warming), sulphur dioxide and nitrogen oxide emissions (acid rain) and ash disposal (heavy metals to landfill). Therefore, high levels of research are being carried out to minimise the environmental effects of coal burn, especially in terms of acid rain mitigation. Efficiency improvements will lead to reduced CO_2 emissions per unit of electricity produced.

Security of Energy Supply

Coal is Europe's largest indigenous energy source, so its use in the past was encouraged as a means of minimising energy imports. However, owing to changes in the world coal trade and cheap imports from other countries, the majority of the European coal industry has closed down during the last fifty years. This has led to a perceived decrease in the importance of coal as a fuel.

A second major impact on the industry has been the termination of the ban on the use of natural gas in electricity generation. As cheap natural gas supplies are plentiful in Europe, there is currently little need for new clean coal plant. However, in the medium to long term, coal is expected to achieve a renaissance, as natural gas becomes a scarcer commodity.

Technology Leadership and Exports

The main rationale behind maintaining a thriving clean coal technology industry is the export market to developing countries, especially in Asia. For Asia, an increase in the market for clean coal is expected over the next few decades. By being technology leaders, European organisations can benefit from their expertise in these markets. The net result is that these exports can be used to develop and test new processes in preparation for penetrating the European electricity generation market itself.

Key Players within the Industry

The players forcing change in clean coal technology are both external and internal to the industry.

The main external influence comes from international and national government funds. For example, the European Commission is a major source of funds encouraging research, development and demonstration of clean coal plant. This is making Europe, with the USA and Japan, a world leader in clean coal technologies, which provides an excellent basis for exports to the rest of the world.

Governments can also force change through regulations and legislation. However, as the majority of the relevant regulations refer to emissions, coal combustion is penalised owing to its high emissions. On the other hand, laws can be introduced requiring long term security of fuel supply. This would help to secure the majority of the remaining deep-mined coal production in Europe, which is currently under threat from cheap imports and competing fuels.

Clean coal technologies will also be affected by the behaviour of utilities. Their aim is to provide the most economic electricity to the consumer within a given set of restrictions imposed by governments. If clean coal developments can make it the best option in certain locations, the long term future for coal combustion in Europe will be assured. Additionally, the faster that technological progress can be made, the better are the prospects for clean coal in the rest of the world.

The final key players are the manufacturers and installers of clean coal technology plant and components. They have a vested interest in the rapid integration of the plant into a first choice technology. Their current markets are in the developing world where experience can be obtained into the operating characteristics of this plant. It is vital that this experience is utilised so that continuing improvements in efficiencies and emissions from clean coal combustion can be made.

Amongst the main industrial actors involved in clean coal developments are Foster-Wheeler Energy, Siemens AG, Babcock Energy International Combustion Ltd, Asea Brown Boveri and many others manufacturers, utilities, and technology integrators.

Key Technology Developments

There are many different clean coal options based upon a small group of technologies. The main ones are pulverised fuel combustion under critical conditions, fluidised bed combustion, and coal gasification. The main technological options arising from these discussed below.

Supercritical and Ultra-Supercritical Coal

Most conventional power stations use sub-critical pulverised fuel (PF) combustion, in which coal is ground into fine particles injected, with air, into a combustion chamber. The particles burn in suspension and heat water tubes in the combustion chamber walls, raising high-pressure, high-temperature steam to drive turbine-generator sets.

Supercritical PF combustion is an evolutionary advance of sub-critical combustion, in which higher steam pressures and temperatures are used (above

221 bar and 540°C) to raise efficiency levels to 44%. Several supercritical plants are in use in Europe, mainly in the Netherlands.

New PF plant in Europe will almost certainly be of supercritical or, in time, ultra-supercritical design, with associated emissions reduction technologies, including selective catalytic and non catalytic reduction (SCR/SNCR), low NOx burners, fuel staging, flue gas desulphurisation (FGD) and dust collection. Equipment manufacturers tend to prefer this technology to more radical new technologies such as IGCC, because it represents an evolutionary advance of the tried and tested conventional steam cycle.

Conventional supercritical coal-fired steam power plant employing elevated, live steam parameters have already been built in past decades. Austenitic materials had to be applied for components operating in the high-temperature at that time. Recent developments in the material properties permit the use of ferritic/martensitic steel in present state-of-the-art hard-coal fired power plants and steam parameters of 221 bar and 540°C. R&D activities for ultra super critical plant are aiming at higher steam parameters of 248 bar and 566°C. For steam generator components, pipes, and turbine components, nickel based materials and methods of manufacturing the named components from these materials have to be developed. However, ultra super critical is still at the development stage, and demonstration plants of this technology can not be expected before 2005-2015.

Pressurised and Atmospheric Fluidised Bed Combustion (PFBC/AFBC)

In fluidised bed combustion, a continuous high-velocity stream of air is injected through a "bed" of mixed inert material, ash and fuel, creating a "fluidised bed" in which the fuel particles move about freely in suspension. In bubbling atmospheric FBC, the bed remains at the bottom of the combustion chamber, with a defined surface level. In circulating FBC, which is more efficient, higher gas velocities are used resulting in a turbulent cloud of solids throughout the combustor.

FBC allows rapid and complete combustion of fuel at temperatures around half those of conventional PF plants. In-bed heat exchangers are used to generate steam for industrial or power plant use. Emissions of NOx are limited because the combustion temperatures are low, and SO_2 can be removed by adding limestone to the bed.

Bubbling AFBC technology is very useful for the efficient combustion of poor quality fuels (e.g. low-grade coal, lignite, peat, wood, biofuels, and waste). The technology is well established for industrial and small power applications and for CHP. Circulating AFBC is generally preferred for large-scale power applications (over 250MWe).

PFBC is based on AFBC but takes place within a large pressure vessel, at pressures around ten times above atmospheric. PFBC can be integrated into a combined-cycle system, where electricity is generated first in a conventional steam turbine-generator set, and then the hot exhaust gases drive a gas turbine to generate additional electricity. More heat is then recovered from the turbine exhaust gas, to give an overall thermal efficiency of about 44%.

There are three commercial-scale bubbling PFBC plants in the EU, and several demonstration plants are operating elsewhere in the world. Second-generation PFBC plants are now being designed, which incorporate fluidised bed gasification followed by char combustion (a hybrid cycle).

PFBC can therefore be considered as commercially available and applicable for a large variety of coals. The pressurised combustion reduces the space requirements and the fluidised bed can be designed to control emissions of SOX (by admixing limestone) and NOx (by air splitting and temperature control in the furnace) below the environmental standards without cost intensive flue gas cleaning. Using cyclones and candle filters, the flue gas particles can be reduced by 99.9 % before entering the gas turbines, which provides sufficient protection of the turbines and complies with most requirements for emission control.

Integrated Gasification with Combined Cycle (IGCC)

In IGCC systems, coal is gasified (by reaction with steam, oxygen or air), and the hot gases are cleaned and burnt in a gas turbine. Exhaust heat from the gas turbine (and from the gas cleaning and gasification processes) is used to raise high-pressure steam for additional electricity generation. IGCC plant can achieve net efficiencies of 45% and can be used for low-grade fuels (e.g. petroleum coke or refinery residues).

IGCC technology is still at the demonstration stage, with demonstration plant in Europe (at Buggenum in the Netherlands and at Puertollano in Spain) and in the USA. It offers the possibility of power generation at very high efficiencies, taking in this manner full advantage of the achievements on gas turbines and combined cycle technology. Decisive characteristics of integrated gasification cycles are the gasification processes (entrained flow, fluidised bed and fixed/moving bed) and oxidant (oxygen- or air-blown). Demonstration plants with an installed capacity of 200-300MW are under construction. The commercial availability of larger units is not expected before 2005.

Integrated Gasification with Humid Air Turbine (IGHAT)

In IGHAT, gasification of coal is followed by combustion in a Humid Air Turbine (HAT) cycle. HATs are based on gas turbines, without a bottoming steam cycle, with two modifications to improve efficiency. First, multi-stage compression of the turbine air with recuperative water cooling between each stage to reduce the compressor work. Second, additional flue gas heat recovery in a recuperator for water and its use for compressed air saturation prior to combustion.

In conventional gas turbines, more than a third of the power generated in the expander is absorbed to compress the working fluid. This loss of energy is increased by the requirement for an excess airflow to allow cooling of the turbine components. Humid air turbines (HAT) allow for lowering this inconvenient disadvantage in two ways: first, they use intercoolers on a multistage compressor, therefore reducing the power requirement of compression and producing low temperature heat. Secondly, moisture is added to the compressed air (20-40%) in a saturator. When HAT is integrated with a coal gasifier (IGHAT), there is a large amount of heat available from the gasifier and other processes (including air separation plant for oxygen blown gasifiers). The hot water is brought into contact with the compressed air in a counter current saturator. Humid air leaving the saturator may have a moisture content of 20% (natural gas HAT) and 40% (IGHAT). This is a key stage of the HAT cycle, since the saturator is a multistage column and the heat exchange approaches reversibility. The humid air is further heated against the hot turbine exhaust in the recuperator. The variable boiling point working fluid of steam and air avoids the large temperature divergence between the water and turbine exhaust in a gas turbine combined cycle. Thus, the moisture addition increases the work output form the turbine, while the intercooler reduces the compressor work requirement. This combines to increase the net power output. The IGHAT cycle can achieve efficiencies higher than those of conventional combined cycles, and is especially adapted for base-load electricity production. An additional feature of this cycle is that, given the high moisture content of the exhaust gas, it is particularly well suited for district heating.

Direct Coal Combined Cycle (DCCC)

DCCC combustion in a gas turbine offers the potential to increase the efficiency of electricity generation from coal by eliminating losses during gasification. However, the main technical barrier is damage to the turbine blades from hot ash particles.

The research on direct coal firing in gas turbines has been carried out for over forty years. Since 1986, Westinghouse and Textron have run a project at Morgantown Energy Technology Centre of the US DoE, with federal funding. The initial difficulties were related to the severe effects of coal ash on turbine blade path components (corrosion, erosion, and deposition). Latter-day effort on direct coal firing (DCF) has concentrated on developing high-pressure (12-16 bar) slagging coal combustors, which allow removal of ash as a liquid prior to entering the turbine. The hot gas clean up must take place above the ash melting temperature (1400-1600°C) and high pressure (at least 18 bar). This high temperature provides additional gains in exergy with respect to other advanced coal cycles, such PFBC or IGCC. The molten ashes accumulate on the edge of the slagging chamber by a centrifugal effect. Nevertheless, there is an efficiency penalty with respect to gas-fuelled natural cycles: the foreseeable efficiency ranks around 44%, and, in general, may be estimated as the efficiency of natural gas combined cycle minus a penalty of around 6 percentage points. Its estimated costs would be twice those of gas-fuelled combined cycle.

It should be kept in mind that DCFCC is not yet a proven technology, its state of development being still in the laboratory phase. Apart from the slagging combustion system and the liquid ash and contaminants removal, all other components of DCFCC (gas turbine, steam cycle, etc.) are commercially available technologies.

A study carried out by Westinghouse/Gilbert-Commonwealth Inc. in 1990 investigated the economics of direct coal fired gas turbines fitted with slagging combustion technology. The study concluded that the cost of electricity from this plant type could be 11-18% cheaper than a comparable 220 MWe pulverised coal boiler fitted with FGD.

Other Options

The Combined Kalina Cycle is an upgrade of the standard Rankine cycle, which, as used in most power generation processes today, uses a single fluid (water) which is expanded in a steam turbine and then condensed and returned to the boiler. The Kalina cycle uses a mixture of two or more fluids, typically ammonia and water, and varies their ratios at different parts of the cycle. This allows optimisation of thermodynamic efficiency, with possible improvements of up to 10%. Conventional steam cycle plant can be adapted to use the Kalina cycle by suitable changes to the condenser.

Other possible future technologies at the R&D stage are Magneto-Hydrodynamics (MHD) and Pressurised Pulverised Coal Combustion (PCCC).

Techno-Economic Performance and Projected Developments

The technical performance of each of the above options is shown in Table 8-2, whereas the economic data of these five technologies are outlined in Table 8-3.

Table 8-2: Technical performance of clean coal technologies (2000).

Technology	Gross capacity	Efficiency (%)		Lifetime
	(MW)	2000	2030	(years)
Supercritical	650	44	49-52	35
PFBC	550	43	44-46	35
IGCC	760	[1]46	50-53	35
IGHAT	400	40	44-46	35
DCCC	40	[1]46	50-52	20

Notes: [1]1990 ECU

Table 8-3: Clean coal technology economic data (2000-2030).

	Capital cost (ECU/kW)		Operation & maintenance costs	
Technology	2000	2030	Fixed (ECU/kW/year)	Variable (m ECU/kWh)
Supercritical	1268	968	45	1.8
PFBC	1030	900	77-67	2.4-2.2
IGCC	[2]1370	900	70-60	1.75-1.5
IGHAT	1300	900	35-30	1.0
DCCC	[2]1200	950	45-30	4.0

Notes: [2]2005

Environmental Impacts

There are three main environmental impacts associated with coal combustion. These are:

- carbon dioxide (CO_2) - global warming;
- sulphur dioxide (SO_2) and nitrogen oxides (NO_x) - acid rain;
- ash disposal issues.

CO_2 emissions are directly related to the carbon content of coal, and thus to the efficiency of the plant. However, the acid rain effect of each technology is much more variable. Emission figures are shown in Table 8-4.

Table 8-4: Clean coal technologies environmental performance-acid rain.

	Efficiency	Emissions (kg/TJ)			
	(%)	SO_2		NOx	
Technology	2000	2000	2030	2000	2030
Supercritical	44	55	30	100	79
PFBC	43	69	65	85	80
IGCC	46	[1]19	17	[1]38	33
IGHAT	40	10	6	20	20
DCCC	46	[1]20	20	[1]55	50

Note: [1]2005

In terms of emissions output, supercritical applications use abatement technologies such as Flue Gas Desulphurisation (FGD) plant, Selected Catalytic Reduction, and low nitrogen oxide (NOx) burners. Emissions abatement for the others is integral within the technologies themselves (i.e. hot gas clean up before combustion, limestone slagging, etc.)

Market Penetration and Prospects for Future Developments

In Europe, the current market penetration of clean coal technologies is extremely small. There are only a few sites, the main ones being in the Netherlands and Spain. However, only super-critical has actually penetrated the market, as the others are demonstration plants.

However, on a world wide basis, there are several clean coal generating stations in the developing world where there is rapid growth in electricity demand, particularly in countries such as China. It is these areas where significant growth in market penetration will be achieved. As experience is gained and the technologies become more economic, the potential for significant penetration in Europe will increase during the next decade.

While there are many reasons for the very small penetration of clean coal plant in the EU, two are dominant. The first was the lifting of the EU directive banning natural gas use in power stations, which was predominantly prompted by the realisation that gas was being wasted by flaring off to get to the underlying oil, rather than by the move towards liberalisation. The second reason was the availability of efficient and relatively cheap Combined Cycle Gas Turbine (CCGT) technology. Consequently, even cheap standard coal plant cannot compete in direct terms with the combined cycle gas turbines currently being constructed.

The other major barrier to the widespread introduction of clean coal is its environmental impact. Even the relatively low emissions from clean coal plant are greater than those from natural gas-fired electricity generation. With climate change commitments requiring the reduction of overall carbon dioxide emissions, the clean coal technologies may only become an option where they replace a 'dirtier' technology, which is old coal or oil plant.

On a world-wide basis, coal will remain a major primary fuel for at least the next two decades. Countries with a large indigenous coal resource and a rapidly growing demand for electricity, such as China and India, will inevitably build significant new coal-fired electricity generating capacity. In these circumstances, however, competition arises from traditional pulverised fuel plant, which is cheap and an established technology. This issue cannot be underestimated.

There are two aspects to be considered when exploring the future potential importance of clean coal. The first relates to the fuel. Coal is the most abundant of the fossil fuels and therefore has a long term future after natural gas and oil reserves become more expensive. Thus, the long term prospects for clean coal are good. However, this assumption is based upon the concept that the reserves are accessible. The European (deep mine) coal industry is shrinking rapidly and is already a tiny proportion of itself 20 years ago. If this decline continues until there is almost no remaining European deep mine industry, many of these current reserves will become inaccessible and will cease to be an effective fuel source.

The second aspect is technology. Coal-fired generation will only achieve its full potential if clean efficient plant is developed and installed. Therefore, it is imperative that the best technological options for clean coal are developed as soon as possible in preparation for its resurgence as the technology of choice.

Economic Competition, Constraints and Costs: "Technology Push" and "Demand Pull"

Both technology push and demand pull have affected the development of clean coal. They both have positive effects for the development of clean coal technologies. In the present European market, there is very little scope for clean coal in the short term. However, the medium and long term prospects for the technologies involved can only be brought forward if demand is stimulated by government.

As the European electricity market becomes more liberalised, clean coal plant will only penetrate to significant levels if it becomes economic against alternative fuels and technologies. This can be achieved by subsidising the capital cost of the plant or by subsidising plant operation. Plant operation can be subsidised by receiving a premium price for electricity generated (similar in principle to the UK Non Fossil

Fuel Obligation) or by reducing operating costs (tax credits or fuel subsidies - for example, the former German KohlPfennig, Spanish electricity subsidies, etc.)

The alternative means of stimulating demand is to force it. This is achieved through regulations requiring the implementation of clean coal plant, or by preventing the market penetration of competing options.

Results of R&D

Research and development in clean can, ultimately, make clean coal technologies more economic, by affecting:

- operating efficiencies;
- emissions (sulphur dioxide, nitrogen oxides, particulates);
- ash management and disposal;
- control of the combustion process.

R&D is also used to build demonstration plants from which the lessons learnt for full-scale plant are fundamental as part of the demonstration process.

In terms of R&D management, there is also the question of combining all of these separate issues for the completed plant, as different areas of research can impinge upon the studies being performed in a different section of the electricity generating process.

Therefore, in order to achieve the best benefits and value, it is vital that research and development is co-ordinated so that all results can be integrated to provide a valuable product.

Conclusions: Prospects for Clean Coal into the 21st Century

On a world wide basis, the prospect for clean coal is extremely good, especially in rapidly developing markets such as Asia. This is especially the case in more remote areas where the availability of alternative fuels is poor. However, in these countries, speed of development is often the primary concern and the traditional pulverised fuel combustion plant is often selected, mainly because the technology is established, easy to construct and cheap. Therefore, it is in these situations that the advantages of clean coal, such as higher efficiencies and lower emissions, need to be emphasised through international agreements, legislation, and regulation.

In Europe, there is little perceived short-term competitive scope for large investment in clean coal technology plant, mainly owing to the low cost of natural gas and the maturity of the market itself with minimal growth expected. However, in the medium to long term, as natural gas becomes a more scarce fuel and prices increase, and in conjunction with further economic improvements in clean coal technologies, clean coal might experience a renaissance as a feasible option for new large scale electricity generating plant.

Another major issue is environmental legislation or taxation related to climate change and acid rain. A measure such as the imposition of a carbon tax could encourage the implementation of clean coal plant as opposed to pulverised fuel plant. However, the overall impact of a carbon tax would be to penalise coal and possibly prevent its widespread market penetration.

In terms of research expenditure, coal is a medium to long term fuel option, which will maintain a significant presence in world wide electricity generation. The future of coal combustion is through the general introduction of clean coal technologies. R&D and demonstration are necessary to achieve economic competitiveness for these technologies, and to enable clean coal to achieve a significant market share in electricity generation. However, it is vital that clean coal research is carefully co-ordinated so that the optimal benefits can be achieved for the minimal cost.

Therefore, it may be concluded that clean coal has the potential to become a major technology for electricity generation. However, there are several risks and possible obstacles in Europe in terms of security of fuel supply, which is less the case in countries that still have large amounts of accessible reserves. Other risks include environmental regulations, technological improvements, cost reduction and market confidence that have to be met or overcome, if clean coal is to achieve its potential.

8.1.5 Fuel Cells for Stationary and Mobile Applications

Fuel cells, by directly converting fuel chemical energy into electric energy, can attain higher yields than that of thermal machines. Apart from their high global performance in the direct production of electricity, fuel cells present some other advantages: silent functioning, small pollutant emissions, relatively small size, good performance in partial charging, modular construction, short manufacturing time, quick replacement and economic maintenance. For all these reasons, fuel cells could represent one of the most promising future technologies.

An Efficient and Clean Technology, but still too Expensive

Since 1990, fuel cells have generated a great deal of interest among a growing number of energy suppliers, manufacturers of components, chemical companies, and R&D organisations.

There is a family of alternative fuel cell technologies corresponding to different electrolytes: Alkaline Fuel Cell (AFC), Phosphoric Acid Fuel Cell (PAFC), Molten Carbonate Fuel Cell (MCFC), Solid Oxide Fuel Cell (SOFC), Polymer

Electrolyte Membrane Fuel Cell (PEMFC). The evolution of this family, or cluster, is characterised by its increasing variety.

Although there are currently close to one hundred demonstration operations in the world (Table 8-5), fuel cells are still at an initial stage of their life cycle:

- phosphoric acid fuel cells (PAFC), which constitute the first generation, are the only ones at a pre-commercial stage;
- the second generation, molten carbonate fuel cells (MCFC) and polymer electrolyte membrane fuel cells (PEMFC), are at a stage of large-scale demonstration prototypes. PEMFC prototypes with a 10–30 kW capacity have been tested in transport and emergency supplying;
- solid oxide fuel cells (SOFC), representing the third generation, are at a research-development stage.

Table 8-5: Fuel cell capacities by technology.

	Existing Units	**Installed Capacities**	**Largest Capacities**
PAFC	160	32	11 MW (IFC, Japan)
PEMFC	11	0.4	120 kW (Ballard, Canada)
MCFC	n.a.	n.a.	2 MW (ERC, USA)
SOFC	n.a.	n.a.	25 kW (Westinghouse, Japan)

Fuel cells present interesting characteristics in the production of electricity:

- 40% to 65% power yield;
- combined heat and power production (co-generation) that allows rising total yields up to 85%;
- great variety of usable fuels (natural gas, propane, butane, naphtha, methanol, carbonic oxide gas, hydrogen, gasified coal, biomass or land-field gas etc.);
- negligible noise and atmospheric pollution;
- moderate power and easily divisible standard size;
- small maintenance cost, due to the reduced number of mobile parts;
- high quality of the steam produced by mcfc and sofc, which can be used for co-generation both in buildings and industry.

They can become competitive from one hundred kWe to several MWe in three segments:

- gas supplied sites of the tertiary sector;
- isolated collective network sites;
- interconnected sites or average power units (from a few MWe to a few tens of MWe) could be implemented at the centre of a consumption area.

The possibility of electric vehicles powered with fuel cells, which would allow autonomy of 250-300 km against the 80 km of the best batteries today, has been seriously considered for some time. Using fuel cells for mobile applications requires the following:

- reasonable weight, linked to the problem of hydrogen supply and storage;
- constant/steady power;
- an almost complete insensitivity to atmospheric temperature;
- rapidly replaceable spares and an excellent degree of reliability. Fuel cells could compete with diesel vehicles in road transport, and could substitute electric vehicles in urban policies aimed at limiting local pollution, which is usually based on promoting clean vehicles.

8.1.6 Drawbacks to Diffusion

There are several obstacles that prevent extensive commercialisation of fuel cells. Considering the very small number of fuel cells produced every year and the materials employed, the cost and the purchase price of fuel cells remains high (see Table 8-6). This is because there are no economies of scale and, at the moment, production is characterised by the absence of standardisation and the high intensity of qualified work.

Table 8-6: Present costs and target costs for fuel cell technologies.

	PEMFC	PAFC	MCFC	SOFC
Present Cost ($/kW)	n.a.	2500-100000 (system)	n.a.	n.a.
Objective ($/kW)	46 (60 kW) mobile 2000 (40 kW stat.) 1000 (250 kW stat.)	250-400 (cell) 1500-2000 (system 200 kW)	1500-200 (system)	215-650 (cell) 1600-3750 (system 50 kW)

As consumers are not familiar with the advantages of fuel cells, they cannot be contributors to the diffusion of fuel cells. Also, the absence of a well-established fuel cells industry contributes to the high risk and cost and to the financing obstacles for their commercialisation.

There is great difference between the results divulged by laboratories and those obtained in practice. Considerable efforts are necessary to overcome the process of change from a single cell, functioning in a controlled environment, to a cell module operating in real world conditions.

For large-scale mobile implementations a number of technical obstacles must be overcome before commercialisation, including the development of increasingly thinner proton exchange membranes, optimisation of electrodes, minimisation of platinum etc. Also, a number of technical and economic problems must be resolved regarding the storage of hydrogen. In the case of the cells destined to be used in electric vehicles, hydrogen can be supplied either by on-vehicle storage of methanol, associated to a reforming device which transforms fuel into hydrogen whenever necessary, or by pure hydrogen storage. The latter can be stored in five different ways: as compressed gas, in its liquid form through intermediate metallic hydrides, in hollow microspheres and in carbon nanofibres. The first option (methanol) is more convenient and cheaper than the second is. In addition, it would be compatible with the existing network of petrol stations. However, this method would result is some pollution, in the form of carbon oxide (CO) and carbon dioxide.

Competitiveness with other Power Production Systems

In order for fuel cells to become competitive with other power production systems (such as the GTCC, steam turbines, diesel units for stationary implementations etc.), especially in co-generation ranging from some hundred kW to some MW capacity, their cost of capital must be reduced to 1500-1600$/kW, depending on the application.

For mobile implementations, fuel cells have the advantages of being much less polluting and of being able to yield much higher efficiencies than current power systems. The fuel cell market covers four market segments: city buses, fleet vehicles, private vehicles and locomotives. Their diffusion by sector is likely to begin with city buses and fleet vehicles before spreading to individual vehicles and, finally, to locomotives. Competitive access of fuel cells is more difficult for mobile applications than for stationary applications. In the case of buses, even if the cost of a PEMFC module were to fall to 300$/kW, a fuel cell bus would still be 30% more expensive than a diesel bus. Mass production economies could reduce costs to less than 150$/kW. For fuel cell private cars to become competitive with electric vehicles (e.g., in local anti-pollution policies based on clean vehicles), the PEMFC system must be reduced to 46-55 $/kW.

Competitiveness between Different Types of Fuel Cell

Several technical challenges remain unresolved and it is still too soon to forecast which type of fuel cell will reach commercial success. PEMFC and SOFC, which only contain solid components, seem to be the most promising, since they avoid liquid electrolyte defects. Thus, SOFC can have simpler designs, cheap ceramic or metallic materials, which could help reduce costs. Tests carried out in laboratory show that SOFCs can have more than a 10-year life span, that is, twice that of liquid electrolyte cells. PEMFCs seem to be more suitable for transport applications: they work at relatively low temperatures (around 80°), providing significant energy from atmospheric temperature, ensuring a short start-up time. They are insensitive to the presence of carbon dioxide gas in the air, unlike "alkaline" fuel cells.

Evolution of Agents' Strategies and Support Policies

Limited information and the complexity of the fuel cell technologies (AFC, PAFC, MCFC, SOFC, and PEMFC) make necessary the pooling of a number of different skills. This has led to co-operation and partnerships between a number of seemingly unrelated entities. Such entities include industrial firms that develop and sell fuel cells, gas and electricity companies, big automobile manufacturers and professional associations. The main industrial firms developing and selling fuel cells include: IFC (US), CLC (Europe), Toshiba (Japan), Ballard (Canada), Fuji Electric (Japan), ERC (US), Melco (Japan), Sanyo (Japan), IHI (Japan), MC-Power (US), BCN (the Netherlands), Hitachi (Japan), Westinghouse (US), MHI (Japan), Allied Signal (US), Dornier, Siemens, ECN (the Netherlands), Sulzer (Switzerland) Ztek (US), De Nora, Ansaldo Ricerche (Italy).

Their objectives for the financing of technical progression towards commercialisation are the following:

a) in order to reduce costs:

- the reduction of surface area and volume per power unit
- the improvement order systems (i.e. intelligent systems)
- the simplification of manufacturing
- the reduction of the cost of auxiliary elements and the increase of their durability: turbo-compressors, filters, catalysts, etc.

b) in order to facilitate their commercial exploitation:

- easy use and maintenance, especially for transport applications
- normalising and modularising individual elements in order to facilitate technology exchange

- fuel cell integration in other equipment
- gas and electricity companies that are major fuel cell buyers include General Electric (US), Fuji Electric Company (Japan), Mitsubishi Electric Company (Japan), Elfraft, ELSAM (Denmark), Sydrakft, Vattenfall, Iberdorla (Spain), ENEL, AEM, Acoser (Italy) -the two latter being council services-, SEP (the Netherlands), Tokyo Gas (Japan), British Gas (UK), Ruhrgas (Germany), SNAM (Italy), Naturgas Syd (Denmark), Imatran Voima (Finland), Autriche Ferngas (Austria), Enagas (Spain), ENI (Italy). These companies consider fuel cells as a potential future opportunity to serve their clients' needs more effectively.

Automobile firms have been devoting a rising amount of investment in the search for a fuel cell powered vehicle. Until recently, the experimental fuel cell vehicle, with the size of a small lorry, was effectively a small factory where almost all the space was occupied by the fuel cell, its compressor and its tank. Progress has been made towards miniaturisation. Daimler-Benz is the first major international vehicle producer to have presented an electric passenger vehicle run on fuel cells. The first prototype model, called "New Electric Car" (NECAR I), was presented in April 1994 by Mercedes-Benz. Daimler-Benz presented the first "standard" monospace run on PEMFC and developed by Ballard Power System, in June 1996. This vehicle can develop 50 kW gross power, with a weight of 6 kilogram per kW system. It reaches a maximum speed of 110 km/h, with a 250-km autonomy. Other producers in this field are General Motors, Ford, Chrysler, Toyota, Honda, BMW and Renault.

Public Policies

Despite substantial efforts at developing this technology, fuel cell commercialisation has been slower than was expected. It has been supported more actively in Japan and the United States, by means of governmental programmes and private companies, than in Europe where interest in fuel cells is more recent.

The present challenge for governments is to help increase the size of the market during the first years of commercialisation, when costs are still high due to the relatively low production levels. Distribution of fuel cell production units should benefit from the reinforcement of environmental measures and the pressure exerted towards some power production decentralisation in Europe.

Cost Perspectives

According to Daimler-Benz researchers, technological improvements will allow a ten fold cost reduction of current prototypes of fuel cells, by 2003. Thus, PEMFC could cost 5,500-6,800 $/kW by 2003. In order to put this in the competitive environment that will be faced by fuel cells, it should be mentioned that the cost of capital of a combined cycle is 600 $/kW, that of a diesel bus 150 $/kW, and that of a private vehicle 55 $/kW. Thus, in a pessimistic scenario built on the basis of learning curve analysis, fuel cells will be competitive with combined cycles after the first half of the coming century. Following the most optimistic scenarios, PEMFCs will be competitive with combined cycle plants around 2020, with diesel powered buses in 2050 and with conventionally fired private cars in 2070.

Perspectives for Market Penetration

Recent views by some researchers of world-wide sales of fuel cells amounting to 1500-2000 MW/year by 2000 seem to be excessively optimistic. As will be seen in later chapters, the degree of penetration of fuel cells even in the longer term is critically dependent on a large number of factors including, technological developments, energy prices etc.

Fuel cell diffusion for stationary applications will probably begin with niche decentralised production, followed by co-generation prior to extending to centralised production, where competitiveness is more difficult to achieve because costs of conventional technologies are rather low. Diffusion of fuel cells into the transportation sector is likely to follow stationary applications.

8.1.7 Wind Power Generation

At the beginning of the eighties new tax policies in the United States favoured the use of renewable energy, and a market for low and medium power turbines was created. Tax incentives together with the regulation concerning the conditions of repurchasing electricity by electric power companies produced an extremely rapid increase in the number of devices installed. This was especially the case in California, which became the "the world's biggest wind centre".[18] Wind power installed in California increased from 10 MW in 1981 to almost 1000 MW in 1986, but this rapid growth was suddenly interrupted as incentives were eliminated in the mid-eighties.

The elimination or reduction of both federal and Californian State tax incentives in 1985, as well as the fall in oil prices, slowed down noticeably the rate of installations in the United States: 400 MW were installed in 1985, 300 MW in 1986, 150 MW in 1987 and 60 MW in 1988. Keeping in mind that the American market of wind power devices was relatively important at that time, the industrial over-capacity that resulted from the collapse of demand, led to extreme competition between producers some of whom have since disappeared.

---

[18] I. Miles et al, 1994.

The limitations of the American market led European producers (and mainly the Danish who had high participation in the United States) to concentrate again on their national markets. Thus, in the Netherlands and in Great Britain, wind power production became really significant at the beginning of the nineties.

It should be noted that wind power generation is not only developed in industrialised countries. India in particular experienced an important growth since the beginning of the nineties, to the point that the Indian market was second, after Germany, in new wind power installations.[19]

At the end of 1996, worldwide wind power installations exceeded 6000 MW and the annual electricity generation amounted to more than 10 TWh. While this is low with respect to the scale of the electricity sector in general, it should not be ignored since the growth rate of the global wind market has reached 20 to 30% during in recent years.

The distribution of worldwide installed windmills has changed significantly in recent years. The United States, which accounted for 55 % of the world's capacity in 1993, represented only 27 % of the market in 1996, whereas Europe's share increased from 41 % to 57 % over the same period.

Introduction to Wind Power Technology, Present Performance and Perspectives

Wind turbines allow the production of electricity for individual applications, such as lighthouses and buoys, or for the electrification of houses or villages in regions which are not easily accessible, especially in developing countries.

Nevertheless, since the beginning of the eighties, the development and implementation of wind power generation has mainly centred on electricity generation to feed the power supply grid. In order to reduce the costs of connection, infrastructure, maintenance etc., wind power devices are more and more systematically grouped to form wind power centres, also called wind farms, which may contain hundreds of turbines (California) and generate power of up to tens of MW.

Currently, in the most favourable sites, from the point of view of wind resources, the costs to produce a kWh from wind are similar to those of traditional coal-fired power stations. However, economic competitiveness is still far from being reached in more conventional sites.

Furthermore, the rotor diameter, which was 10 to 12 m in the early eighties, now commonly approaches 40 m, corresponding to 500 kW nominal power (as opposed to 50 kW with a 10 m rotor), with 40 m high masts. 750 kW windmills

---

[19] B.T. Madsen, 1995.

are currently being built and commercialised, and 1 500 kW devices with 60 m rotors are also being tested. These much larger dimensions of windmills are justified by economy-of-scale effect: more favourable distribution of very high windmills, higher productivity with larger rotors, because of comparably lower maintenance costs, better exploitation of the available sites, etc.

The average productivity of Danish windmills (450 kW model) reaches 900 kWh/m2 whereas the first models (55kW) produced approximately 400 kWh/m2. Similarly, in California the average productivity increased from 500 kWh/m2 in 1985 to 850 kWh/m2 at the end of the eighties, and some wind farms with particularly good wind conditions even reached 1450 kWh/m2 (San Gorgonio Farms - USA) or 1300 - 1400 kWh/m2 with the 10 best Danish devices.[20]

The costs of wind power generation today are of about 0.05 up to 0.10 ECU/kWh.[21] It is obvious that this is still far from the 0,02 $/kWh stated by the DOE at the beginning of the nineties, and it is doubtful that this level may ever be reached.[22] Nevertheless, there is significant progress with respect to the costs of the first Californian installations in 1981 and 1982 which were near 0,50 $/kWh.

Turbine installation cost almost 3-4000 $/kW at the beginning of the eighties. Now the turbine manufacturing cost amounts approximately to 1200 $/kW[23] or 1100 $/kW[24] for medium power devices (4-500 kW). The costs of installation, foundation, road access, connection to the power supply system, the purchase of the site, etc. have to be added, constituting 25 % to 50 % of the turbine cost. According to an AIE analysis in the principal wind power developing countries average manufacturing costs were of 780 - 1200 $/kW in 1995, i.e. an average of 1000 $/kW which means a slight decrease compared to previous years.[25]

Operation and maintenance costs have decreased notably during the last 15 years, and, at present, may amount to 0.01 $/kW in the most favoured sites. Nevertheless, these costs, like those of installation, are totally dependent on the site, the number and the dimensions of the installed windmills.

---

[20] P. Gipe, 1995.

[21] TERES II, 1996.

[22] K. Karas, 1992, quoted by P. Gipe.

[23] P. Gipe, 1995.

[24] Braun and Smith, 1992.

[25] AIE, 1995.

Environmental Impacts

Atmospheric emissions caused by wind power generation are insignificant compared to those produced by fossil energies. Usually they result from the energy contained by materials or used to construct, to install or to maintain the devices.

The noisiness of windmills is often mentioned as a nuisance. While it should not be totally neglected, the noise level near windmills is rather limited: a modern windmill of 300 kW operating at a wind speed of 8 m/s produces 45 dB at a distance of 200 m, which is "noisier than a bedroom at night but quieter than a house during the day".[26]

The visual impact of windmills, and even more so that of wind farms, often in nature reserves, is probably the main environmental disadvantage of this new type of energy. The first Californian installations, in particular, have significantly contributed to form a pejorative image of wind power.

Perspectives and Expected Progress with Respect to Performance

The energetic performance of modern windmills, which is already relatively high (80% of the theoretically available energy is extracted by the turbine at an optimum velocity[27]) can be only slightly increased.

Furthermore, to improve the performance of the devices, their dimensions have to be increased in order to profit from the much higher wind speeds at a greater altitude and to enlarge the swept rotor surface. This method, in use since the beginning of the 1970s, allows the average power of commercialised windmills to become 10 times higher. The 500 kW threshold, reached in the mid 1990s, is being surpassed by new 750 kW devices now on the market.

Present pilot studies with offshore wind power generation seem to be quite interesting. By means of offshore production it is possible to profit from the much higher wind speeds and better location, avoiding major visual impact by installing the wind farms at a certain distance from the shore. Nevertheless, the obtained increase in productivity does not compensate for the extra costs of foundations, electric cables and maintenance, so that an offshore produced kWh is 30 to 60 % more expensive than that of the equivalent overland configurations.[28]

The European Wind Energy Association (EWEA), which groups the companies involved in the sector, aims at covering 10 % of the energy requirements of

---

[26] EWEA, 1991.

[27] M. Grubb, 1996.

[28] M. Grubb, 1996.

Europe in 2030 by means of wind power. This would require the installation of some 100 000 MW of wind power.[29] Such a target would imply growth rates as high as those observed in recent years in Europe over long periods of time. Taking into account the enduring economic limitations and the difficulties involved in implanting new sites in some countries, this objective is probably very difficult to reach.

The Importance of the Resource and the Limitations of its Availability

A study, carried out in Denmark, evaluates the resource at 780 TWh. It is based on the hypothesis of the systematic and evenly distributed installation of 1 MW windmills evenly distributed over the all areas.

The introduction of new limitations, visual impact, noise, impact on the fauna, etc. lowers this potential by some 10 TWh. This potential might be complemented by offshore production. The same study carried out on a European scale (12 countries) led to a potential of 620 TWh, i.e. 40 % of the total electricity generation.[30]

However, the problem of irregularity is more limiting. As the availability of the resource is uncertain, wind power-stations can not guarantee their maximum installed power at a given moment, and thus, the installed wind power generation capacity should not be counted at 100 % for annual peaks and it could weaken the reliability of electric systems.

Most experts agree that a relative 5 to 10 % wind power contribution would not produce any weakening of the power supply system.[31] However, it might be inadvisable for wind to contribute at levels over 25 to 45 %, depending on the actual power supply system.[32]

R&D Actions

After the petrol crisis the renewed interest in wind power resulted in the promotion of vast research programmes in industrialised countries. These programmes are mainly oriented towards the development of large-sized devices. Research into the economies of scale effects seems to be the most appropriate and rapid method to reduce turbine costs. Therefore, first the Americans and then the Europeans established research programmes, implicating aerospace industries (Boeing, British Aerospace, Messerschmidt, etc.) or other organisations involved

---

[29] EWEA, 1991.

[30] M. Grubb and N. Meyer, 1993.

[31] AIE, 1993 quoted by M. Kliman.

[32] M. Grubb and N. Meyer, 1993.

in the sector like NASA, in order to develop wind power turbines capable of producing several MW. Some of these devices have reached 3 MW with rotor diameters of almost 100 m, as for example: WEG LS-1, 3MW (GB), Mod-5B, 3,2 MW (USA), WTS-4 4 MW (USA), Growian 3MW (D).

These programmes were doomed to failure. This is made clear by the fact that, at the same time, the Danish were developing their own technological trajectory based on low power windmills (7-30 kW) and on a tried and tested idea. This trajectory has been boosted by incremental improvements to the initial concept and to operation apprenticeships. These have been encouraged by market enlargement policies ("market incentives") established by the Danish government to favour small independent producers (especially co-operatives). This is how the present technical standard has been reached. The 500 – 700 kW commercial devices currently available clearly show their derivation from the first Danish windmills of some ten kW.

The prevailing trajectory has consisted in progressively increasing the power of windmills based on a very classical conception, but without apparent limits so that certain companies are getting ready to commercialise 1 MW devices. The advances on the power (and size) scale have been achieved by the validation of previous steps and the incorporation of accumulated knowledge. This approach, which may not seem very ambitious, allowed a significant progress to take place. This progress was evident both in terms of productivity of the devices (250 – 300 kWh/m2 in 1980 to 600 kWh/m2 in 1988 and 800 kWh/m2 in the mid-nineties) and of production costs (4000 $/kW at the beginning of the eighties to 1250 $/kW in the mid-nineties).[33] At the same time the reliability and availability of turbines was considerably improved.

Consequently, technological progress in the wind power sector seems now to confirm a unique standard constituted by a horizontal-axis windmill with two or three blades mounted on a tubular tower about 50 m high, which generates 500 to 750 kW. Present R&D programmes aim to pursue this trajectory incorporating improvements which are basically incremental (variable velocity, compound materials, easily adaptable towers, etc.), but, in general, technological issues are not the only ones to be considered in the definition of research programmes. In this aspect, the orientations given by ETSU in Britain for wind power research programme are very instructive.

The Stimulation of the Demand

Most countries resorted to investment aid in order to facilitate the appearance of wind power generation, which was established by the PURPA Law in the United

[33] See EWEA, 1991 and P. Gipe, 1995.

States. Investment subsidies like tax-deductible credits have significantly contributed to the explosion of the Californian market. Similarly, the first Danish support programmes for the wind power markets were based on a 30 % subsidy of the investment costs. This subsidy was then progressively reduced, then abolished in 1989.

Fixed price systems oblige electricity companies to buy the electricity generated by independent producers. In general, they are calculated on the basis of average domestic sale prices or of the avoided costs, independent of supply guaranties.

Finally, these price incentive measures may be completed by the creation of reserved markets corresponding to a fixed proportion of electricity production to be generated by windmills, or renewable energies in general. In Great Britain, for example, 1500 MW are reserved to be produced by renewable sources till 2000, within the framework of the NFFO (Non Fossil Fuel Obligation). In France, continual invitations to tender for EOLE 2005 should provide the means to reach the 250 to 500 MW objective.

The Industrial Participants

The wind power generation industry has advanced considerably during the last fifteen years. Europe occupies the first place in the world's wind power generation sector. Danish constructors are the European leaders and hold a very important place in the international market: in 1994 Danish exports represented 50 % of the global market. German and Dutch constructors have also followed the development of their national markets, strongly boosted by public policies, but up to now they have been less present in the international market. Similarly, the American constructors with the exception of Kenetech, handicapped by the fact that wind power technology is mainly increasing in Europe, are relatively insignificant in the global market.

Table 8-7: Major construction companies of wind turbines.

Constructor	Accumulated store in MW	Accumulated store number of devices
Vestas (DK)	623	4 500
Kenetech (USA)	400	3 900
Micon (DK)	265	>2 000
Bonus (DK)	>250	2 000
Nordtank (DK)	253	2 123
Enercon (D)	250	nd
Mitsubishi (J)	>210	370
HMZ Windmaster (B)	> 95	nd
Nedwind (NL)	> 55	nd
Wondworld (DK)	> 50	nd

Source: F. Armand, 1995.

8.1.8 Photovoltaic Electricity

The photovoltaic cell (PV) consists of semi-conducting electronic components. The cell constitutes an elementary power generator. The resulting current is a function of the light intensity and the cell's surface and yield. A 10cm x 10cm cell generates energy of the order of 1.5 Wc under a 0.5 V-tension. Several elementary cells are therefore connected to each other in a module, in order to provide a greater amount of energy and tension, i.e. 35 Wc under 12 V. The final product commercialised by the PV industry is the module, which can be assembled in the shape of panels regrouping several modules.

The major relevance of PV energy stems from its functioning principle: the electricity is produced without moving parts from a renewable energy, solar light. PV energy is, at least in theory, reliable and durable and its functioning does not produce any emissions or noise. Furthermore, the PV can be separated into modules. A PV generator can be limited to a module of some tens of watts or produce several hundreds of kW, or tens of MW.

Technological Aspects

Silicon represents 97% of the world's PV production: crystalline silicon (mono- and multi-) and amorphous silicon. The development of other technologies is still marginal but they could play a more important role if the PV technology were to develop further.

Crystalline Silicon

The basic raw material used in the industry is the silicon from the wastes of the electronic industry. This material is recast to obtain ingots that are later cut in layers several hundred microns thick to form PV cells.

Multicrystalline silicon presents several advantages compared to monocrystalline silicon. It demands lower quality raw materials, the production technique of the ingots is faster and less energy- consuming and cells are directly produced in a squared shape that allows better module refilling rates. These advantages are in part compensated by the lower crystallographic quality of the material, which implies an inferior photoelectric yield to that of monocrystalline silicon.

The highest current yield of industrial multicrystalline cells is 13%, which corresponds to the yield of the functioning modules of the order of 10%. The theoretical yield limit for crystalline cells is close to 30%. Crystalline silicon presents the best compromise at present between yield and cost of production.

Amorphous Silicon

Production processes of crystalline photocells are not compatible with continuous production processes that would allow an important reduction of production costs. Thin-layer elaboration processes go beyond these constraints by depositing very thin layers of semi-conducting material on a glass substratum. The main thin-layer technology, and the only one currently industrialised, is the amorphous silicon (a-Si).

Production costs are relatively low and economies of scale could lead to a further reduction. The main a-Si shortcoming stems from its current yield, which is significantly inferior to the yield of crystalline cells: the a-Si yield is unstable in solar rays and can fall to 4-5%-stabilised yield in real conditions of use.

Other Components of PV Systems

A PV system also includes the module assembly structures, in order to ensure mechanical resistance and to orient the PV, as well as electric connections and, possibly, a storage system and electronic order system. In the event of very powerful systems, the acquisition of a site, its preparation, the impact studies, the network connections, etc., contribute to the overall cost of the installation. Specific batteries were developed for the storage of electricity some years ago for isolated PV systems, allowing for high discharges which automobile batteries could not admit.

Concerning metallic structures, one of the main perspectives for development is module integration in ceilings or façades. This latter possibility is being seriously considered by the PV industry since the cost of the PV modules is only slightly above 50% of the total cost of a PV installation. With no symmetrical progress of the costs of all the components, the improvement of the cell yields or the decrease of their production cost will only exert a small influence over the cost of the PV kWh.

Investment, Functioning and Maintenance Costs

The cost of a PV installation can be systematically divided into two parts: the PV element stricto senso (the modules) and the non PV components. Currently, the production cost of the modules is estimated to be 420 ECU/m2,[34] Estimates for the costs of non PV components are an average of 3500 ECU/kW (but without batteries).[35] In general, some 40 to 60% of the total cost of the system originates from the non PV components.

In order to assess the cost of PV equipment, distinctions should be made according to the range of power envisaged and the applications regarded (several W or tens of them, versus some MW, autonomous systems supplied with stock versus network systems).

The total cost of PV equipment is rather variable, depending on the capacity and the nature of the service to be rendered (small-size autonomous system vs. electric plant connected to the network). In Europe, the cost of a power production plant connected to the network ranges from 6000-8000 ECU/kW, whereas the cost of a system integrated in a façade or roof may be lower.[36] For smaller-size installations or smaller volume orders, costs can be rather high. For example, small units destined to ensure lighting in rural areas of non electrified developing countries (an average 50 Wp) can reach or surpass 20 $/Wp (16000 ECU/kWp), including the costs of the final equipment.

Present estimations for PV power production costs range from 25 to 300 cents (1990) / kWh, 25 to 250 cents / kWh for isolated systems, but 30 to 40 cents (0.24 to 0.32 ECU/kWh) for PV plants.

Environmental Impact

The environmental impact of PV-cell power production is small. Surface occupation is clearly one of the main environmental limitations of PV energy. The issue of toxic wastes follows the use of potentially dangerous materials in the construction of PVs. Nevertheless, the amounts at stake are small compared to

---

[34] EPIA, 1996.

[35] Wrixon, 1993.

[36] Wrixon, 1993.

other industrial sectors and recuperation and treatment techniques should reduce the scope of the problem.

Economic Perspectives

PV producers estimate that crystalline silicon yields will reach 15% return at the beginning of the 20th century. The yield could increase to 18%, through further improvements of existing processes. The production of amorphous silicon tandems should reduce instability problems and ensure stable yields of around 10% or more towards 2005-2010.

At the same time, the production costs of PV modules is likely to continue to decrease. The cost of the production of crystalline PV modules could evolve from 350 to 250 ECU/m2, that is, a maximum cost of 1.8 ECU/Wc by 2010. The cost of production of the a-Si modules could reach 0.8 ECU/Wc (80 ECU/m2). Parallel improvements are envisaged on non PV components resulting from successive learning stages. The total cost of a PV plant with non PV components could, therefore, descend to 1000 ECU/kW (from a 3000 ECU/kW minimum today) for an ideal configuration based on modern technologies.

World Market

World-wide sales of PV in 1994 amounted to almost 70 MWc. The total installed capacity all over the world is estimated at 350 MWc, excluding non energy applications.

Traditionally, four main categories of PV energy applications have been distinguished:

- professional applications: power supply for isolated or difficult-access areas;
- interior or leisure applications;
- isolated systems for domestic or collective use: isolated systems are used in mountain huts, villages or rooms not connected to the grid;
- systems connected to the grid: this category covers both "powerful" power PV plants (beyond 1 MWc) and small individual equipment (several kWc of the "solar roof" type.

By prolonging the rate of growth observed in the period 1984-94 (15%/year), world-wide annual sales in 2010 would rise to 630MWc, i.e. an accumulated installed capacity of about 4000MWc.[37]

---

[37] EPIA, 1996.

There is consensus among experts that a significant increase will take place in the isolated PV equipment market over the coming years, and, particularly, in the equipment destined for rural power supply of developing countries.

8.1.9 PV Cell Producers

The production of PV modules takes place mostly in OECD countries: Europe (essentially Germany, France, Italy, Netherlands and the United Kingdom), Japan, USA. Japan and the USA dominated the market to a great extent until 1987 but Europe has gradually improved its position and today occupies the second position, following the United States. Module production also takes place in India, China, Brazil, Argentina etc., but is of smaller scale.[38]

About twenty companies represented more than 80% of the volume in the world market in 1994. The largest five companies accounted for more than 50%. Siemens is the world leader with a 20% market share, having overtaken the American firm ARCO Solar. Most of these firms belong to larger groups, usually to companies from the oil, chemical or electronic sectors. The participation of electric companies is much more limited and relatively recent. The rate of use of the production capacities for x-Si are an average of 75% (84% for USA and 60% for Europe). On the other hand, the rate of use of the capacities for a-Si is inferior to 50% (36% for USA).

Public Policies

Public policies in favour of PV energy are divided between programmes aimed at improving the commercial performance, at developing new technologies, at experimenting with new applications, etc., and actions to support PV markets (investment grants, protective tariff policies, etc.).

R&D policies of IEA countries in the field of renewable energies seem to be influenced by the political and economic context. Stimulated by the oil crisis of the 1970s, spending on renewables R&D suffered following the decrease of the oil price and of the constraints of public budgets.

Over this period (1984-94), the PV absorbed a third of R&D public credits and its share increased even more in 1993 and 1994, when it was granted more than half of the total R&D budget. 85% of the R&D credits assigned to PV originate from only to four countries: Japan, the United States, Germany and Italy.

In Europe, an important contribution to R&D is provided by the Joule programme (I & II). A total 32-million ECU have been devoted to PV research (Joule I & II)

---

[38] Production is assigned to the country of location of the company's headquarters .

and, in relation to PV, to research on improving x-Si performances, exploring the potentialities of new materials and developing thin-layered cells.

The difference between demonstration and diffusion, in the case of PV applications, is not always obvious since the number of units involved usually remains relatively limited (particularly concerning large production installations). However, programmes carried out by several countries (USA, Germany, France, and Japan) are mainly concerned with diffusion, insofar as specific and permanent incentive mechanisms are put into practice to encourage the adoption of PV power production systems. Other parallel mechanisms help its diffusion in export markets and, particularly, in the electrification of rural areas in developing countries. Certain countries have carried out specific aid programmes in this area.

9 Technology Scenarios to 2030: Baseline and Alternative Technology Scenarios[39]

Technological development is a complex phenomenon that has drawn the interest of many scholars. Understanding the nature of the forces behind the process of scientific discovery, technological innovation and know-how dissemination is a challenge for which many theories have been formulated and are still under discussion. It is customary to distinguish between basic research, often supported by public funding, and applied research, whose outcome is protected by intellectual property rights under some legal scheme. It is probably impossible to forecast completely the nature and stage of development of the technologies that will be relevant for power generation in 2030. Indeed, at least for the time being, there are no large scale energy models, which even try to model technology progress endogenously.[40]

The difficulties of forecasting over such a long period are based not only on the nature of technological discoveries and breakthroughs, which are almost by definition unexpected, but also on the dependence of the technology of 2030 on the actions taken by economic agents between now and then. Future power generation technologies will also be determined, by the market structures, which are currently changing extremely rapidly.

Consequently, the methodology adopted in this chapter is to construct a number of possible technology scenarios for the power generation system over the next 30 years. It is hoped that this approach will stimulate debate on the likely benefits of these alternative technology futures and on the likely priorities for public funding.

---

[39] The principal author of this chapter was Antonio Soria of IPTS. The application of the technology scenarios were carried out by Patrick Criqui of IEPE and Niko Kouvaritakis of ECOSIM, at the world level, and by Pantelis Capros and Leonidas Mantzos of NTUA.

[40] A number of efforts to address this weakness are currently under way, including the one by the EU funded TEEM project.

These scenarios were based on a number of *feasible* and *plausible energy technology trajectories* that have been constructed around a limited number of technology clusters, which were discussed in the previous chapter. One of the reasons for this approach was the necessity of identifying the benefits of a breakthrough « other things being equal », as well as the possible synergies within each cluster. Indeed, some of the key technologies within each cluster could experience significant progress, and therefore induce important cost reductions in the related technology families which, in turn, could lead to much higher penetration of these technologies.

Another key defining characteristic of the scenarios presented below is the degree of market concentration. As it has already been mentioned, some technologies, because of their capital requirements, size of optimal plant etc., are more likely to be associated with decentralised market structures.

9.1.1 Energy Technology Baseline Projection

In order to analyse the impact of likely future energy technology trends, a reference technology deployment trajectory has been explored within this research. It corresponds to a BAU (Business As Usual) scenario, for whose numbers a consensus has been achieved after an in-depth look in the literature survey.

The electricity generating costs for some selected technologies as they are characterised in the technology database and corresponding to years 1990 and 2030 are reported in the following graphs. For these cases it was assumed that, during the period considered, the price of coal would increase by 5% in real terms, lignite price would remain stable, whereas oil and natural gas prices would experience substantial increase, by 50% and 80%, respectively. These assumptions are in conformity with the results of the POLES model on international energy prices.

Table 9-1: 1990 Electricity generation costs by load for some selected technologies.

	Hours per year							
Electricity cost (mECU/kWh)	**0.50**	**1.00**	**2.00**	**3.00**	**4.00**	**5.00**	**6.00**	**7.00**
Hard Coal 200-500 MW	435	224	119	84	67	56	49	44
Hard Coal>500 MW	338	176	95	68	55	46	41	37
Lignite + FGD (200-500 MW)	403	206	107	74	58	48	41	37
GTCC 200-350 MW	213	114	65	49	40	36	32	30
Oil Boiler (200-500 MW)	383	203	112	82	67			
Nuclear 1000-1500 MW LWR	691	346	173	115	87	69	58	50
Large Hydro	651	326	163	109	82	65	55	47
Supercritical Coal	401	207	110	77	61	51	45	40
IGCC>500 MW	532	272	142	98	77	64	55	49
PFBC>500 MW	410	211	112	79	63	53	46	42
Integrated Oil Gasif. Comb. Cycle 200-500 MW	434	226	122	87	70	60	53	48
New Nuclear design	720	360	180	120	90	72	60	51
Advanced Coal Cycle	307	160	86	61	49	42	37	33
Small Hydro>2 MW	465	233	116	78	58			
Wind on shore > 0.5 MW	383	192						
Low Temp. Solar	330	165	83					
Biomass	635	319	161	108	82	66	55	48
Fuel Cell Stationary	515	278	159	119	99	87	80	74

Table 9-2: 2030 (BAU) Electricity generation costs by load for some selected technologies.

	Hours per year							
Electricity cost (mECU/kWh)	**0.50**	**1.00**	**2.00**	**3.00**	**4.00**	**5.00**	**6.00**	**7.00**
Hard Coal 200-500 MW	392	203	108	77	61	51	45	41
Hard Coal>500 MW	336	175	94	67	54	46	40	37
Lignite + FGD (200-500 MW)	363	185	96	67	52	43	37	33
GTCC 200-350 MW	176	95	54	41	34	30	27	25
Oil Boiler (200-500 MW)	351	186	103	75	62			
Nuclear 1000-1500 MW LWR	731	366	183	122	92	73	61	53
Large Hydro	520	260	130	87	65	52	44	37
Supercritical Coal	319	165	88	63	50	42	37	33
IGCC>500 MW	434	222	117	81	64	53	46	41
PFBC>500 MW	370	191	102	72	57	49	43	38
Integrated Oil Gasif. Comb. Cycle 200-500 MW	374	196	106	77	62	53	47	43
New Nuclear design	645	323	161	108	81	65	54	46
Advanced Coal Cycle	232	121	65	47	38	32	28	26
Small Hydro>2 MW	426	213	106	71	53			
Wind on shore > 0.5 MW	312	156	78					
Low Temp. Solar	188	94	47					
Biomass	544	273	138	93	70	57	48	41
Fuel Cell Stationary	515	278	159	119	99	87	80	74

9.1.2 World Energy Technology Scenarios

The scenarios presented below have been constructed on the basis of the present state and future potential of each technology cluster, as discussed in Chapter 10. It should be underlined again that, although the scenarios are supposed to capture a complete, coherent and self-consistent hypothesis on the whole energy system, they have been developed only for the power generation technologies.

The technologies involved in scenarios 1 and 2 are based on the assumption of centralised electricity production and distribution system. On the contrary, scenarios 3, 4 and 5 are related to decentralised power generation systems.

9.1.2.1 Centralised Electricity Production: the Nuclear Renaissance and Incremental Innovations

Under this hypothesis, when the time for massive decommissioning of nuclear power plants installed in OECD countries will arrive (around the year 2010), no other reliable and carbon free alternative will be available for baseload electricity production. In this case, one could forecast the substitution of old nuclear power plants by new, safer (possibly smaller) nuclear devices. Nevertheless, the possible downsizing of this technology would never reach a degree in which the power generation could be considered as decentralised. The typical power of these new, advanced reactors could lie around the 600 MW, i. e. close to the typical size of coal-baseload power plants (which ranges around 300-600 MW). The possibility of disruptive innovations in this field (such as the *energy amplifier* concept) should be considered also within this scenario.

The applicability of this scenario is mainly limited to the OECD countries, which already have nuclear power, the possibility for new economies to shift towards this scheme *ex nihilo* being lower. Nevertheless, some emerging economies could also follow this way (India, Pakistan, and China). The implications that this may have in terms of global geostrategy are beyond the scope of our discussion, but maybe however very important. However, in terms of nuclear weapon proliferation, the choice for massive nuclear-based power generation in a limited number of these countries would not dramatically change the composition of the world nuclear club of nations. In this scenario, new and renewable energy technologies would operate in the fringes and progress at a relatively low speed.

Basically, the global picture that has to be retained within this scenario is:

- the *power sector* is supposed to rely on the reengineering of nuclear power plants, with progress in safety and waste management and the finding of socially acceptable solutions to decommissioning and retrofitting. In a longer time horizon, passive reactors and/or energy amplifiers, focusing on intrinsic safety, enhanced operation flexibility and smaller size would allow for the massive replacement of nuclear by nuclear, and for the partial nuclearisation of some emerging economies. At the European scale, this nuclearisation (or re-nuclearisation process) may take place first on a reduced cluster of countries (France, Belgium, Germany, and Sweden). A second group could then follow, including the UK, Spain, the Netherlands and Italy;

- the *transport sector* would be supposed to stay relying on oil-based fuels, but allowing for a progressive penetration of hybrid cars first, and then electric battery cars;
- from the *natural resource capital stock* point of view, for this scenario to occur, one should probably assume also high primary fossil fuel prices and reduced reserves, especially of natural gas.

The technology characterisation supplied to the models under these assumptions were the following:

- all the nuclear capacity dismantled will be replaced by new nuclear in OECD countries. To achieve this, one has to be *really optimistic* for what concerns the projected costs of nuclear, and quite pessimistic on the future fossil fuel prices;
- a significant share of baseload electricity production in the largest emerging electricity demanders (China and India) will be nuclear-based;
- the rest of the technologies could be assumed to follow a non constrained but conservative path.

9.1.2.2 The Cleaner Fossil Fuel Based Baseload Electricity Production

Under this scenario, nuclear power might be progressively substituted by cleaner fossil-fuel-based power generation schemes for baseload production. The degree of diffusion of these technologies may depend on the restrictions applying to a given world zone: the developed world is likely to shift toward GTCC schemes for medium and peak load production. These technologies still have a good potential for efficiency improvement, provided a smooth evolution of the gas prices. However, new large consumer nations sitting on huge coal reserves (China, India, South Africa etc.) are likely to use these resources. The potential of advanced clean coal technologies is to be found mainly there.

The possible technological trajectory along this scenario would include as a first step the adoption of fluidised bed combustion power plants (in all its possible variants), to be progressively replaced by supercritical and ultra-supercritical power plants that may reach thermal conversion efficiencies far beyond 50%. The utilisation of ceramic filters and advanced inox steel will allow for noticeable operation improvements and simultaneous specific pollutant emissions reductions. Marginal improvements to the today predominant Rankine thermodynamic cycle may also be obtained by the use of modified thermodynamic fluids. Further on, new coal-based technologies (Integrated Gasification, possibly in combination with Humid Air Turbines, Direct Coal Fired Combined Cycle, Magneto-hydrodynamics Cycle, etc.) have to prove their full potential beyond 2010.

Therefore, in international terms, the picture may be associated to *the hypothesis of a joint implementation success story*, within a basically non changing power sector structure. The limitations imposed by the baseload coal-dominated sector would in this case preclude (at least at the first stages of this scenario) a massive decentralisation of the grids.

The sector-wise picture that could describe this scenario is as follows:

- the *power sector* is supposed to rely on the technology innovations concerning clean coal technologies and the gasification of solids. The Rankine-cycle technological trajectory would start from traditional coal boilers and fluidised bed combustion plants, which are supposed to be rapidly replaced by supercritical and ultra-supercritical power plants. This would be due to significant progress in material science, allowing for the development of steels able to withstand the extremely severe enthalpy conditions of the main steam in these plants. Simultaneously, developments in the gasification technologies may take place, allowing for biomass-based (either waste or crops) techniques, both based on the gas turbine combined cycle thermodynamic approach;
- ;he *transport sector* would also in this case stay relying on oil-based fuels. Oil should more and more become a specialised fuel for transportation applications. The expected technological progress in this field would be incremental, allowing for a progressive penetration of low-consumption cars (5 l/100km and less), giving less space for the penetration of the electric car. This because no cheap baseload electricity would be available, contrary to what could happen in the first scenario;
- for what concerns the *heat market* within this scenario, the expected changes with respect to the present situation should be of minor importance. The diffusion of cogeneration, although important, may have a limited potential ceiling due to the limits on natural gas and the difficulties in implementing such schemes for non utility purposes at lower scale;
- from the *natural resource capital stock* point of view, within this scenario, one should simultaneously assume high natural gas prices (or low reserves) *and* low coal prices (or high reserves).

The foreseen technology characterisation supplied to the models under these assumptions is the following:

- the nuclear will be replaced by nuclear in OECD countries only in a relatively reduced share (new reactors may be introduced for experimental purposes);
- baseload electricity production will be progressively occupied by clean-coal technologies, first in OECD, and immediately also in the emerging economies, who might benefit from joint implementation schemes. Natural gas is expected

to enter in the power sector very intensively, but operating mostly in the peak hours.

9.1.2.3 The Gas Induced Decentralised Power Generation System

This technology path corresponds to a more radical change within the organisational structure of the energy system. Under this hypothesis, it is assumed that today's economies of scale in producing and distributing energy (and more precisely, electricity) will progressively disappear, being substituted by technologies operating with *diseconomies of scale* (in other words, with *increasing marginal costs*). It is also supposed that, for this scenario to develop, natural gas supply should be abundant and relatively cheap.

As a first expected result, the optimum production-unit size would be lower. Consequently, large, high voltage linear transport lines will loss their importance and will be substituted by local, radial, low-voltage grids, locally operated. The interconnection between different generators, necessary to ensure an adequate grid reliability, would therefore be multiplied at a very reduced scale. This may facilitate the economic competition between independent power producers, as well as a significant degree of technology diversification. However, it is expected that this decentralisation in power generation would not be strong enough so as to induce dramatic changes in the distribution system.

The most salient technology that may be effectively downsized without significant efficiency loss is the gas turbine (either in single or combined cycle). As this scenario would be almost entirely dominated at the beginning by natural gas, an appropriate network of gas pipes or any other transporting system is mandatory in this case. Long-distance electric lines would be partly replaced by gas pipes. The generation and distribution of electricity would more and more take place at a local level.

This system may be complemented by some traditional centralised baseload electricity generation, decentralised co-generation to satisfy heat demand, as well as some intermittent renewable electricity production operating in their market niches.

The sector-wise picture is the thus the following:

- the *power sector* would be supposed to rely on the innovations concerning gas turbine-related technologies. Solid fuels and some nuclear capacity would also be present for a share in baseload electricity generation. The former would be used under IGCC schemes. To summarise, further advances in small GTCC and coal power plants may be foreseen, as well as on-site conversion of wastes and biomass residues;

- the presumed abundance of natural gas and the pressure to limit carbon emissions would induce some changes within the *transport sector* along this scenario: one may assume that the oil-fired explosion engine vehicle may progressively shift towards compressed natural gas (CNG) vehicles. The potential market share of these vehicles would in the end be determined by the price paths of oil and natural gas. The use of bio-methanol from biomass pyrolysis as a transportation fuel may also be foreseen;
- the *heat market* within this scenario may develop according to the general decentralisation trend, by a significant penetration of co-generation schemes within the industrial sector. The end-use efficiency increase in this sector would therefore be noticeable;
- the *demand-side* target technologies would have a comparatively minor importance, since the emerging generation technologies would be able to accommodate both to peakload and baseload. Significant improvements are to be assumed for what concerns the development of CNG vehicles;
- concerning the *natural resource capital stock* issue, one should assume optimistic hypotheses on the global reserves of natural gas, since it would occupy a progressively higher role within the world's primary energy mix.

As a consequence, the pressure on natural gas prices would be progressively higher. As natural gas prices increase, coal gasification and direct coal combustion may start to be more and more attractive, provided that, in the meantime, the corresponding capital costs had already undergone already a significant decrease.

To summarise, the fundamental technological features of this scenario are:

- nuclear power is replaced by nuclear only in a very limited share;
- the availability of natural gas at low price would contribute for massive installation of small scale GTCC power plants;
- coal resources would partially be channelled to power production through gas-fuelled Brayton turbines: IGCC, DCCT (direct coal combustion turbine) or similar technologies which allow for more flexible, load-tracking devices: optimistic assumptions are made on the costs of these technologies.

9.1.2.4 The Energy Efficient Decentralised Power Generation System

This scenario would represent a still more radical change within the organisational structure of the energy system. It may be considered as a technologically enhanced version of the previous one. Indeed, in the short run, its development would overlap with scenario 3: gas turbines would allow for a first move towards decentralisation with lower size the production units. But as other alternative technologies will be available, they would enter progressively into the market. In particular, natural-gas fuel cells for decentralised electricity production would be the natural sequel to the initial small-size gas turbine capital equipment.

Consequently, large, high voltage linear transport lines will loss their importance and will be substituted by local, radial, low-voltage grids, locally operated. The interconnection between different generators, necessary to ensure an adequate grid reliability, would therefore multiply at a very reduced scale. This may facilitate the economic competition between independent power producers, as well as a significant degree of technology diversification. The existence of an appropriate network of gas pipes or any other transporting system is also a requisite for this scenario to develop. The traditional centralised baseload electricity generation would remain important only during the first transient period, to progressively loose importance afterwards. Decentralised co-generation would also contribute to satisfy heat demand. The technology push induced by fuel cells would allow for a more intense expansion of intermittent renewable electricity expansion production, possibly using hydrogen as an energy carrier.

The sector-wise picture is the following:

- the natural gas is supposed to dominate a more and more decentralised *power sector*, first via gas turbine technologies, then with the even more efficient energy conversion fuel cells. As in the previous scenario, solid fuels and some nuclear capacity would also be present for a share in baseload electricity generation. In a second step, fuel cells and new systems for independent power producers will represent a disruptive innovation (although still gas-based);
- the emergence of fuel cells for static applications is supposed to spread to electric fuel cell automobile. This might imply a dramatic change in the *transport sector*. The availability of a cheap primary fuel (natural gas in this case) is a prerequisite for this scheme to emerge. Later on, the issue of alternative fuels (hydrogen from renewables) may depend on the long-term behaviour of the fossil fuel reserves, as well as on the prices of these conversion technologies (biomass distillation, renewable-based water electrolysis, etc.);
- the generalised decentralisation of the energy system would foster the outbreak of energy service companies, as well as *local markets for heat and power*;
- as in scenario 3, the *demand-side* target technologies would have a comparatively minor importance, because of the stronger links between supply and demand, thanks to the decentralised generation technologies;
- as in scenario 3, one should also assume optimistic hypotheses on the global reserves of natural gas within this scenario, and, therefore, low gas prices at least in the first phase.

As a consequence, the pressure on natural gas price would be progressively higher. As natural gas price increases, coal gasification and direct coal combustion may start to be more and more attractive, provided that, in the meantime, the corresponding capital costs had undergone already a significant decrease.

To summarise, the technological characterisation would look like this:

- the role of centralised electricity generation is significantly reduced and nuclear power is replaced by other types of baseload generation units;
- the availability of natural gas at low price would contribute for massive installation of small scale GTCC power plants in the first phase of the period of analysis and of low-cost fuel-cells in the second phase;
- in order to properly simulate this scenario, one has thus to assume *really optimistic* capital costs decrease for fuel cells beyond 2005-2010;
- the fuel cell development would facilitate the use of coal and biomass resources through gasification schemes.

9.1.2.5 A Future of Renewable Energy Technologies

The picture selected within this hypothesis is maybe the less likely in the short and medium term. Its early stages may be coincident with scenarios 2 and 3. Further in advance, one could devise a world situation in which, either due to supply-side shocks from the fossil-fuel market, or to a dramatic decrease of renewable energy technologies (RET) costs, or to both reasons, RET would reach a commercial status. The industrial organisation within this scenario may combine aspects from the fully centralised electricity generation and transmission scheme, and those of the decentralised downsizing scheme.

Among the first, we may outline the possibility for long-distance, baseload electricity hydro resources that may be put in exploitation if the adequate conditions in transmission technology apply (super conductor lines, etc.). Concerning the second, home-produced electricity or heat (via photovoltaic, or low temperature solar thermal devices, whose market penetration may be induced by using the appropriate building regulation) may be introduced.

The fundamental feature that would characterise this scenario is the necessity to find appropriate technological solutions to provide a sufficient degree of flexibility in the system, in order to accommodate to a significant share of intermittent power generation. The key technologies would therefore concern energy storage. This may take place at the power source, by accumulation of energy under several forms: thermal (molten salt reservoirs), mechanical (pump storage), chemical (synthetic fuels) or electrical (batteries), to be delivered to the grid when the original source is not available. The energy storage can also take place at the final consumption level. In addition, demand-side management technologies would have a key role in achieving and maintaining the above-mentioned system flexibility.

This scenario could be described, by sector, as follows:

- the fossil fuels are supposed to diminish their predominant role within the *power sector*. This does not mean that they would disappear, but rather that they will be confined to captive markets. The standard electric networks will be reinforced by new long distance transmission technologies able to put in exploitation large scale renewable power generation schemes: large hydro, on-shore and off-shore wind power, as well as solar thermal power plants. Large-scale electricity storage could cope with the problem of intermittence of renewables. Simultaneously, small scale decentralised production, based on solar heat and building-integrated PV systems, small wind generators and battery storage would allow for massive decentralisation;
- the emergence of renewables will also imply significant changes in the transport sector. The elaboration of synthetic fuels using renewable energy technologies would certainly induce significant changes in the environmental external costs of transport. The changes may or may not be based on technological disruptions: synthetic methanol could, for instance be used within conventional (although more efficient) internal explosion engine cars, or used in electric fuel cell cars. The energy carrier could also be hydrogen. The adoption of a given technological paradigm will certainly depend on the cost and the opportunity factors;
- as in scenario 4, the decentralisation of the energy system would encourage the appearance of energy service companies operating in local markets for heat and power;
- contrary to what was said for scenarios 3 and 4, a necessary condition for this scenario to occur is that the energy system would move onto a situation with high primary fossil fuel prices (either due to shortage in reserves or to full internalisation of the corresponding external costs).

The main features of this scenario would be:

- nuclear power is replaced by decentralised fossil fuel based electricity generation first, then by baseload, long-distance renewable energy supply;
- the electric grid evolves towards a more decentralised one, combined however with some long-distance, large capacity transport lines;
- the fraction of baseload power occupied by renewables increases as a consequence of developments in energy storage technologies.

9.1.3 Technical Note: Running the Scenarios

Following the definition of the five Technology Scenarios, their simulation required the construction of a *coherent set of input data*. The two models used, POLES and PRIMES, have different features and therefore special care had to be taken, in order to combine suitable hypothesis for technology deployment. POLES, for instance, requires particular assumptions for technology costs and performances in different world zones. This problem is not present when dealing with the European energy markets in PRIMES, although the characterisation of different national energy systems within this model also implies the introduction of country-specific information. The degree of technological disaggregation is also different between the two models. The current version of POLES includes basically 12 conventional electricity generation technologies, as well as 12 new and renewable energy technologies, whereas the number of energy technologies retained in PRIMES is significantly greater. The conventional electricity generation technologies considered in POLES are:

- conventional, large-size hydro
- conventional nuclear LWR
- new nuclear design
- super-critical Coal Units
- integrated gasification combined cycle
- advanced thermodynamic cycle (coal powered)
- lignite-powered conventional
- coal-powered conventional
- oil-powered conventional
- gas-powered conventional
- oil-powered gas turbine combined cycle
- gas-powered gas turbine combined cycle

The important point to note is that in the current state of the models, part of the variable costs (variable operation & maintenance, but not fuel costs) and the fixed costs (investment and fixed operation & maintenance) are exogenous. It has been therefore necessary to modify the set of data (eventually including some capacity constraints) in order to set up the different technology scenarios. The technology database that has been developed in the project was the basis for the identification of a consistent set of technological paths. For each model, the list of 64 technologies revised by the Technology Working Group under the project, have been used to select the most appropriate technologies to be included within each model, in order to capture as adequately as possible the technological development path.

The corresponding data have also been cross-checked by looking at the confidence intervals provided in some of the literature that was used to elaborate the original database.

It should be underlined that modifying the technology data is not the only way in which different energy technology scenarios can be simulated and analysed. Indeed, the modification of some of the structural constants of a given model could be the preferred way to shift from one scenario to another, with the same set of tech.-economic data. Increasing the market share elasticities (or substitution parameters) may cause small perturbations in the costs to have amplified effects on the market shares, as the absolute value of the substitution parameter increases. This would induce a faster penetration of slightly cheaper technologies. This is why the issue of model calibration and design has to be regarded in close connection to the data provided for the scenario definition.

While the NREN module of POLES includes a learning curve function for each technology, this module has not been used in the simulation of the five technology scenarios in order to maintain the consistency of the comparisons. The endogeny of technical progress may have however a capital importance for the modelling and the simulation of the dynamics of the diffusion of renewables. It would allow the integration of the institutional R&D expenditures as an explaining factor of the technological change instead of the more ambiguous concept of technological exogenous progress. Under this new approach, the technical progress, specified as a technological policy decision variable and then subjected to structural and short-term fluctuations, may have a great influence on the diffusion of the renewables.

Some other issues could, if taken into account in further research, lead to the definition of technological policies for the diffusion of renewables. These include:

- the modelling of emerging energy technologies does not currently assume any annual *capacity constraint* for the diffusion of the renewables. These constraints could be added to the models in order to improve the simulation of scenarios corresponding to different degrees of commitment of the industry and the states or to different CO_2 emissions reduction policies;
- moreover, by integrating R&D, it would be possible to build different assumptions on the future support of the authorities towards the diffusion of renewables: specifically, it could involve the possible substitution of the institutional R&D by private R&D expenditures;
- this institutional behaviour issue should be compatible with the calibration of the substitution parameter in the market-share allocation of electricity

generation for each technology. Indeed, within the present structure, *no real competition* is allowed between traditional electricity generation schemes and renewables: whereas the former are installed according to the demand forecast, the capital and other various variable costs, the latter diffuse according to an almost independent, Fischer-Pry-like scheme.

Nevertheless, and bearing in mind the significant effort that has been made to upgrade the structure of the electricity sector within POLES and PRIMES, the above remarks should be kept in mind only as possible future themes for further development.

9.2 Summary of Assumptions

The following tables refer to the summary of assumptions.

Table 9-3: Nuclear technologies alternative scenarios.

ECU90/KW	1995	2030 reference case	2030 scenario
SPECIFIC CAPITAL COSTS			
Nuclear 1000-1500 MM MW LWR LWR	2000	2150	1500
New Nuclear Design	2217	2042	1400
Supercritical Coal		968	
Gas Turbine Combined Cycle	570	470	
FIXED OM COSTS			
Nuclear 1000-1500 MM MW LWR LWR	76	76	50
New Nuclear Design	73	59	30
Supercritical Coal	40	40	
Gas Turbine Combined Cycle	29	24	
FUEL EFFICIENCY %			
Nuclear 1000-1500 MM MW LWR LWR	34	34	34
New Nuclear Design	31	39	39
Supercritical Coal	44	49	
Gas Turbine Combined Cycle	50	59	

Table 9-4: Clean coal technologies alternative scenarios.

ECU90/KW	1995	2030 no tech. progress	2030 reference case	2030 scenario
SPECIFIC CAPITAL COSTS				
Hard Coal (Conven.)	928	928	835	760
Supercritical Coal		1268	968	750
Integr.Gasif.Combined Cycle		1370	1090	820
Advanced Coal Cycle (Direct Comb.)		1200	1000	750
Gas Turbine Combined Cycle	570		470	
FIXED OM COSTS				
Hard Coal (Conven.)	46	46	46	46
Supercritical Coal		40	40	40
Integr.Gasif.Combined Cycle		70	60	60
Advanced Coal Cycle (Direct Comb.)		45	35	30
Gas Turbine Combined Cycle	29		24	
FUEL EFFICIENCY %				
Hard Coal (Conven.)	38	38	39,6	39,6
Supercritical Coal		44	49	49
Integr.Gasif.Combined Cycle		46	49,8	54
Advanced Coal Cycle (Direct Comb.)		46	50	52
Gas Turbine Combined Cycle	50		59	

Table 9-5: Enhanced gas technologies alternative scenarios.

ECU90/KW	1995	2030 reference case	2030 scenario
SPECIFIC CAPITAL COSTS			
Gas Turbines in Combined Cycle 200-350 MW	570	470	410
CHP	900	814	640
Adv. Coal Cycle		1000	780
Supercritical Coal		968	

Table 9-5: Continued.

FIXED OM COSTS			
Gas Turbines in Combined Cycle 200-350 MW	29	24	10
CHP	45	40	33
Adv. Coal Cycle		35	30
Supercritical Coal	40	40	
FUEL EFFICIENCY %			
Gas Turbines in Combined Cycle 200-350 MW	50	59	63
CHP (electricity)	28	40	42
CHP (steam)	45	45	47
Adv. Coal Cycle		50	52
Supercritical Coal	44	49	

Table 9-6: Fuel cells technologies alternative scenarios.

ECU90/KW		**2000**	**2030 reference case**	**2030 scenario**
SPECIFIC CAPITAL COSTS				
Fuel Cell Cars (Hydrogen)	per system	180000	46000	15000
Solid Oxyde (fix-cogen)		1700	800	600
Proton Exchange Membran FC (fix)		3455	864	259
FIXED OM COSTS				
Fuel Cell Cars (Hydrogen)	(ECU90/carKM)	0,07	0,01	0,01
Solid Oxyd FC (fix)		40	30	10
Proton Exchange Membran FC (fix)		50	30	10
FUEL EFFICIENCY %				
Fuel Cell Cars (Hydrogen)	(carKm/kWh)	4	7	9
Solid Oxyd FC (electricity)		60	70	70
Solid Oxyd FC (steam)		20	20	20
Proton Exchange Membran FC (fix)		55	60	63

Table 9-7: Renewable technologies alternative scenarios.

ECU90/KW	2000	2030 reference case	2030 scenario
SPECIFIC CAPITAL COSTS			
Biomass Gasification < 25 MW	1767	1767	960
Photovoltaics in Buildings	12000	4000	2000
Solar Thermal Low Temp.	1200	750	500
Small Hydro > 2 MW	2000	1850	1000
Molten Salt Tower Plant & Storage	2500	1603	1131
Photovoltaics Rural Electr.	13100	10000	5000
Wind on Shore	1300	1100	460
FIXED OM COSTS			
Biomass Gasification < 25 MW	50	30	20
Photovoltaics in buildings	100	50	25
Solar Thermal Low Temp.	10	5	3
Small Hydro > 2 Mw	20	16	16
Molten Salt Tower Plant & Storage	64	34	25
Photovoltaics Rural Electr.	250	180	160
Wind on Shore	16	14	14
AVAILABILITY FACTOR %			
Photovoltaics in Buildings	15	17	17
Solar Thermal Low Temp.	27	31	33
Small Hydro > 2 MW	57	57	57
Molten Salt Tower Plant & Storage	27	32	34
Photovoltaics Rural Electr.	9	9	14
Wind on Shore	20	24	32

10 Scenario Impacts on World Energy and Emissions[41]

This chapter describes the impact of the technological scenarios, which were presented in the previous chapter on the global power generation system. In order facilitate the evaluation of the impacts of the changes in assumptions, the baseline power generation trends are presented first.

10.1 Baseline Electricity Trends

The technologies retained in this study include those considered in the central electricity production module of POLES, as well as some other included in the new and renewable module.

The baseline world electricity mix for 2030 is summarised in Table 10-1 to Table 10-4. At world level, Table 10-1 shows the remarkable success of the gas turbine in combined cycle technology: by the end of the period, this technology represents almost one fourth of total world electricity generation, one third of thermal production. Amongst the baseload technologies, supercritical coal seems to expand rapidly its market share, although IGCC also retains a significant degree of competitiveness due to its flexibility in using fuels of varied quality.

---

41 The principal authors of this chapter are Patrick Criqui of IEPE and Niko Kouvaritakis of ECOSIM.

Table 10-1: POLES world electricity projection 1992-2030.

	1992	2000	2010	2020	2030	1992/ 2000	2000/ 2010	2010/ 2020	2020/ 2030
Electricity Generation in TWh	12670	14998	22377	30764	40769	2.1%	4.1%	3.2%	2.9%
Thermal	7787	9495	15964	23113	31655	2.5%	5.3%	3.8%	3.2%
of which:									
Clean Coal	108	130	566	2007	3348	2.4%	15.8%	13.5%	5.3%
Gas Turbines	271	1040	3840	6410	8931	18.3%	13.9%	5.3%	3.4%
Biomass	138	174	227	283	348	2.9%	2.7%	2.2%	2.1%
Nuclear	2126	2304	2567	2978	3607	1.0%	1.1%	1.5%	1.9%
Hydro+Geoth	2219	2584	3238	4000	4796	1.9%	2.3%	2.1%	1.8%
Solar	0.95	1.75	6.30	20.33	35.81	7.9%	13.7%	12.4%	5.8%
Wind	4.66	9.76	14.86	29.18	60.37	9.7%	4.3%	7.0%	7.5%
Small Hydro	125.3	168.6	183.8	213.4	250.0	3.8%	0.9%	1.5%	1.6%
CHP	405.9	433.2	403.4	408.7	363.7	0.8%	-0.7%	0.1%	-1.2%

Table 10-2: Baseline world electricity mix (TWh) by 2030.

Large Hydro	4795	**Oil Powered Gas Turbine**	1455
Nuclear 1000-1500 MW LWR	2538	**Gas Turbine in Combined Cycle 200-350 MW**	8873
New Nuclear Design	1042	**Gas Turbine in Combined Cycle & CHP**	364
Supercrititcal Coal	5940	**Small Hydro**	249
Integrated Coal Gasification with Combined Cycle	2978	**Wind on Shore**	60
Advanced Coal Cycle	2784	**Solar Power Plant**	31
Lignite & FGD	2761	**Waste Incineration CHP**	81
Hard Coal 200-500 MW	5593	**Biomass Gasification**	268
Oil Boiler Monovalent	414	**Proton Exchange Membran Fuel Cell**	101
Gas Powered Trad.Turbine	556	**Solid Oxyde Fuel Cell**	7

In the OECD countries (see Table 10-3), the growth of total electricity consumption and generation remains very moderate. Gas turbines progress rapidly by the beginning of the projection period and at a slower pace by the end, when this technology represents one third of thermal production. Clean coal technologies also show high growth rates, which remain so until the end of the projection: they represent one tenth of thermal production in 2030. With a very slow rate of development, nuclear continuously loses market shares in the electricity production, excepted during the last decade, where the rate of growth of nuclear is again very slightly superior to the average. In spite of high growth rates - initially for wind power and by the end of the projection for solar power - the contribution of the new renewable technologies to power generation remains very limited in volume.

Table 10-3: POLES OECD electricity projection 1992-2030.

	1992	2000	2010	2020	2030	1992/ 2000	2000/ 2010	2010/ 2020	2020/ 2030
Electricity Generation in TWh	7688	8583	10924	13002	14976	1.4%	2.4%	1.8%	1.4%
Thermal	4359	5078	7268	9012	10589	1.9%	3.6%	2.2%	1.6%
of which:									
Clean Coal	64	74	256	740	1036	1.9%	13.1%	11.2%	3.4%
Gas Turbines	271	794	1950	2707	3205	14.4%	9.4%	3.3%	1.7%
Biomass	84	105	117	144	173	2.8%	1.0%	2.1%	1.9%
Nuclear	1744	1822	1908	2133	2486	0.5%	0.5%	1.1%	1.5%
Hydro+Geoth	1136	1199	1281	1350	1398	0.7%	0.7%	0.5%	0.3%
Solar	0.89	1.69	6.20	20.17	35.38	8.3%	13.9%	12.5%	5.8%
Wind	4.55	8.88	12.86	24.21	43.28	8.7%	3.8%	6.5%	6.0%
Small Hydro	65.06	89.51	92.88	93.64	95.06	4.1%	0.4%	0.1%	0.2%
CHP	376.85	383.23	354.69	368.26	327.58	0.2%	-0.8%	0.4%	-1.2%

In the developing Asia region (see Table 10-4), the total electricity growth rate is very much higher: from 8 % at the beginning of the projection it however regularly declines, to 4 % by the end. Reflecting the primary resource endowments of the region, gas turbines progress rapidly but achieve a much lower share of thermal production in 2030 than in the OECD region. Conversely, clean coal technologies achieve a much higher share than in OECD, as their contribution is in volume comparable to that of gas turbines. As opposed to the respective dynamics of wind and solar power in the OECD, wind technologies show in Asia a much larger potential than that of solar power: the growth rate of wind power in that region might remain high, between 8 an 10 %/yr., all along the projection period.

Table 10-4: POLES Asia electricity projection 1992-2030.

	1992	2000	2010	2020	2030	1992/ 00	2000/ 10	2010/ 20	2020/ 30
Electricity Generation in TWh	1712	2927	6545	10253	14726	6.9%	8.4%	4.6%	3.7%
Thermal	1251	2203	5475	8752	12742	7.3%	9.5%	4.8%	3.8%
of which:									
Clean Coal	7.40	6.64	200.30	997.76	1813	-1.4%	40.6%	17.4%	6.2%
Gas Turbines	0	146	905	1413	2218	NA	20.0%	4.6%	4.6%
Biomass	7.41	6.65	4.83	4.12	7.84	-1.3%	-3.1%	-1.6%	6.6%
Nuclear	97	190	296	401	560	8.7%	4.5%	3.1%	3.4%
Hydro+Geoth	309	467	711	1016	1308	5.3%	4.3%	3.6%	2.6%
Solar	0.02	0.03	0.07	0.06	0.05	1.5%	10.8%	-1.9%	-2.2%
Wind	0.10	0.81	1.73	3.73	9.63	30.2%	7.8%	8.0%	9.9%
Small Hydro	48.66	58.61	55.52	74.16	101.59	2.4%	-0.5%	2.9%	3.2%
CHP	5.22	6.26	5.19	5.39	3.63	2.3%	-1.8%	0.4%	-3.9%

The five technology stories presented above have been simulated and compared to this baseline scenario, in order to identify the quantitative impacts at world level and by main regions of some major technological breakthroughs in the future.

10.1.1 The Nuclear Scenario

This scenario has been implemented in POLES by altering the tech.-economic characteristics of two types of nuclear plants:

- a standard large LWR, which in the reference case saw its capital, cost slightly increasing over time was assumed to be about 30 % cheaper compared to reference in 2030. Furthermore fixed operation and maintenance costs were assumed to be about 35 % lower;
- a new evolutionary nuclear design assumed to be introduced gradually in the reference case and finishing costing about 5 % less to construct and about 25 % less to operate than the LWR by to 2030 gained a substantial share of the nuclear share (over 25 %) thanks primarily to its inherent safety characteristics. For the nuclear scenario this type of plant was assumed to be 30 % cheaper to construct and 50 % to operate than in the reference case.

These changes have had a substantial impact on the competitiveness of these plants in terms of generating costs: whereas in the reference case they were not competitive in most regions of the world even for the higher annual loads they become generally cost-attractive vis-à-vis combined cycle gas turbines [42] in the region of 5-6 thousand hours per year and vis-à-vis supercritical coal technologies [43] in the region of 4-5 thousand hours per year. Such enhanced competitiveness is particularly marked for the new nuclear design. It is worth noting however that the changes were phased in gradually and power plants but especially nuclear ones are characterised by a slow turnover.

Table 10-5: The nuclear scenario: change in global electricity generation from reference scenario, TWh, in 2030.

Large Hydro	-31	Oil Powered Gas Turbine	-90
Nuclear 1000-1500 MW LWR	1104	Gas Turbine in Combined Cycle 200-350 MW	-1163
New Nuclear Design	4770	Gas Turbine in Combined Cycle & CHP	-54
Supercrititcal Coal	-691	Small Hydro	-13
Integrated Coal Gasification with Combined Cycle	-389	Wind on Shore	-14
Advanced Coal Cycle	-229	Solar Power Plant	-7
Lignite & FGD	-823	Waste Incineration CHP	-10
Hard Coal 200-500 MW	-534	Biomass Gasification	-56
Oil Boiler Monovalent	-41	Proton Exchange Membran Fuel Cell	-23
Gas Powered Trad.Turbine	11	Solid Oxyde Fuel Cell	-1

[42] These are basically the “winning” technologies in the reference case.

[43] These are basically the “winning” technologies in the reference case.

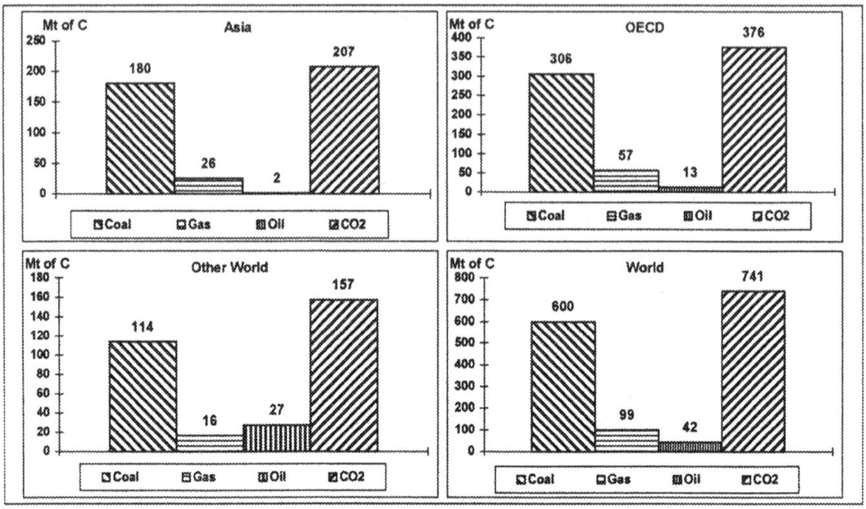

Figure 10-1: The nuclear scenario by zones: comparison with the baseline (2030).

The overall effect of the scenario is a worldwide reduction of CO_2 emissions of 5.3 % (8.6 % in the OECD). Naturally this effect is produced almost exclusively within the power generating sector. World-wide: the new nuclear design plant passes from about 1000 TWh to 5800 TWh and the more conventional LWR from 2500 TWh to 3600 TWh overall nuclear contribution increases from under 9 % to over 22 % (from 16 % to 37 % in the OECD). Clearly nuclear power which suffers from heavy costs in the reference case erodes into the high to medium annual loads displacing about 3000 TWh; of coal fired and 1200 TWh of gas fired electricity production. The most severely affected technology is large scale brown coal fired plants (-30 %) while IGCC and gas turbine CC are both 13 % lower and the remainder coal technologies (8-12%) lower. The relatively important reduction in gas turbine CC is a measure of its big contribution in relative high loads in the reference case. Combined Heat and Power (-15%) and Biomass gasification Combined Cycle (-21 %) suffer from the lower baseload electricity prices resulting from the cheaper nuclear power. Wholesale coal prices are 5-10 % lower while gas prices 3-8 % lower. World oil prices stand virtually unaffected (-1 %) since most of the changes implied by the scenario occur within the electricity sector where petroleum plays a relatively small role.

10.1.2 The Clean Coal Scenario

For the purpose of this scenario three new clean coal technologies have been retained:

1. supercritical coal which in the reference case achieved 49% efficiency and average specific capital cost of around 970 ECU/KW with low operating and maintenance costs (by coal fired power plant standards) was the "winning" coal technology gaining about 14 % of world total central generation and 30 % of world coal generation by 2030. For the purpose of the scenario the efficiency was increased to 55 % and the capital cost brought to 750 ECU/KW by 2030 while a 12 % reductions in O&M costs was also introduced;

2. an IGCC type plant reaching about 50 % efficiency and costing 1100 ECU/KW by 2030 with still relatively high O&M costs of about 60 ECU/KW achieved a penetration of about half the importance of supercritical coal in the reference case. For the scenario the tech.-economic performance of this type of plant was substantially improved to reach 54 % efficiency by 2030 while achieving 25 % reductions in both capital and fixed O&M costs;

3. an Advanced Thermodynamic cycle (direct coal fixed combined cycle) plant costing 1000 ECU/KW and reaching 50 % thermal efficiency by 2030 with relatively low operating costs achieved a penetration comparable to the IGCC plant in the reference case. A capital cost of 780 ECU/KW and an efficiency of 52 % was retained for the scenario resulting in cost performances similar to those of the IGCC plant.

The changes in the aggregated world electricity output by 2030 are summarised in Table 10-6.

Table 10-6: The clean coal scenario: change in global electricity generation from reference scenario, TWh, in 2030.

Large Hydro	-51	**Oil Powered Gas Turbine**	-115
Nuclear 1000-1500 MW LWR	-147	**Gas Turbine in Combined Cycle 200-350 MW**	-1376
New Nuclear Design	350	**Gas Turbine in Combined Cycle & CHP**	-21
Supercrititcal Coal	2111	**Small Hydro**	-38
Integrated Coal Gasification with Combined Cycle	1719	**Wind on Shore**	-18
Advanced Coal Cycle	2254	**Solar Power Plant**	-9
Lignite & FGD	82	**Waste Incineration CHP**	-9
Hard Coal 200-500 MW	-2303	**Biomass Gasification**	-49
Oil Boiler Monovalent	-78	**Proton Exchange Membran Fuel Cell**	-6
Gas Powered Trad.Turbine	19	**Solid Oxyde Fuel Cell**	0

It is worth noting that for regions with access to reasonably cheap gas only the supercritical coal plant, and for loads higher than 6500 hours per year, had a clear advantage over Combined Cycle gas plants. In the scenario under otherwise similar circumstances supercritical coal and Advanced Thermodynamic cycle plants become attractive for loads higher than around 3000 hours/year while the IGCC for loads higher than 4500 hours/year. Some improvements (though clearly on a smaller scale) were assumed especially in terms of capital cost reductions for more conventional coal technologies which were justified as knock on effects arising from the improvements in the new clear coal technologies.

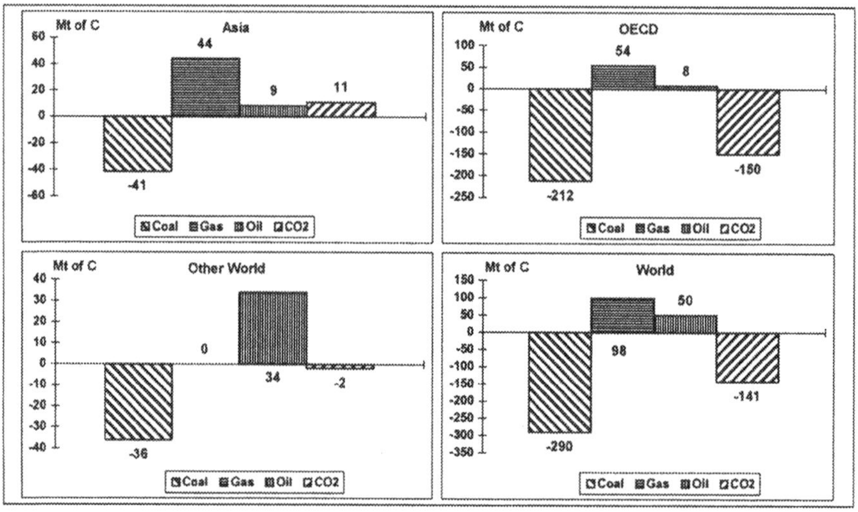

Figure 10-2: The clean coal scenario by zones: comparison with the baseline (2030).

This scenario results in a net increase in world CO_2 emissions (+1 %) which conceals some regional differences: a slight decreases in Asia (-0.2 %) an increase (3.5 %) in the OECD and stagnation in the rest of the world. This rather ambiguous result stems from the fact that while the clean coal plants are more efficient and therefore produce considerably less CO_2 per kWh generated they also become more cost attractive and are often chosen in preference to nuclear and gas fired plants which produce even less CO_2 per kWh: Coal consumption in power stations increases by 7.5 % while gas consumption decreases by 14%, oil by 10 %, nuclear generation by 13 % and hydro generation by 1%. Of the new coal technologies the most dramatically affected is the advanced thermodynamic cycle (+81 %) which sees its share of world power generation go up from under 7 % to around 12 % while the increase in IGCC generation is of the order of 58 % (share up from 7.3 % to 11 %). The performance of the supercritical coal plants is less spectacular + 35 % and a share increase from 15 % to 19 % both because of its relatively deep penetration already in the reference case but also due to the more modest additional tech.-economic improvements implied by the scenario.

The above new coal technologies replace very substantially conventional coal generation (-2300 TWh) and to a lesser extent gas turbine C.C. (-1380 TWh). They act as a break for the development of new nuclear design power plants, which reach only 690 TWh down from 1040 TWh in the reference case. The more conventional LWR are less affected (-150 TWh). Also severely affected is the Biomass gasification CC (-18 %) and especially wind power generation (-30 %) mainly due to the lower baseload electricity prices (nearly 10 % lower). Coal prices in the world and regional markets are 5-7 % higher which is clearly not sufficient to reduce significantly the competitiveness of the new coal technologies.

10.1.3 The Gas Technology Scenario

This scenario is composite and its implementation in POLES involved, apart from modifications in the tech.-economic characteristics of some key technologies, an expansion of world ultimate resources in gas. This latter was carried out by following the 5 % probability upper estimate for undiscovered resources contained in the "World Petroleum Assessment and Analysis" of the U.S. Geological Survey National Centre (1991). Special care was taken to ensure that correlations and statistical independence for undiscovered gas in different regions was reflected in a manner consistent with the above survey. On the other hand no attempt was made to link the possible higher resource base for gas with eventual higher occurrence of petroleum resources. The latter have hence remained at their reference case values.

The geographical distribution of these additional gas resources has been uneven about half of them occurring in the Former Soviet Union with large additions also recorded for Canada, Iran and China.

In terms of the technological side of the scenario the following main gas powered technologies have been treated.

- Gas turbine combined cycle is already the most successful power generating technology in the reference case contributing about 22 % of total generation by the year 2030. This gain in share achieved in the face of substantial increases in natural gas prices is mainly due to the major improvements in tech.-economic performance of this already attractive technology incorporated in the reference case (i.e. an 18 % decrease in capital costs mainly due to an increase in efficiency from around 50 % at present to 59 %). For the scenario capital costs were reduced by a further 12 %, specific fixed operations costs reduced by more than half and efficiencies raised to 63 %. These improvements were assumed to occur mainly by a learning curve effect associated with the greater availability of cheaper gas itself a major component in the structure of generating costs for these technologies.
- Gas Turbine combined Cycle Combined Heat and Power, which is another turbine dependent technology, saw sluggish performance in the reference case to higher gas prices. For the scenario apart from the lower prices this technology is assisted by a two point increase in both electric and steam conversion efficiency but also by a 20 % reduction in specific capital cost, fixed and variable operating costs.

Apart from these gas technologies, the scenario implies improved tech.-economic performance albeit on a smaller scale for two coal technologies: the advanced thermodynamic cycle and especially the IGCC, which benefit from gas turbine improvements.

The technology shift with respect to the baseline projection in terms of global electric power production is outlined in Table 10-7.

Table 10-7: The gas scenario: change in global electricity generation from reference scenario, TWh, in 2030.

Large Hydro	-108	**Oil Powered Gas Turbine**	-1246
Nuclear 1000-1500 MW LWR	-213	**Gas Turbine in Combined Cycle 200-350 MW**	9242
New Nuclear Design	-531	**Gas Turbine in Combined Cycle & CHP**	618
Supercrititcal Coal	-2542	**Small Hydro**	-53
Integrated Coal Gasification with Combined Cycle	1444	**Wind on Shore**	-33
Advanced Coal Cycle	-124	**Solar Power Plant**	-18
Lignite & FGD	-1093	**Waste Incineration CHP**	-21
Hard Coal 200-500 MW	-1635	**Biomass Gasification**	-123
Oil Boiler Monovalent	-255	**Proton Exchange Membran Fuel Cell**	14
Gas Powered Trad.Turbine	-286	**Solid Oxyde Fuel Cell**	0

This scenario represents a radical alternative in terms of world energy market configurations due to the enhanced natural gas resource base: Gas in place is assumed to be 45 % higher than in the reference case. Introducing this assumption in POLES results in new discoveries (from the present to 2030) twice as large in the scenario than in the reference case and world gas reserves. This implies that in spite of much higher cumulative production world reserves stand are more than one and a half times their reference level in 2030 resulting at much lower bulk gas prices (in North America less than half Europe -27% and Asia -20%).

These assumptions have yielded an alternative oil and gas market evolution, which is summarised in the following.

Table 10-8: POLES oil market outlook.

Oil Production (mbd)	**1973**	**1980**	**1990**	**2000**	**2010**	**2020**	**2030**	**1980/ 1973**	**1990/ 1980**	**2000/ 1990**	**2010/ 2000**	**2020/ 2010**	**2030/ 2020**
Gulf Conventional	21.0	18.8	16.3	20.6	33.3	42.0	49.3	-1.6%	-1.4%	2.4%	4.9%	2.3%	1.6%
OECD Conventional	13.3	14.5	15.2	18.0	15.4	14.3	10.6	1.2%	0.5%	1.7%	-1.6%	-0.7%	-3.0%
Other Conventional	24.3	30.2	33.7	32.8	36.0	37.3	35.8	3.1%	1.1%	-0.3%	0.9%	0.3%	-0.4%
World Non Conventional	0.0	0.0	0.3	0.7	0.7	5.7	19.7	NA	NA	7.0%	0.0%	23.9%	13.3%
Total	58.6	63.5	65.2	72.1	85.3	99.2	115.4	1.1%	0.3%	1.0%	1.7%	1.5%	1.5%

Table 10-8: Continued.

Oil Reserves (thousand mbd)	1973	1980	1990	2000	2010	2020	2030	1980/ 1973	1990/ 1980	2000/ 1990	2010/ 2000	2020/ 2010	2030/ 2020
Gulf	305	338	592	592	567	498	401	1.5%	5.8%	0.0%	-0.4%	-1.3%	-2.1%
OECD	58	52	59	52	39	34	35	-1.4%	1.1%	-1.2%	-2.9%	-1.2%	0.4%
Other	173	216	315	443	395	333	262	3.3%	3.8%	3.5%	-1.1%	-1.7%	-2.4%
World Conventional	0	0	917	1086	1000	865	699	NA	NA	1.7%	-0.8%	-1.4%	-2.1%
Non Conventional	0	0	48	130	187	291	399	NA	NA	10.4 %	3.7%	4.5%	3.2%
Total	524	601	966	1216	1187	1157	1097	2.0%	4.9%	2.3%	-0.2%	-0.3%	-0.5%

	1973	1980	1990	2000	2010	2020	2030	1980/ 1973	1990/ 1980	2000/ 1990	2010/ 2000	2020/ 2010	2030/ 2020
World crude oil price	9.8	41.4	20.0	18.1	23.9	30.1	32.7	22.8%	-7.0%	-1.0%	2.8%	2.3%	0.8%
World RP ratio	24.5	25.9	40.6	46.2	38.1	31.9	26.1	0.8%	4.6%	1.3%	-1.9%	-1.8%	-2.0%
Average Recovery Rate (%)	80.0	80.0	80.0	31.0	36.8	43.3	49.8	0.0%	0.0%	-9.0%	1.7%	1.7%	1.4%

Table 10-9: POLES gas market outlook.

Gas Production (billion m3)	1973	1980	1990	2000	2010	2020	2030	1980/ 1973	1990/ 1980	2000/ 1990	2010/ 2000	2020/ 2010	2030/ 2020
OECD	728	0	774	1115	1474	1836	2302	-100.0 %	NA	3.7%	2.8%	2.2%	2.3%
of which N. America	568	582	572	776	1007	1314	1753	0.5%	-0.2%	3.1%	2.6%	2.7%	2.9%
Former Soviet Union	269	404	742	666	716	979	1339	8.5%	6.3%	-1.1%	0.7%	3.2%	3.2%

Table 10-9: Continued.

Middle East	37	40	94	186	411	940	1490	1.5%	8.9%	7.0%	8.3%	8.6%	4.7%
Asia	24	49	116	241	385	515	628	15.8%	8.9%	7.5%	4.8%	2.9%	2.0%
Other	0	0	187	293	482	813	1150	NA	NA	4.6%	5.1%	5.4%	3.5%
World	1149	1397	1914	2501	3469	5084	6909	4.0%	3.2%	2.7%	3.3%	3.9%	3.1%

Gas Reserves (Trillion m3)	**1975**	**1980**	**1990**	**2000**	**2010**	**2020**	**2030**	**1980/ 1973**	**1990/ 1980**	**2000/ 1990**	**2010/ 2000**	**2020/ 2010**	**2030/ 2020**
OECD	15	15	18	18	20	27	40	-0.5%	1.9%	0.0%	1.5%	2.7%	4.0%
of which N. America	9	9	8	7	10	16	28	-0.7%	-0.7%	-1.7%	3.0%	5.4%	5.8%
Economies in transition	23	29	50	79	83	90	97	5.2%	5.4%	4.6%	0.6%	0.8%	0.7%
Latin America	2	5	8	10	13	15	16	15.5%	4.2%	2.4%	2.8%	1.7%	0.5%
Asia	2	5	9	12	13	13	12	21.0%	5.4%	3.2%	0.8%	-0.4%	-0.8%
Other	24	28	51	70	85	96	98	2.5%	6.4%	3.2%	1.9%	1.3%	0.3%
World	67	82	135	188	215	241	263	4.1%	5.2%	3.4%	1.3%	1.2%	0.9%

RP ratios	**1973**	**1980**	**1990**	**2000**	**2010**	**2020**	**2030**	**1980/ 1973**	**1990/ 1980**	**2000/ 1990**	**2010/ 2000**	**2020/ 2010**	**2030/ 2020**
N. American market	14.9	16.1	15.9	9.9	9.3	10.4	12.6	1.5%	-0.1%	-4.6%	-0.6%	1.1%	1.9%
European market	68.0	58.1	56.7	82.6	70.2	55.1	47.1	-3.1%	-0.3%	3.8%	-1.6%	-2.4%	-1.5%
Asian market	161.7	90.6	75.1	52.2	46.1	41.7	32.4	-10.9%	-1.9%	-3.6%	-1.2%	-1.0%	-2.5%
World			70.6	75.4	61.9	47.5	38.1			0.7%	-1.9%	-2.6%	-2.2%

Table 10-9: Continued.

Import prices ($90/boe)													
N. American market	20.4	30.0	10.8	14.4	15.1	13.4	10.6	8.1%	-9.7%	2.9%	0.5%	-1.1%	-2.4%
European market	11.3	21.4	15.2	12.5	15.0	20.0	23.6	13.6%	-3.4%	-1.9%	1.8%	3.0%	1.7%
Asian market	19.2	36.5	20.6	22.8	26.9	27.3	32.4	13.6%	-5.6%	1.0%	1.6%	0.2%	1.7%

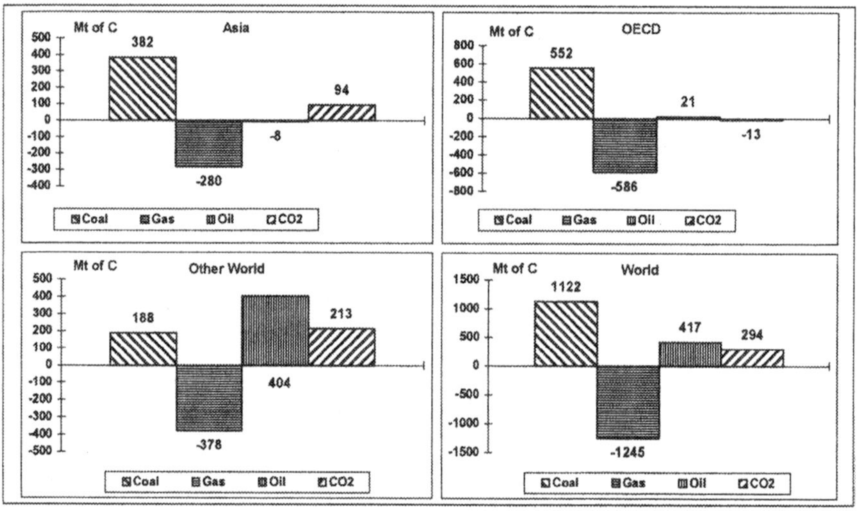

Figure 10-3: The gas technology scenario by zones: comparison with the baseline (2030).

At these levels gas becomes competitive not only vis-à-vis oil (which it was already) but also for a number of substantial market segments vis-à-vis coal. Consequently oil prices are dragged down (by 9%) and coal prices fall sharply from already low levels (-20% in North America and -10% elsewhere). Accompanying these changes is a substantial increase in international gas trade (for example in China gas imports double despite a 50% increase in domestic production). Yet the impact of the scenario on CO_2 emissions is a paltry -2.1% in 2030 (+0.3% in the OECD). The reason for this apparent paradox are numerous and are summarised below.

Generally low gas prices and lower energy prices have meant that final consumption of gas has increased (by 27%) but lower prices of other fuels have meant that their consumption has not fallen sufficiently to compensate for gas increases. In the case of electricity the combination of low fuel costs and improved power plant tech.-economic characteristics has produced lower generating costs and hence lower electricity prices.

This has been particularly striking for low to medium loads (residential/commercial electricity) where due to the suitability of GTCC to meet such demand, electricity prices have fallen on average woldwide by nearly 20%. This naturally led to higher electricity consumption (up by 8.5 %) which the accompanying multiplier effect associated with transformation losses.

Within the power-generating sector, many striking changes occur woldwide. Generation from gas turbines combined cycle. Increase by 9250 TWh (more than double the reference case) bringing their share to 42% of total generation instead of the 22% of the reference case. This increase is achieved at the expense of super critical coal (-2550 TWh) conventional coal (-1640 TWh) oil (-1500 TWh), brown coal (-1100 TWh) but also nuclear (-750 TWh, 70 % of which the new nuclear design which finds little room to develop). Furthermore, one clean coal technology (the IGCC) because of the improvements associated with turbines finds itself reinforced in the scenario (+1450 TWh worldwide). Three quarters of this latter increase takes place in Asia where the benefits of the enlarged gas supplies are less obvious as most of the increment has to come from distant sources involving costly transport thus leaving coal competitive over a large part of the load curve. Combined Heat and power c.c. sees its contribution increasing by 170% to nearly 1000 TWh and in many parts of the world it comes close to saturation imposed by the industrial constraints to its potential. On the other hand, many renewable sources of electricity see their prospects slashed by half (notably biomass gasification combined cycle, Wind power plants and Solar thermal). Even small hydro plants and waste incineration see their contribution reduced by around 20%.

10.1.4 The Fuel Cell Scenario

This is not strictly speaking an autonomous exercise but an extension of the Gas technology scenario presented above. It retains the enhanced gas resources with all their ramifications and incorporates the same improvements relating to gas turbine combined cycles, CHP, as well as the turbine related clean coal technologies. In addition, three fuel cell technologies are given much improved tech.-economic characteristics.

- The proton exchange membrane fuel cell for fixed applications which in the reference case saw fuel efficiency rising from 55% in 2000 to 60% in 2030 has it rising further to 63% for that latter year. Likewise, specific capital costs which fell from 3450 ECU/kW in 2000 to 864 ECU/kW in 2030 in the reference case are allowed to reach one third of that value and the fixed operating costs for the scenario are also 1/3 of their reference value while specific variable operating costs are allowed to halve.

- The Solid Oxyde fuel cell with cogeneration which is also a stationary application had efficiencies rising from 60% to 70% between 2000 and 2030 this latter efficiency was deemed to represent an upper limit and was therefore kept unchanged in the scenario. Specific capital costs which reached 800 ECU/kW in the reference case were further reduced to 600 ECU/kW for the scenario while fixed and variable operating costs by 2030 represent in the scenario one third of their respective values in the reference case. Variable costs reductions are mostly due to large stack cost reductions.
- A hydrogen fuel cell car was included on the rationale that stationary fuel cell technology improvements as outlined above would also have an impact on smaller mobile forms required for vehicles. Although important improvements have been assumed relating to cost per kW, stack costs and car Km/kWh performance and technical lifetime they have proved insufficient for producing a significant impact: in the best of cases i.e. in the scenario by the year 2030 the propulsion system alone costs an additional 20000 ($90) for a relatively small 40 kW car.

As the fuel cell scenario is additional to the gas technology scenario and therefore embodies most of the features of the latter, a comparison with the reference case would entail to a large extent a repetition. A more interesting comparison can be made between the fuel cell and gas technology scenarios.

Table 10-10: The fuel cell scenario: change in global electricity generation from reference scenario, TWh, in 2030.

Large Hydro	-123	Oil Powered Gas Turbine	-1252
Nuclear 1000-1500 MW LWR	-230	Gas Turbine in Combined Cycle 200-350 MW	8454
New Nuclear Design	-625	Gas Turbine in Combined Cycle & CHP	624
Supercrititcal Coal	-2601	Small Hydro	-51
Integrated Coal Gasification with Combined Cycle	1368	Wind on Shore	-32
Advanced Coal Cycle	-181	Solar Power Plant	-18
Lignite & FGD	-1167	Waste Incineration CHP	-20
Hard Coal 200-500 MW	-1680	Biomass Gasification	-123
Oil Boiler Monovalent	-260	Proton Exchange Membran Fuel Cell	927
Gas Powered Trad.Turbine	-300	Solid Oxyde Fuel Cell	16

The proton exchange membrane fuel cell technology which in the gas technology case had reached the relatively small contribution of 115 TWh worldwide by 2030 makes significant inroads reaching 1030 TWh. On the other hand the Solid Oxyde fuel cell though trebling in importance still does not account for more than 24 TWh worldwide. Fuel cell vehicles for the reasons mentioned above do not go beyond a demonstration curiosity.

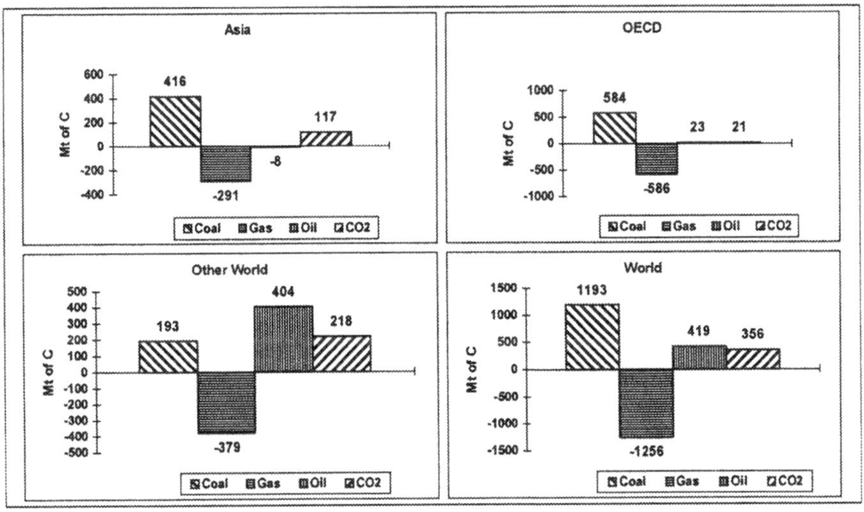

Figure 10-4: The fuel cell scenario by zones: comparison with the baseline (2030).

As the improvements in stationary fuel cells are spread over time and the power generating sector is characterised by slow turnover, one suspects that the 1050 TWh of fuel cell contribution would not be the end of the story had the horizon been extended beyond 2030. This caveat notwithstanding, it is interesting to remark that the fuel cell penetration is accompanied by a massive reduction of 790 TWh in gas turbine combined cycle generation which is hardly surprising given the overwhelming pre-eminence of this technology in new power plant markets which characterised the gas technology scenario. Other significant losers are the IGCC (-75 TWh) for very much the same reasons affecting the GCC and the new nuclear design losing a further 95 TWh as the market for new power plants gets even more saturated with new power technologies.

Under these circumstances it is not surprising that world CO_2 emissions decrease only by a further 0.4-0.5% making a 2.5% difference from reference. It seems that there are strict limits as to what can be achieved by adding attractive technologies to one another when essentially they are competing for the same patch.

10.1.5 The Renewable Energy Technology Scenario

This Scenario is different from the ones previously examined in that it is not organised around a single technological breakthrough or a generically linked cluster of technological developments. It is rather representative of a situation where a major R&D effort would be directed on decentralised renewable technologies producing drastic improvements in the technico-economic characteristics of a number of them which are otherwise technologically heterogeneous.The main technologies affected in this scenario are:

- *biomass gasification* for electricity production in small scale (less than 25 MW) combined cycle plants. In the reference case this technology saw no improvement in specific capital costs but a 40 % reduction in fixed operating costs and an increase in efficiency from about 35 % at present to 43 %. Under these circumstances the contribution of this technology slightly more than doubled by 2030 (registering a very low average growth of under 2.5 % p.a.).

For the technology scenario specific capital costs were assumed to halve by 2030 with some further improvements in efficiency and operating costs. The impact of these changes was a near seven fold increase in contribution between the present, and 2030, albeit with an inflexion of growth towards the end of the period due to increasing costs of the biomass itself as the cheaper sources are gradually exhausted.

- *Photovoltaics in buildings,* involving mostly cells incorporated in window panels saw their capital cost plunge from about 12,000 ECU/ kW to 5,000 ECU/kW by 2010 and 4,000 ECU/kW by 2030 in the reference case. Specific fixed O& M costs were halved between the present and 2030. Impressive as these reductions are they left the cost of electricity produced at a generally uncompetitive level and the technology contributed a relatively symbolic, 4,6 Twh world-wide by 2030. The technology scenario implies a halving of 2030 reference costs this implies a reduction of the cost of the Kwh delivered to around 0.17 ECU which though still not generally competitive with network electricity allows at least the development of niche markets and leads to a contribution of around 32 Twh world-wide by 2030. Of these only about 6 Thw are produced in developing countries despite the fact that on average they enjoy better insolation conditions. This discrepancy underlines the "luxury" nature of this technology even with the additional improvements.

- *Molten Salt Tower Solar plant* with storage was the main solar thermal power technology retained with a capital cost reduced to 1600 ECU/ kW fixed O & M costs of 34 ECU/kW p.a. and a capacity factor of 36 % by 2030. In the reference case it achieved a contribution of 31 Twh word-wide. In the technology scenario with these figures reaching 1130 ECU/kW 25 ECU/kW. p.a. and 38 % respectively its contribution reaches 114 Twh world-wide. It is worth noting that though this latter figure is relatively modest it does not represent the full potential impact of the scenario as by 2030 this technology is still in the stage of vertical take off in industrialised countries and has not even entered that stage in the developing countries where most of the physical potential exists.

Table 10-11: Renewables technological alternative scenarios.

ECU90/KW	2000	2030 reference case	2030 scenario
SPECIFIC CAPITAL COSTS			
Biomass Gasification < 25 MW	1767	1767	960
Photovoltaics in Buildings	12000	4000	2000
Solar Thermal Low Temp.	1200	750	500
Small Hydro > 2 MW	2000	1850	1000
Molten Salt Tower Plant & Storage	2500	1603	1131
Photovoltaics Rural Electr.	13100	10000	5000
Wind on Shore	1300	1100	460
FIXED OM COSTS			
Biomass Gasification < 25 MW	50	30	20
Photovoltaics in buildings	100	50	25
Solar Thermal Low Temp.	10	5	3
Small Hydro > 2 Mw	20	16	16
Molten Salt Tower Plant & Storage	64	34	25
Photovoltaics Rural Electr.	250	180	160
Wind on Shore	16	14	14
AVAILABILITY FACTOR %			
Photovoltaics in Buildings	15	17	17
Solar Thermal Low Temp.	27	31	33
Small Hydro > 2 MW	57	57	57
Molten Salt Tower Plant & Storage	27	32	34
Photovoltaics Rural Electr.	9	9	14
Wind on Shore	20	24	32

- *Small hydro* which in the reference case was assumed to be a mature technology registering insignificant gains over the projection period sees its capital cost halved for the scenario and hence its contribution double from 250 Twh to 505 Twh word-wide by 2030. However by the end of the period the technology displays clear signs of saturation as the best available sites are gradually exhausted.

- *On shore wind turbine* (over 500 kW capacity) offers the highest contrast between reference and scenario. In the reference case no significant reduction in capital costs was assumed all improvements being concentrated in a relatively modest increase in capacity factors to a range of 12-30 % depending on site. The technology scenario by contrast implies a reduction of capital costs to one third of their present level and further significant increases (+ 33 %) in capacity factors.

These developments render wind power highly competitive, its intermittent character notwithstanding and lead to massive development world-wide (nearly 1600 Twh by 2030). This development is furthermore fairly evenly distributed about half occurring in industrialised countries.

The projected differences in terms of world power output by 2030 with respect to the baseline scenario are showed in Table 10-12.

Table 10-12: The renewables scenario: world power production by technologies with respect to the reference scenario (2030).

Large Hydro	-14	Oil Powered Gas Turbine	-148
Nuclear 1000-1500 MW LWR	-58	Gas Turbine in Combined Cycle 200-350 MW	-907
New Nuclear Design	-112	Gas Turbine in Combined Cycle & CHP	54
Supercrititcal Coal	-412	Small Hydro	256
Integrated Coal Gasification with Combined Cycle	-224	Wind on Shore	1500
Advanced Coal Cycle	-157	Solar Power Plant	83
Lignite & FGD	-387	Waste Incineration CHP	1
Hard Coal 200-500 MW	-410	Biomass Gasification	510
Oil Boiler Monovalent	-30	Proton Exchange Membran Fuel Cell	4
Gas Powered Trad.Turbine	-43	Solid Oxyde Fuel Cell	0

In addition to the above the renewable scenario included three more technologies where the improvements implied were insufficient to produce a significant impact. These were low temperature passive solar which due to insufficient cost reductions remained a niche option, rural photovaltaic which becomes uncompetitive with network connection and therefore faces a shrinking market potential as electrification proceeds and methanol from biomass for use in vehicles which even with the improvements assumed in the technology scenario remained more costly than conventional oil and failed to penetrate significantly in the absence of preferential fiscal treatment.

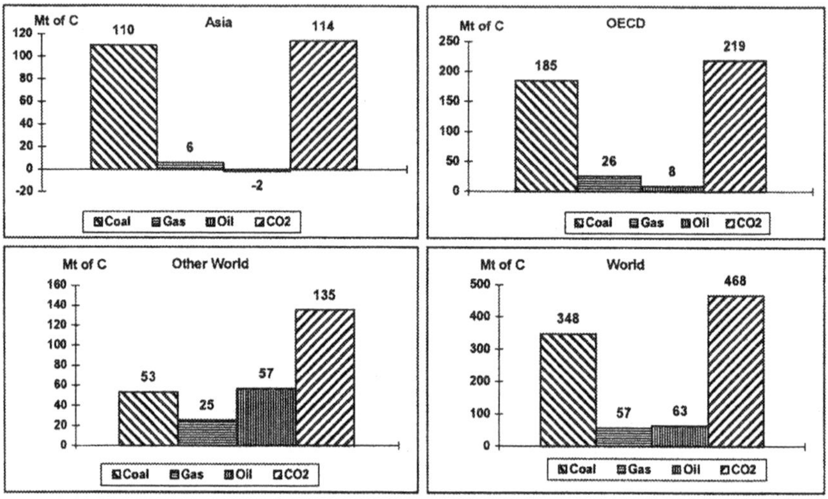

Figure 10-5: The renewable breakthrough scenario by zones: comparison with the baseline (2030).

The net result of the scenario is a 3.3 % reduction in world-wide emissions (2 % in Asia and 5 % in OECD). This is primarily achieved by an across the board reduction in centrally produced electricity (-1600 Twh of coal fired -950 Twh of gas fired and -180 Twh of nuclear generated electricity world-wide).

World oil prices are little affected (-1.3 %) but wholesale gas and coal prices are more substantially reduced (around - 5 %) allowing for some increase of the consumption of these fuels outside the power generation (especially in industry) which, partially offsets the substantial reductions in power generation. Generally the scenario has major implications for the field of new and renewable technologies but fails to produce a major impact on the CO_2 problem because these technologies are mostly applicable to the power generating sector, often (being mostly intermittent) even a limited section of this sector displaying also marked geographical "niches". Another reason fot the relatively subdued overall impact of this scenario is that the most important technologies become truly competitive only towards the end of the period thus having little time to penetrate to their full potential.

10.2 Conclusion

The main findings of the previous comparison exercise between the 5 considered energy technology scenarios are briefly summarised in Table 10-13.

Table 10-13: Summary table of the impact of technology scenarios, world in 2030, and comparison with reference case.

Scenario	Coal	Oil	Gas	Nuclear	Hydro	CO_2 Emissions
Nuclear	-94%	-0.8%	-4.0%	142%	-0.7%	-5.3%
Clean Coal	4.5%	-1.0%	-3.9%	-13.1%	-1.1%	1.0%
Gas Technology	-17.5%	-8.1%	49.7%	-19.6%	-2.2%	-2.1%
Gas and Fuel Cells	-18.6%	-8.2%	50.1%	-22.5%	-2.6%	-2.5%
Renewable	-5.4%	0.1%	-2.3%	-4.5%	-0.3%	-3.3%

A cursory glance at the table above is sufficient to realise that the technology scenarios as defined in the current work do not offer panaceas for the global CO_2 emission problem. The reasons for this are inherent in the way the scenarios were constructed as well as energy market structures and the way their functioning is represented in the POLES model. In lieu of conclusions some brief points are proposed here for discussion.

- The technology scenarios have been defined in terms of clusters of technological breakthroughs affecting only a part of the energy market and hence global CO_2 emissions. They have almost exclusively concentrated on power generation and often essentially on mere segments of the load curve. Important though the power generating sector may be, it still accounts for about a one third of world CO_2 emissions projected to increase to 40% by 2030. By simple arithmetic technologies addressing only this sector risk at the best of times to have an impact, which seems 2.5 to 3 times too small.
- Scenarios involving major tech.-economic improvements in fossil fuel technologies produce weakened impacts on CO_2 emissions because at the same time as they reduce specific emissions they make these technologies economically attractive, not only vis-à-vis more polluting technologies but also more polluting ones. Choices are therefore influenced in a far from unambiguous manner as far as CO_2 emissions are concerned. In this context it is worth noting that all supply-side technological improvements result one way

or another in a reduction in the cost of consuming energy and hence potentially cause consumption to increase. In any case, it is hardly surprising that the scenarios involving non fossil energy (the nuclear and renewable scenarios) have produced markedly better outcomes as far as CO_2 emissions are concerned.

- The two scenarios involving increased gas availability have, in view of the major restructuring they imply in world energy markets, produced surprisingly weak results as far as CO_2 emissions is concerned. This is due partly to the uneven geographical distribution of the enhanced resources resulting in the gas not being available at sufficiently cheap prices where it could have made the biggest impact (the big Asian coal users: China and India) but also due to a series of secondary market effects which brought fuel prices down and caused demand firmness especially in electricity and gas. All this, notwithstanding the indisputable fact that these scenarios represent a more comfortable energy market situation with all its concomitant virtues of security and a propitious environment for economic development.
- There is a clear need for energy saving technology breakthroughs, as these are likely to suffer less from the ambiguities and secondary effects associated with supply technologies. It is, however, very difficult to identify a cluster of technologies which would be homogeneous enough to be convincing as an identifiable alternative while at the same time addressing a wide spectrum of energy demand. A possible area where such a cluster could be convincingly found is in the transport sector where perhaps the fuel cell vehicle could be revisited with more drastic improvements in its characteristics.
- It is hard to see how clusters of energy technologies could by themselves make a major impact on the global CO_2 problem unaccompanied by major policy initiatives albeit market related ones. Combining technology breakthroughs with internalisation of external costs through taxation or tradable permits would magnify the impact and tend to neutralise some of the more ambiguous side effects quite apart from the fact that these policy instruments can by definition act on a much wider front.

11 Technology Stories with PRIMES2 for the European Union; an Analysis for Demand Side, Power and Steam Generation Sector[44]

11.1 Introduction

This chapter explores the implications for the European Union of accelerating the progress of key energy technologies in the demand and supply sectors over the period 2010-2030.

The chapter does not consider the driving forces or policies (e.g. R&D support) that would induce contribute in obtaining the assumed technology progress. The analysis aims at deriving the benefits arising from such a progress, in terms of system costs and pollution. The use of a system model also allows estimating the likely size of the market for the technologies and the side effects on the energy system.

This chapter builds on the results of PRIMES V.2 model including the interactions between demand and supply.

The baseline scenario is the one presented in other chapters and concerns the EU member states. The scenario is consistent with the world energy projections made with the model POLES, in particular regarding the world energy prices evaluated by POLES as a result of demand and supply interaction in the worldwide markets.

For sensitivity analysis purposes, the approach considers two baseline scenarios, which for the EU differ, only in the dynamic evolution of world energy prices. The baseline scenario involving relatively low rates of increase of gas and oil border prices to the EU is called "low price baseline", while the one assuming high growth of gas and oil prices in the long run is called "high price baseline". Both scenarios assume stability for the price of imported coal.

---

[44] This chapter presents the most updated version of the activity on technology stories carried out with PRIMES. The work involved Pantelis Capros, Leonidas Mantzos and Kostas Delkis. The results were presented in successive workshops jointly organised by the IEA/OECD and the European Commission.

For analytical purposes, the chapter presents the results separately for each cluster of technologies, for which accelerated technology progress occurs. Such a set of assumptions lead to a scenario called "technology story". One such story concerns the demand-side and six concern the supply-side (in fact the power and steam generation technologies). The chapter also presents the combined effects from the demand and the supply-sides.

The following matrix summarises the definition of the model runs.

Table 11-1: Definition of technology stories.

<table>
<tr><td colspan="2"></td><td>Low prices for oil and gas</td><td>High prices for oil and gas</td></tr>
<tr><td rowspan="2">Demand side</td><td>Baseline</td><td>Low price baseline</td><td>High price baseline</td></tr>
<tr><td>Demand-side story</td><td>Low price and demand-side story</td><td>High price and demand-side story</td></tr>
<tr><td rowspan="2">Supply side
(6 cases)</td><td>Demand-side progress as in the baseline</td><td>Low price and supply-side stories without demand effects</td><td>High price and supply-side stories without demand effects</td></tr>
<tr><td>Demand-side progress as in the demand-side story</td><td>Low price and supply-side stories with demand effects</td><td>High price and supply-side stories with demand effects</td></tr>
<tr><td colspan="2">Number of scenarios for the EU</td><td>14</td><td>14</td></tr>
</table>

11.2 Definition of Technology Progress Stories

A technology story assumes a different from baseline evolution of technological progress affecting a defined group of energy technologies. The assumptions correspond to an assertion that a cluster of technologies, grouping techniques that are based on similar technological components, will improve over time at a rate higher than other technological groups.

Assumptions underlying the technology stories are described in detail in chapter nine. The demand-side story and a pessimistic supply-side story have been added.

In summary, the technology stories are defined as follows:

- the *Nuclear story*, involving considerable progress in the nuclear plant technologies (conventional and new emerging technology) regarding capital cost, flexibility of dispatching, variable and fixed operation costs;
- the *Clean Coal story*, involving significant reduction of investment costs of key clean coal technologies (supercritical, IGCC and PFBC plant types) and improvement of thermal efficiencies;
- the *Gas story*, involving continuous improvement of the technical-economic characteristics of gas turbine combined cycle plants and other small gas turbines plants, accompanied by higher resource availability of natural gas and consequently lower (than in the corresponding baseline) supply prices for natural gas;
- the *Fuel Cells story*, involving, in addition to the assumptions of the gas story, significant cost and performance progress in fuel cell technologies for power and steam generation, that are considered as a result of, commercialisation, hence maturity and economies of scale. This story is also accompanied by high resource availability of natural gas, as in the gas story;
- the *Renewable story*, involving significant improvement of the technical-economic characteristics of new renewable technologies for power generation, including technologies that involve intermittent supply of power and technologies that use biomass and wastes;
- the *Pessimistic (supply-side) story* which assumes that the technical-economic characteristics of power and steam generation technologies do not change over time and remain frozen at their levels as in 2000;
- the *Demand-side story* which assumes that the capital cost of new end-use technologies (except steam generation and cogeneration) decreases gradually up to 35% compared to baseline. This concerns those technologies that are improved or advanced compared to ordinary end-use techniques.

Technology progress in cost and efficiency terms begins from 2010, accelerates between 2010-2020 and reach maturity by 2030. The generic time pattern of the assumed evolution of technology is shown in figure.

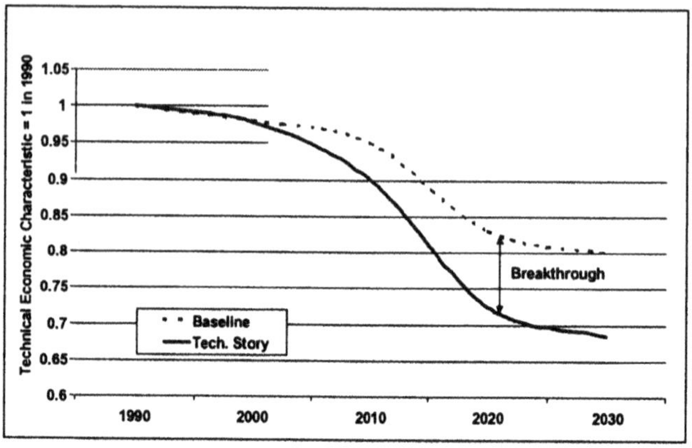

Figure 11-1: Pattern of technology progress.

11.3 The Baseline Scenario

11.3.1 World Energy Prices

As mentioned, the border prices of energy imports to the EU change in the future as a result of interactions between demand and supply at a world level. The POLES model simulates those interactions.

In the baseline scenarios, the oil and gas prices bear relatively significant increases in the longer run, reflecting progressively binding supply and resource constraints along with high growth of world energy demand. On the contrary, the abundance of coal resources induces stability of coal prices.

The low price baseline projects an average increase of oil and gas prices, in real terms, of the order of 2.1% per year for the period 2000-2010 and 1.5% for the period 2010-2030. The corresponding assumptions of the high price baseline are 3% (2000-2010) and 2.1% (2010-2030). In the low price baseline, this leads to a price (for oil) of 19$'90/bbl in 2030, while in the high price baseline the same price becomes 33$'90/bbl in 2030. Similarly, the border price for natural gas becomes equal to 4.9$'90/Mbtu in 2030 in the low price baseline, while becoming 8.5$'90/Mbtu in 2030 in the high price baseline.

The supply-side stories that involve progress in technologies that use natural gas for power generation, also assume that world gas resources prove to be more abundant than in the baseline. In this case, the gas price decreases by 13% in 2030 compared to the low gas baseline, and by 22% in the case of high price baseline.

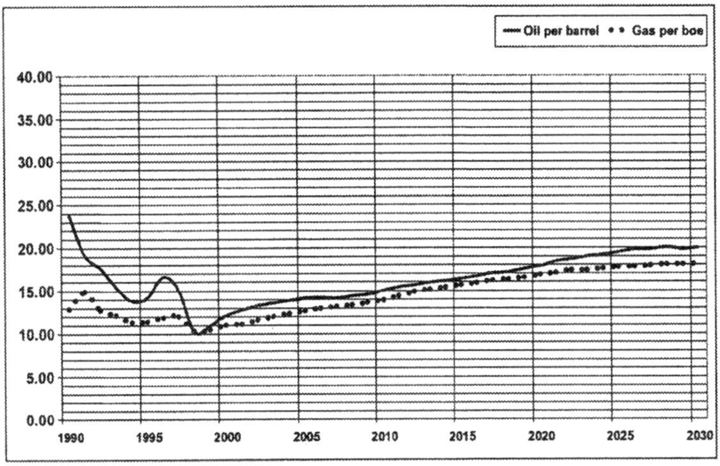

Figure 11-2: EU border prices in $90 (low prices).

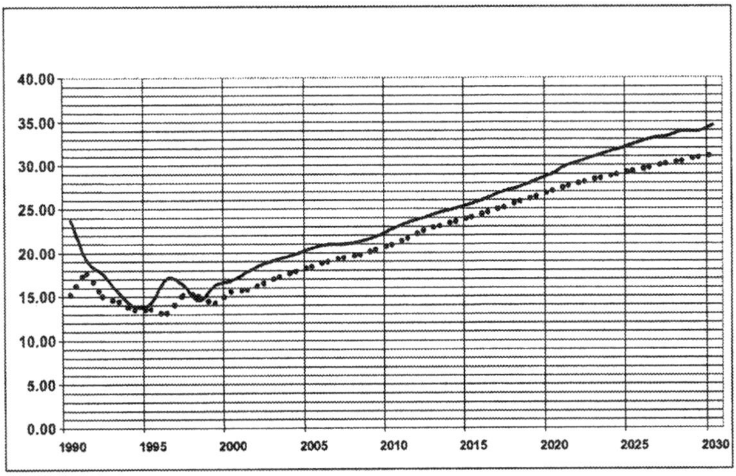

Figure 11-3: EU border prices in $90 (high prices).

11.3.2 The Baseline Scenario

The baseline scenario for the EU assumes continuation of current trends and policies, including technology progress in both demand and supply, as well as market liberalisation in power and steam. Global environmental constraints do not influence the baseline projection.

Continuous efficiency gains prevail in the baseline trends. The overall energy intensity (primary energy requirements over GDP) improves at a rate of 1.5% per year. Final energy requirements increase with activity and income, but at rates that differ by sector. Activity in heavy industrial processes, as well as energy needs, slowdown over time. On the contrary, energy for transportation rises substantially, however at rates progressively slower than in the past. Electricity demand increases more than total final energy, mainly through the accelerated adoption of electricity-based technologies and the emergence of new uses of electricity in the tertiary and domestic sectors. The end-use technological system improves, as mentioned before, but there is no profound change in the consumption patterns or the nature of technology. For example, fuel cell cars (or electric cars), renewables in final energy and new energy carriers (like hydrogen) do not emerge. The evolution pattern in the supply-side is similar. There is continuous improvement in the efficiency of technologies, but there is no structural change other than that induced by electricity market liberalisation. The process of liberalisation, accompanied with a technological progress that allows small size power and steam plants to gain competitiveness relatively to the traditional large scale plants, induces a decentralisation of generation and the rapid penetration of power and steam cogeneration. The resulting change is mainly based on gas technologies (GTCC and small gas machines), that exhibit higher efficiency rates. Benefiting from low oil and gas prices, this trend continuous to develop even in the last decade (2020-2030) when massive nuclear decommissioning takes place (lifetime duration of 40 years is generally assumed for existing nuclear plants).

The penetration of natural gas in power and steam generation implies improvement of the carbon intensity, especially in the short-medium term. In the long term, the decommissioning of nuclear plants contributes to deterioration of carbon intensity. The combined effect for CO_2 result in a continuous growth of CO_2 emissions, at rates substantially lower than GDP growth. The rising trend slowdown towards 2020, but afterwards increases again substantially. Transportation and at a lesser degree the tertiary sector, are responsible for the increase in CO_2 emissions in the medium run, while power generation is mainly responsible in the longer run.

The PRIMES model incorporates differentiation of capital cost and performance according to the size of the plant. Independent power producers, because of the size of their applications, can select from a menu of technologies that generally include lower-sized plants compared to the menu for utilities. Lower-sized plants are at present significantly more costly than larger plants. However, there is a strong trend in some technologies (like the gas turbines, for example) in favour of reducing the unit cost differences between small independent generators and utilities. Such a progress evidently enables decentralisation of generation, depending also on demand growth and fuel availability.

Technology progress in relation to economies of scale is introduced in the baseline scenario (and all technology stories) independently of the main technology evolution by cluster of techniques. This explains, among others, that the scenarios project relatively high penetration of cogeneration.

For example, in all scenarios CHP electrical capacity as a percentage of total thermal capacity exceeds 36% in 2030, while being about 23% in 2000. This share depends upon the nature of technology story but remains a dominant factor in the analysis. Consequently, generation decentralisation also increases. For example, while the share of utilities in power generation was more than 95 % in 1995, this share is projected to steadily grow reaching 65 % by 2030 in the baseline scenario.

The nuclear policy question practically arises beyond 2020, as decommissioning of existing capacities is considered beyond 40 years of plant lifetime. Even in the context of relatively high growth of oil and gas prices, nuclear energy is not any more competitive The baseline scenarios lead to almost full abolishment of nuclear energy by 2030.

In the baseline scenario, the gas turbine combined cycle plants and the small gas turbines and machines are very competitive in the short-medium term. Their massive deployment characterises the evolution of the power and steam generation system. In the context of the low price baseline, the domination of gas-based generation prevails in the longer run, as well. The share of electricity produced from gas rises up to 58% in 2030. According to the baseline projection, a total of about 600 GW of power and CHP generation plants will be constructed in the EU, in the period 2010-2030. The gas based small and large plants will represent 75% of incremental constructions, while traditional technologies and clean coal will represent only 17%.

In the context of high price baseline scenario, gas based generation becomes less competitive than coal-fired plants in the period beyond 2020. This is a consequence of relative fuel prices, but it is also accelerated by the technology progress of clean coal. In this case, the share of gas-based electricity generation drops to 38% in 2030 and its market share in total investment in 2010-2030 becomes 57%. In the context of high oil and gas prices, clean coal takes a significant part of nuclear decommissioned: about 150 GW of clean coal technologies are invested in the 2010-2030 and the total share of conventional and clean coal investments becomes 34%.

The high price baseline scenario is more efficient than the low-price baseline in the demand-side (improvement by 1% per year) as a result of higher prices. During the last decade of the projection, coal penetrating more than gas in power and steam generation, implies a further increase of CO_2 emissions and a decline of average thermal efficiency. The combined effects from demand and supply-sides lead to a deterioration of CO_2 emissions in the high price baseline, in 2030, compared to the low price baseline (6.5% more emissions).

Apart from coal and natural gas, there is no significant fluctuation in the use of other fuels during the projection period. The use of Lignite and peat remains high in some countries such as Greece and Germany, which have high domestic production and low domestic extraction costs. The use of biomass increases after 2000, getting its biggest share by 2020. All these fuels are constrained from limited availability in their supply.

Wind, photovoltaic and other renewable technologies lose a small part of their share at the end of the period. Concerning wind energy and photovoltaics, there is slowdown of new investment beyond the short-medium run. In case of low price baseline, renewables represent 6.5% of gross inland consumption in 2030, while that share increases in the high price baseline (7.1%).

The dynamic evolution of efficiency indicators by sector in the demand-side is continuously improving. In the industrial sectors, the efficiency indicator (energy per unit of physical production) improves at an average annual rate of about 0.45% (2000-2030). Restructuring away from traditional processes (like integrated steelworks and basic chemicals) and in favour of high value added activities (including acceleration of recycling of materials) contribute to further improvement of the overall industrial efficiency indicator. The progress in the tertiary and domestic sectors is more pronounced, being higher than –1% per year (computed as a ratio of energy demand per unit of income of households or value added in services). The changes in the transport sector are more complex, since technology progress combines with structural shifts in the allocation of mobility by transport mode. While mobility grows at a rate similar to that of GDP, energy use becomes more efficient as activity shifts more into faster and more collective transport modes. The engineering efficiency of vehicles improves at an annual rate of about 1%, while the ratio of energy for transportation over GDP improves at a rate of 1.5% per year.

The case of high prices (in the baseline scenario) has rather small effects on final energy demand. The technology improvement accelerates in most sectors. However, in some processes or sectors that decline, as a consequence of higher prices, technology progress slowdown, since capital turnover also slowdown in these sectors. The overall efficiency in the demand-side improves (energy demand decreases), accompanied with higher electricity penetration in the end-use sectors.

The following tables summarise the results for the low-price and the high-price baseline scenarios.

Table 11-2: Summary results of the low price baseline scenario.

Baseline Scenario - Low Fuel prices			Mtoe				*annual growth rate*		
EUROPE 14	**1995**	**2000**	**2010**	**2020**	**2030**	***95-00***	***00-10***	***10-20***	***20-30***
Gross Inland Consumption	*1362*	*1450*	*1576*	*1629*	*1600*	1.26%	0.84%	0.33%	-0.18%
Solid Fuels	237	220	186	167	173	-1.45%	-1.68%	-1.09%	0.37%
Liquid Fuels	574	624	673	704	700	1.71%	0.76%	0.44%	-0.05%
Natural Gas	273	313	417	472	578	2.74%	2.91%	1.24%	2.04%
Nuclear	205	212	216	191	40	0.67%	0.19%	-1.23%	-14.53%
Electricity	1	1	1	2	3	-9.98%	-2.06%	13.47%	4.72%
Renewable Energy Sources	71	79	82	94	107	2.08%	0.38%	1.33%	1.32%

Baseline Scenario - Low Fuel prices			Mtoe				*annual growth rate*		
EUROPE 14	**1995**	**2000**	**2010**	**2020**	**2030**	***95-00***	***00-10***	***10-20***	***20-30***
Electricity Generation in TWh	*2306*	*2543*	*3000*	*3342*	*3652*	1.97%	1.67%	1.09%	0.89%
Nuclear	810	845	862	765	174	0.84%	0.20%	-1.19%	-13.77%
Hydro - Renewables	293	356	396	425	453	3.97%	1.06%	0.71%	0.64%
Thermal (incl. biomass)	1203	1342	1742	2152	3025	2.21%	2.64%	2.14%	3.46%
Steam Generation in TWh $_{th}$	*1034*	*1123*	*1288*	*1391*	*1472*	1.67%	1.38%	0.77%	0.57%
CHP plants	323	503	692	1022	1290	9.29%	3.25%	3.97%	2.36%
Boilers	711	620	596	369	181	-2.70%	-0.40%	-4.68%	-6.85%
Generation Capacity in GWe	*569*	*623*	*714*	*804*	*880*	1.83%	1.37%	1.20%	0.91%
Nuclear	132	136	134	116	23	0.68%	-0.18%	-1.46%	-14.92%
Hydro (without pumping)	106	109	110	110	106	0.58%	0.07%	0.00%	-0.33%
Thermal (incl. biomass)	328	368	448	546	711	2.29%	1.99%	2.02%	2.66%
of which CHP	64	85	141	214	255	5.79%	5.20%	4.28%	1.79%
Solar, wind, geothermal	3	10	22	32	40	25.73%	8.23%	3.66%	2.30%
Average Load Factor %	*46.2*	*46.6*	*48.0*	*47.4*	*47.4*	0.14%	0.30%	-0.11%	-0.02%
Boilers (GWth)	*239*	*209*	*157*	*114*	*85*	-2.60%	-2.85%	-3.11%	-2.95%
Fuel Inputs for Power and Steam Gen.	*364*	*386*	*443*	*499*	*608*	1.18%	1.37%	1.20%	2.00%
Solids	171	159	133	122	140	-1.49%	-1.75%	-0.87%	1.41%
Oil	88	93	99	115	106	1.03%	0.62%	1.52%	-0.85%
Gas	83	110	184	224	314	5.85%	5.25%	1.99%	3.45%
Biomass / Waste	22	24	27	38	48	2.20%	1.05%	3.54%	2.43%
Hydrogen	0	0	0	0	0	-	-	-	-
Average Thermal Efficiency %	*52.8*	*54.9*	*58.9*	*61.1*	*63.6*	0.78%	0.70%	0.37%	0.40%

Baseline Scenario - Low Fuel prices			Mtoe				*annual growth rate*		
EUROPE 14	**1995**	**2000**	**2010**	**2020**	**2030**	***95-00***	***00-10***	***10-20***	***20-30***
Final Energy Demand	*883*	*946*	*1041*	*1089*	*1125*	1.39%	0.96%	0.45%	0.33%
by fuel									
Solid Fuels	43	38	33	28	21	-2.44%	-1.52%	-1.39%	-3.00%
Liquid Fuels	404	434	467	476	479	1.46%	0.75%	0.18%	0.07%
Natural Gas	177	193	216	226	233	1.73%	1.12%	0.45%	0.30%
Steam	67	73	85	93	100	1.69%	1.54%	0.90%	0.69%
Electricity	169	186	221	249	275	1.95%	1.73%	1.18%	1.00%
New fuels	0	0	0	2	5	-	4.79%	15.17%	10.10%
Renewable Energy Sources	23	21	18	15	13	-1.09%	-1.57%	-1.77%	-1.31%
by sector									
Industry	244	259	287	304	314	1.20%	1.01%	0.59%	0.33%
Residential	242	251	261	264	265	0.75%	0.42%	0.11%	0.03%
Tertiary	123	136	158	173	185	2.15%	1.48%	0.89%	0.70%
Transports	275	300	335	348	361	1.77%	1.12%	0.38%	0.37%

Baseline Scenario - Low Fuel prices			Mtoe				*annual growth rate*		
EUROPE 14	**1995**	**2000**	**2010**	**2020**	**2030**	***95-00***	***00-10***	***10-20***	***20-30***
CO2 Emissions	*3029*	*3169*	*3384*	*3505*	*3740*	0.91%	0.66%	0.35%	0.65%
industry	381	392	401	395	370	0.57%	0.21%	-0.13%	-0.65%
tertiary	200	215	232	240	243	1.48%	0.78%	0.32%	0.15%
households	430	439	447	450	446	0.42%	0.18%	0.07%	-0.10%
transports	800	871	966	988	1004	1.72%	1.04%	0.23%	0.16%
supply side	1219	1252	1338	1431	1677	0.54%	0.67%	0.67%	1.60%

Baseline Scenario - Low Fuel prices							*annual growth rate*		
EUROPE 14	**1995**	**2000**	**2010**	**2020**	**2030**	***95-00***	***00-10***	***10-20***	***20-30***
Indicators									
Energy intensity (toe/1990MECU)	240	225	192	166	138	-1.33%	-1.54%	-1.47%	-1.82%
Carbon Intensity (tn of CO2/toe)	2.224	2.186	2.148	2.151	2.338	-0.35%	-0.17%	0.01%	0.84%
Import Dependency (percent)	47.6	49.3	57.5	65.1	76.6	0.72%	1.55%	1.26%	1.63%

Table 11-3: Summary results of the high price baseline scenario.

Baseline Scenario - High Fuel prices			Mtoe				*annual growth rate*		
EUROPE 14	**1995**	**2000**	**2010**	**2020**	**2030**	***95-00***	***00-10***	***10-20***	***20-30***
Gross Inland Consumption	*1362*	*1448*	*1577*	*1618*	*1602*	1.23%	0.85%	0.26%	-0.10%
Solid Fuels	237	228	220	231	346	-0.76%	-0.37%	0.48%	4.15%
Liquid Fuels	574	615	647	645	636	1.38%	0.52%	-0.03%	-0.15%
Natural Gas	273	311	399	443	464	2.59%	2.53%	1.04%	0.48%
Nuclear	205	213	220	195	40	0.70%	0.37%	-1.24%	-14.72%
Electricity	1	1	1	2	3	-8.25%	-1.20%	12.51%	4.85%
Renewable Energy Sources	71	81	90	103	114	2.65%	0.98%	1.43%	0.95%

Baseline Scenario - High Fuel prices			Mtoe				*annual growth rate*		
EUROPE 14	**1995**	**2000**	**2010**	**2020**	**2030**	***95-00***	***00-10***	***10-20***	***20-30***
Electricity Generation in TWh	*2306*	*2584*	*3058*	*3404*	*3717*	2.30%	1.70%	1.08%	0.88%
Nuclear	810	845	878	778	173	0.86%	0.38%	-1.20%	-13.95%
Hydro - Renewables	293	359	400	454	513	4.11%	1.09%	1.28%	1.22%
Thermal (incl. biomass)	1203	1379	1779	2171	3031	2.77%	2.58%	2.01%	3.39%
Steam Generation in TWh_{th}	*1034*	*1136*	*1301*	*1400*	*1480*	1.90%	1.36%	0.74%	0.56%
CHP plants	322	510	770	1147	1415	9.61%	4.21%	4.07%	2.12%
Boilers	711	626	530	253	65	-2.54%	-1.64%	-7.15%	-12.68%
Generation Capacity in GWe	*569*	*626*	*726*	*830*	*910*	1.90%	1.50%	1.34%	0.93%
Nuclear	132	137	134	116	23	0.72%	-0.20%	-1.46%	-14.92%
Hydro (without pumping)	106	109	110	110	107	0.58%	0.07%	0.00%	-0.23%
Thermal (incl. biomass)	328	369	459	563	721	2.38%	2.20%	2.06%	2.51%
of which CHP	64	85	147	231	286	5.99%	5.60%	4.58%	2.17%
Solar, wind, geothermal	3	11	23	41	59	26.88%	8.11%	5.99%	3.55%
Average Load Factor %	*46.2*	*47.1*	*48.1*	*46.8*	*46.6*	0.38%	0.19%	-0.26%	-0.04%
Boilers (GWth)	*239*	*211*	*158*	*85*	*52*	-2.47%	-2.82%	-6.03%	-4.77%
Fuel Inputs for Power and Steam Gen.	*364*	*396*	*454*	*501*	*626*	1.69%	1.38%	0.99%	2.24%
Solids	171	166	166	184	311	-0.61%	-0.02%	1.06%	5.38%
Oil	88	88	80	67	54	0.01%	-0.98%	-1.72%	-2.20%
Gas	83	116	175	206	211	6.98%	4.22%	1.61%	0.24%
Biomass / Waste	22	26	33	44	49	3.34%	2.56%	2.92%	1.02%
Hydrogen	0	0	0	0	1	-	-	-	-
Average Thermal Efficiency %	*52.8*	*54.6*	*58.3*	*61.3*	*62.0*	0.67%	0.66%	0.50%	0.12%

Baseline Scenario - High Fuel prices			Mtoe				*annual growth rate*		
EUROPE 14	**1995**	**2000**	**2010**	**2020**	**2030**	***95-00***	***00-10***	***10-20***	***20-30***
Final Energy Demand	*883*	*940*	*1035*	*1080*	*1116*	1.26%	0.96%	0.43%	0.33%
by fuel									
Solid Fuels	43	40	34	30	23	-1.76%	-1.40%	-1.27%	-2.73%
Liquid Fuels	404	430	463	470	473	1.29%	0.75%	0.15%	0.05%
Natural Gas	177	184	205	214	221	0.80%	1.08%	0.42%	0.33%
Steam	67	74	86	94	101	2.04%	1.54%	0.88%	0.68%
Electricity	169	189	226	253	279	2.28%	1.76%	1.17%	0.97%
New fuels	0	0	0	2	5	-	5.32%	15.74%	10.11%
Renewable Energy Sources	23	22	19	16	14	-0.48%	-1.37%	-1.64%	-1.17%
by sector									
Industry	244	256	283	300	309	0.99%	0.98%	0.58%	0.33%
Residential	242	250	262	265	267	0.73%	0.46%	0.12%	0.04%
Tertiary	123	136	157	171	183	2.08%	1.47%	0.82%	0.72%
Transports	275	297	332	345	357	1.59%	1.13%	0.36%	0.34%

Baseline Scenario - High Fuel prices			Mtoe				*annual growth rate*		
EUROPE 14	**1995**	**2000**	**2010**	**2020**	**2030**	***95-00***	***00-10***	***10-20***	***20-30***
CO2 Emissions	*3029*	*3170*	*3402*	*3519*	*3977*	0.91%	0.71%	0.34%	1.23%
industry	381	380	385	379	355	-0.06%	0.13%	-0.16%	-0.64%
tertiary	200	212	227	231	236	1.19%	0.70%	0.18%	0.19%
households	430	437	446	449	446	0.30%	0.21%	0.07%	-0.07%
transports	800	863	957	976	988	1.53%	1.04%	0.20%	0.12%
supply side	1219	1279	1387	1483	1952	0.97%	0.81%	0.67%	2.79%

Baseline Scenario - High Fuel prices							*annual growth rate*		
EUROPE 14	**1995**	**2000**	**2010**	**2020**	**2030**	***95-00***	***00-10***	***10-20***	***20-30***
Indicators									
Energy intensity (toe/1990MECU)	240	224	192	165	138	-1.36%	-1.52%	-1.54%	-1.74%
Carbon Intensity (tn of CO2/toe)	2.224	2.189	2.158	2.174	2.482	-0.32%	-0.14%	0.08%	1.33%
Import Dependency (percent)	47.6	49.1	56.3	63.3	75.0	0.64%	1.38%	1.18%	1.71%

11.4 The Supply Side Stories

11.4.1 The Nuclear Story

In this scenario, we assume significant improvement in the technical-economic characteristics of the conventional nuclear technology, as well as the emergence of a new nuclear design.

In particular:

- for a standard large LWR, it is assumed that whereas in the baseline case the capital cost was slightly increasing over time, in the nuclear story the cost of capital becomes about 30 % cheaper by 2030. Furthermore fixed operation and maintenance costs were assumed to become about 35 % lower;
- a new evolutionary nuclear design with inherent safety characteristics is introduced after 2020 with capital cost about 13 % and operating costs about 25 % less than the LWR. For the nuclear scenario this type of plant was assumed to be 30 % cheaper to construct and 50 % to operate than in the baseline case.

The above changes make nuclear plants more competitive in terms of generating costs. In the baseline case they were not competitive even for the higher annual loads. In the nuclear story they become generally cost-attractive, particularly in some countries like France. In the last decade, whereas in the baseline only 23 GW of nuclear plants remain in operation in the EU, in the nuclear story the technology re-emerges leading to additional 26 GW mainly of new nuclear design type.

The additional constructions of nuclear plants are nevertheless rather limited. The penetration of nuclear is effected to the detriment of GTCC plants mainly, since they both compete for the base load. Other plant types are not affected. Cogeneration and renewables remain almost unchanged from the baseline scenario.

CO_2 emissions are significantly lower in the nuclear story, compared to baseline. By 2030 the emissions from power and steam generation become lower by –5.6% compared to the Baseline case. The implications on CO_2 emissions appear in particular at the last decade of the projection. For the total energy system, CO_2 emission reduction is about 2.5% in 2030. The positive effect on CO_2 from nuclear energy is less than expected mainly because it displaces gas-fired plants in the base load.

In the case of high price baseline, nuclear energy becomes slightly more attractive than in the case of low prices. Additional nuclear capacities amount to 42 GW.

The main competitor is in this case clean coal, which is mainly displaced by nuclear in the base load. This results into bigger positive effects for CO_2 emissions, which reduce by 4.5% in 2030 in the energy system (8.8% in power and steam generation).

Finally, the nuclear story leads to a reduction of average power and steam generation costs by about 1% in 2030 compared to the Baseline scenario.

11.4.2 The Clean Coal Story

For the purpose of this scenario three new clean coal technologies have been retained:

- supercritical coal having a rate of 44 % efficiency by 2030, improve in terms of average specific capital cost (863 ECU/kW instead of 1114 ECU/kW in Baseline) and fixed maintenance costs (20 ECU/kW);
- IGCC plants improve in terms of thermal efficiency reaching by 2030 a rate of 50% (46 % in Baseline), and capital costs (1003 ECU/kW instead of 1333 ECU/kW in Baseline);
- PFBC plants also improve in terms of capital costs (780 ECU/kW instead of 1040 ECU/kW in Baseline) and thermal efficiency reaching a rate of 44 % by 2030.

These improvements combined with low coal prices make clean coal technologies very attractive in comparison to natural gas plants mainly for base load operation. The winner within the family of clean coal technologies seems to be PFBC. In total, the clean coal story involves the construction of about 130 GW on top of the baseline scenario. This acts to the detriment of GTCC. Small gas plants, renewables and nuclear are not affected. The clean coal story induces higher penetration of cogeneration, than in the baseline (30 GW more), implying also a slightly higher degree of production decentralisation.

The evolution within the clean coal story has substantial adverse effects on CO_2 emissions, which rise by 10 % at the end of the period from baseline (24% in power and steam generation).

In the case of high prices, clean coal has already penetrated in the baseline scenario. On top of that, additional constructions of clean coal plants amount to 90 GW, again displacing GTCC plants in the base load. The adverse effects on CO_2 are again significant.

Average power and steam generation costs decrease by about 4% in this story.

11.4.3 The Gas Story

GTCC is a mature technology and investment activity is widespread around the world. In this gas story scenario we assume a significant improvement of this technology over time, both in terms of thermal efficiency (63 % compared to 59 % in Baseline) and capital cost (460 ECU/kW instead of 528 ECU/kW in Baseline. In the gas story assuming abundance of natural gas supply, reflecting lower gas prices than in the baseline, further facilitates the penetration of GTCC and smaller gas machines.

The combination of these changes provides high potential to GTCC plants for electricity and also for cogeneration. The considerable improvement of GTCC plants contributes to narrow the gap in generation cost terms between large utilities and small producers. Independent producers penetrate more easily in the market, further leading to higher decentralisation of generation (47% in 2030) than in the baseline scenario (41%). The increased penetration of GTCC operates to the detriment of clean coal and conventional thermal technologies. Cogeneration also penetrates more, mainly through GTCC cogeneration plants, allowing in the same time for lowering the need for peak load plants. Other power technologies, like nuclear and renewables, are slightly negatively affected.

CO_2 emissions are reduced by -6 % in 2030 compared to baseline (-14% in power generation), as a consequence of the use of gas, the decline of coal use and the spectacular improvement in average thermal efficiency and cogeneration.

The gas story also involves an important decrease of average generation costs (-15%), as a consequence of lower fuel prices but also the low and improved capital costs associated with GTCC plants.

In the case of high price baseline, the benefits for CO_2 emissions are even more spectacular (-12% in 2030), since in the corresponding baseline clean coal penetrates more than in case of low fuel prices. About 190 GW of GTCC plants are constructed on top of those built in the baseline, displacing clean coal and conventional thermal technologies.

11.4.4 The Fuel Cells Story

The fuel cells story is considered to be an extension of the Gas story, in the sense that Fuel Cells use also natural gas, since the scenario has not considered as economic the development of alternative fuels for fuel cells, such as methanol and hydrogen. The fuel cells story considers that the progress is enabled through the development of SOFC (Solid Oxide Fuel Cells) which allow for high heat to power generation rates.

The main assumptions are similar to those of the Gas story, including the assumption that GTCC plants are becoming very competitive and the abundance of natural gas keeping prices low. In addition, the fuel cells story assumes an important improvement of the technical-economic characteristics of Fuel Cells. For example, regarding fuel cells we assume:

- the electrical efficiency raises up to 70 % by 2030 (65 % by 2015);
- capital cost becomes 25 % less than in Baseline case;
- O&M costs are 66.7 % lower in this story.

The evolution of technology over time is such that Fuel Cells becomes competitive after 2020 starting from a negligible initial share. Within this limited timeframe of the study, the share of fuel cells rises up by 2030 to 4% only. The system is again based on natural gas technologies (mainly GTCC), which represents about 64 % in terms of power generation.

The deployment of fuel cells allow for further decentralisation in generation (50%) as the cost differences due to economies of scale are the smallest among the generation technologies. In addition they allow for higher penetration of cogeneration. The use of solid fuels declines over time reaching the lowest value among the technology stories.

The fuel cells story is very efficient in terms of CO_2 emissions, which are found to be reduced by -8% compared to the Baseline scenario (-21% in power and steam generation). CO_2 emissions in 2030 increase only by 9% from 2000, while this rate was double in the baseline scenario.

In terms of average generation costs, the fuel cells story achieves the lowest level (-15%) among the technology stories, even slightly lower than the gas story.

The case of high price baseline confirms these results. Clean coal is massively displaced by fuel cells, decentralisation and cogeneration that augment substantially. The benefits in terms of CO_2 emissions are more significant: -14% from baseline.

11.4.5 The Renewable Story

The renewable story involves a considerable technological progress in intermittent power production technologies and a decrease of costs associated to the use of biomass. In particular:

- for wind turbines the capital cost decreases from 849 ECU/kW (in the Baseline scenario) to 355 ECU/kW in 2030;

- for photovoltaics used in buildings, the capital cost reaches 1235 ECU/kW compared to 2470 ECU/kW in the Baseline scenario;
- for small hydro plants the capital cost is reduced to 1137 ECU/kW (2104 ECU/kW in the Baseline scenario);
- for tidal-ocean techniques having capital cost that decrease from 1831 ECU/kW (in the Baseline scenario) to 990 ECU/kW;
- for geothermal plants reaching a capital cost equal to 1042 ECU/kW (1928 ECU/kW in the Baseline scenario).

The O&M costs of the above technologies are in this story also lower by 50% compared to the Baseline scenario.

Biomass gasification is assumed also to progress, in terms of capital cost (from 946 ECU/kW in the Baseline scenario to 513 ECU/kW in 2030) and O&M costs (from 29.7 ECU/kW to 19.8 ECU/kW).

The technology improvements are enough to make renewable power plant considerably more attractive. The scenario involves additional constructions of about 120 GW contributing by 35% in power and steam generation in 2030 (to be compared to 15% in 2000 and 18% in 2030 in the baseline). Renewables represent 13.5% of Gross inland consumption in 2030 (7% in the baseline).

The deployment of renewables operates mainly to the detriment of GTCC plants and has little effects on other power and steam technologies. Also, the expansion of renewables seems to develop to the detriment of decentralised gas-based generation. This explains why the system is found to be slightly more centralised than in the baseline and the penetration of cogeneration is smaller than in the baseline.

As a result of these developments CO_2 emissions are significantly reduced (by –7% compared to baseline in 2030, -17% in power and steam generation) despite the adverse effects from the reduction of gas use. The average production cost reduces by 3-4%.

In the case of high price baseline, the penetration of renewables is of the same order of magnitude as in the case of low price baseline. In case of higher prices, renewables displace clean coal and conventional technologies, leading to a more spectacular decrease of CO_2 emissions (-10%).

11.4.6 The Pessimistic Supply Side Story

This story assumes that the technical economic characteristics of power plants remain frozen at their level of 2000. In other terms, the technology progress incorporated in the baseline scenario is cancelled.

The conventional thermal technologies keep their competitive place in the market, so over time they are displaced by new technologies at rates less than in the baseline. Mainly cogeneration plants and GTCC are negatively affected by the frozen progress. Clean coal also suffers. Nuclear and renewables are almost not affected. Adverse affects are also observed for generation decentralisation (30% only in 2030).

The effects on CO_2 emissions are significant: total CO_2 emissions increase by 5% from baseline (11% in power generation). Average generation costs increase by 10% in 2030 from baseline. Similar are the results for the case of high price baseline. The following figures summarise the findings from supply-side stories compared to low price baseline scenario.

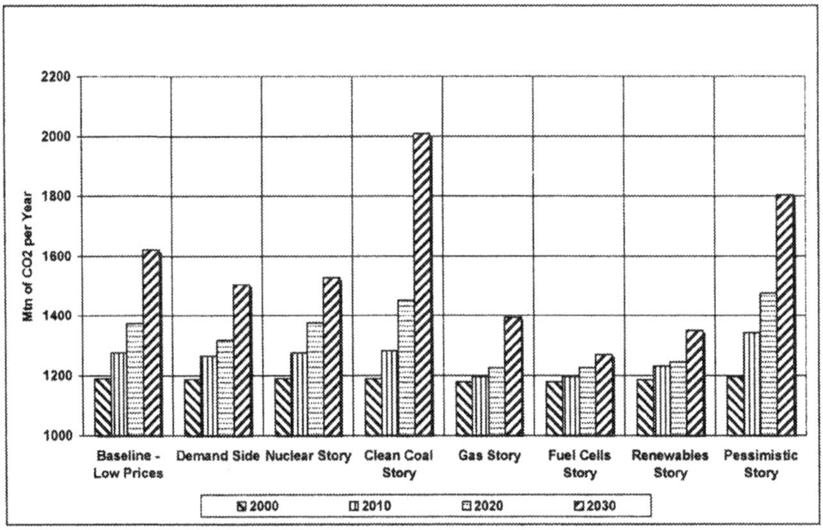

Figure 11-4: CO_2 emissions from the power and steam generation sector of the EU (scenario with low prices in the baseline).

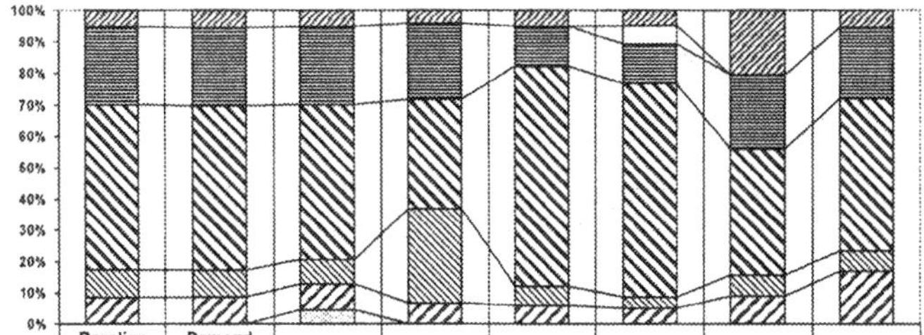

GW	Baseline scenario with low fuel	Demand Side scenario	Nuclear story	Clean Coal story	Gas story	Fuel cells story	Renewables story	Pessimistic story
Hydro-Renewables	31	30	31	24	29	29	127	32
Fuel Cells	0	0	0	0	0	33	0	0
Small size Gas plants	147	136	147	144	72	70	146	133
GTCC	315	287	293	208	398	387	250	285
Clean Coal	53	48	47	179	35	20	42	38
Conv. Thermal	51	47	50	40	34	29	56	100
Nuclear	0	0	26	0	0	0	0	0

Figure 11-5: EU-14, % structure of cumulative new capacities 2010-2030 (scenario with low prices in the baseline).

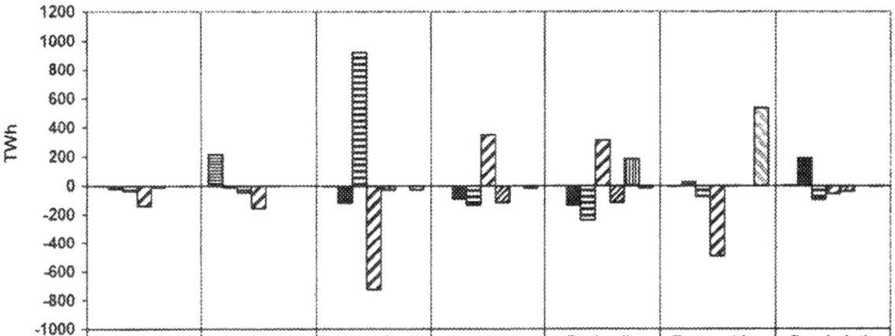

	Demand side story	Nuclear story	Clean Coal story	Gas story	Fuel cells story	Renewables story	Pessimistic story
Nuclear	-2	217	-2	1	2	-7	3
Conv. Thermal	-21	-12	-119	-91	-136	23	191
Clean Coal	-35	-46	921	-132	-240	-78	-95
GTCC	-142	-160	-725	350	314	-494	-56
Small size Gas plants	-12	-1	-26	-117	-119	-4	-39
Fuel Cells	0	0	0	0	185	0	0
Hydro-Renewables	-2	-1	-28	-16	-17	533	-7

Figure 11-6: EU-14, change of electricity production by plant type in 2030 (scenario with low prices in the baseline).

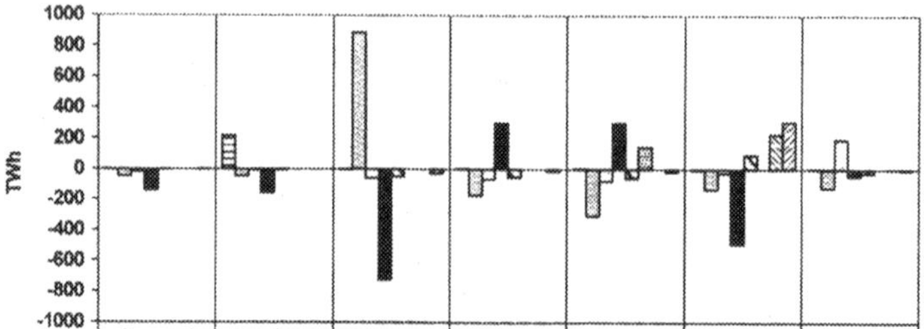

	Demand side story	Nuclear story	Clean Coal story	Gas story	Fuel cells story	Renewables story	Pessimistic story
Nuclear	-2	217	-2	1	2	-7	3
Solids	-49	-46	883	-170	-306	-133	-121
Oil except refinery gas	-19	-6	-57	-65	-78	-26	191
Gas incl. refinery gas	-139	-159	-726	299	300	-487	-44
Biomass and Waste	-4	-7	-50	-54	-59	93	-24
New Fuels	0	0	0	0	146	0	0
Large Hydro	-1	0	-1	0	-1	228	0
Small Hydro-Renewables	-2	-1	-27	-15	-17	306	-7

Figure 11-7: EU-14, change of electricity production by energy form in 2030 (scenario with low prices in the baseline).

The following figures summarise the findings from supply-side stories compared to high price baseline scenario.

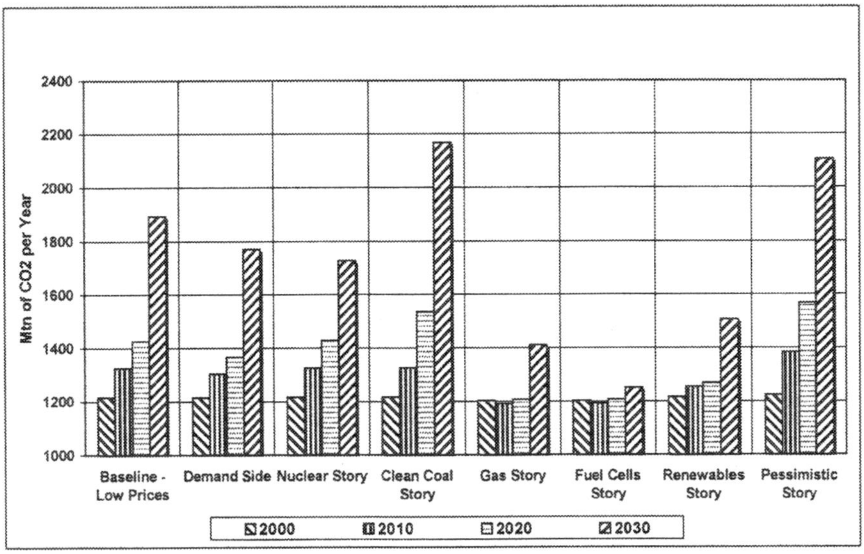

Figure 11-8: CO_2 emissions from lower and steam generation sector of the EU (scenario with high prices in the baseline).

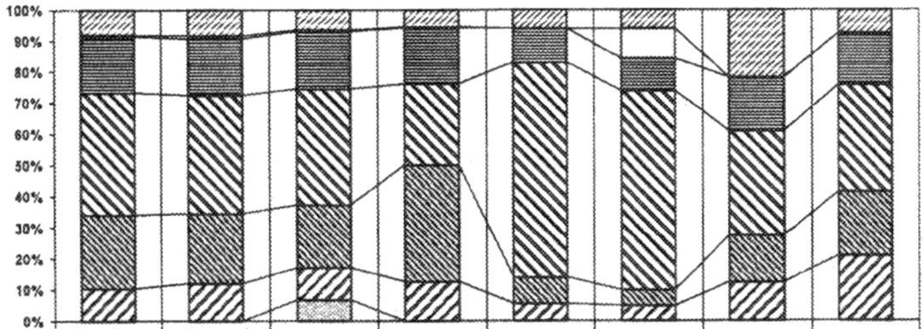

GW	Baseline scenario with low fuel	Demand Side scenario	Nuclear story	Clean Coal story	Gas story	Fuel cells story	Renewables story	Pessimistic story
Hydro-Renewables	50	49	40	34	35	35	140	46
Fuel Cells	3	5	3	2	0	56	1	0
Small size Gas plants	114	108	114	112	65	61	109	103
GTCC	244	222	229	160	401	371	214	208
Clean Coal	147	132	124	231	49	29	96	125
Conv. Thermal	65	71	63	78	33	28	78	127
Nuclear	0	0	42	0	0	0	0	0

Figure 11-9: EU-14, % structure of cumulative new capacities 2010-2030 (scenario with high prices in the baseline).

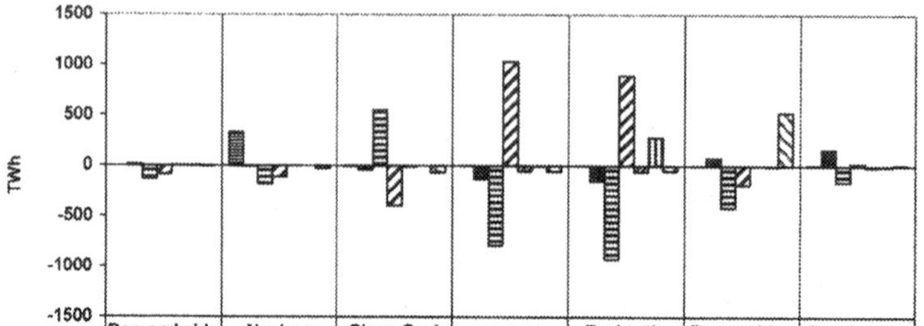

	Demand side story	Nuclear story	Clean Coal story	Gas story	Fuel cells story	Renewables story	Pessimistic story
Nuclear	0	331	-8	-1	-1	-7	2
Conv. Thermal	13	-11	-41	-133	-153	72	159
Clean Coal	-135	-179	549	-789	-927	-417	-162
GTCC	-86	-112	-402	1031	892	-191	18
Small size Gas plants	2	-1	-13	-57	-62	-3	-12
Fuel Cells	4	0	0	-12	279	-9	-12
Hydro-Renewables	-8	-29	-69	-56	-55	522	7

Figure 11-10: EU-14, change of electricity production by plant type in 2030 (scenario with high prices in the baseline).

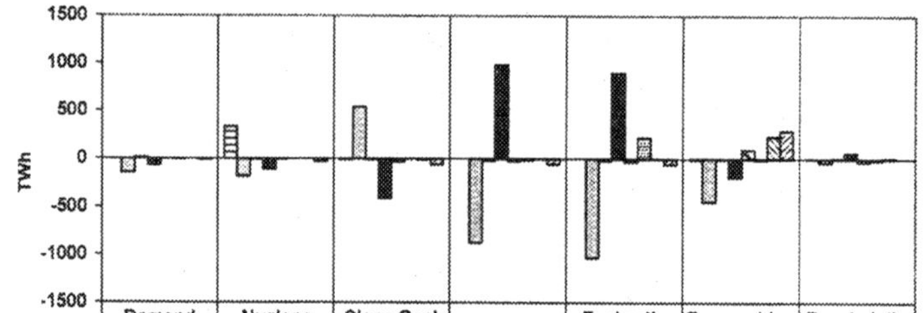

	Demand side story	Nuclear story	Clean Coal story	Gas story	Fuel cells story	Renewables story	Pessimistic story
Nuclear	0	331	-8	-1	-1	-7	2
Solids	-143	-185	542	-878	-1033	-445	-40
Oil except refinery gas	14	-3	-7	-23	-23	-5	7
Gas incl. refinery gas	-71	-110	-411	981	896	-189	65
Biomass and Waste	1	-5	-32	-26	-31	99	-30
New Fuels	-4	0	0	-12	219	-9	-12
Large Hydro	0	0	-6	1	1	233	8
Small Hydro-Renewables	-7	-29	-63	-56	-56	289	-1

Figure 11-11: EU-14, change of electricity production by energy form in 2030 (scenario with high prices in the baseline).

11.5 The Demand Side Story

11.5.1 Effects from Demand Side Progress

The demand-side story is defined in a rather simple way. It is assumed that all new technologies in end-uses become cheaper by 35% compared to the baseline. This is assumed to be achieved gradually over time, allowing for a gradual acceleration of the penetration of improved and advanced technologies in all sectors. However, the assumptions are such that none of the new technologies ever becomes cheaper than the ordinary technology.

The demand-side sub-models of PRIMES V.2 have a uniform structure. Each sub-model represents a sector that is further decomposed into sub-sectors and then into energy uses. A technology operates at the level of an energy use and utilises energy forms (fuels). The following graphic illustrates the hierarchical decomposition of the demand-side models.

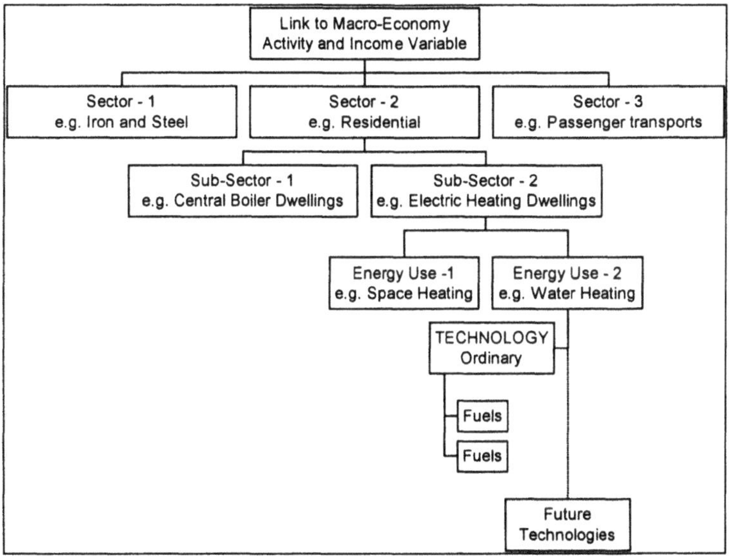

Figure 11-12: Decomposition structure of the demand-side models.

The fact that future technologies become cheaper than in the baseline induces acceleration of their penetration. This has multiple indirect effects in the energy system. First, by means of learning by doing and market maturity mechanisms, the technology that starts penetrating more is further facilitated to do so. Then, because of differentiated structure of technology mix and cost structures by energy use and sub-sector, substitutions among uses and sub-sectors also occur. For example, the demand-side technology progress also affect the material recycling rates and the degree of adopting new industrial processes in some sectors. Those sub-sectors that decline bear slowdown of technical progress along with slowdown of capital replacement schedule. Other sub-sectors that develop more rapidly also benefit from accelerated capital turnover and the embedded technological improvement.

The demand-side sectors are also affected by indirect effect coming from the supply-side. Efficiency gains in demand-side induce changes in the demand for electricity and steam addressed to the supply-side. The adjustment of the supply sectors also implies changes in emissions, fuel mix and selling prices. The latter influence the choice of fuel mix in the demand-side.

The following table shows the improvement of energy efficiency indicators at the level of the model's sectors.

Table 11-4: Energy efficiency indicators: average rate of change per year (2000-2030).

Scenarios with low prices in the Baseline					
	Baseline	Demand Side Story		Baseline	Demand Side Story
Industry					
integrated steelworks	-0.16%	-0.34%	cement	-0.33%	-0.48%
electric processing	-0.32%	-0.47%	ceramics	-0.90%	-0.97%
primary aluminum	-0.26%	-0.33%	glass	-0.58%	-0.64%
secondary aluminum	-0.35%	-0.48%	glass recycled	-0.86%	-0.96%
copper	-0.48%	-0.61%	other building materials	-0.43%	-0.53%
zinc	-0.37%	-0.53%	pulp	-0.55%	-0.59%
lead	-0.11%	-0.18%	paper	-0.77%	-0.84%
other non ferrous products	-0.33%	-0.43%	food, drink, tobacco	-0.42%	-0.48%
fertilisers	-0.37%	-0.43%	engineering	-0.46%	-0.56%
petrochemical	-0.44%	-0.50%	textiles	-0.44%	-0.48%
inorganic chemicals	-0.43%	-0.55%	other industries	-0.38%	-0.47%
low energy chemicals	-0.42%	-0.51%			
tertiary	-1.06%	-1.30%	households	-1.70%	-2.15%
passenger transports (income related)	-1.12%	-1.50%	goods transports (GDP related)	-1.72%	-1.78%
Vehicle efficiency					
passenger transports (toe/Mpkm)	-1.00%	-1.38%	goods transports (toe/Mtkm)	-1.27%	-1.34%
Gross Inl Cons / GDP (toe/1990MECU)	-1.61%	-1.84%			

The effects of demand-side technology progress as defined in this scenario induces significant (but not dramatic) effects in favour of energy efficiency improvement. The effects are more pronounced in the domestic sector and in transports. The overall effect, in terms of energy intensity, is substantial, leading to a yearly rate of improvement that is well beyond the historical average.

Final energy demand in 2030 (compared to the baseline) decreases by 8% (3% in industry, 13% in residential, 7.5% in tertiary and 8.5% in transports). Liquid fuels (-9%) and natural gas (-9.5%) are more affected than other energy forms. On the contrary, electricity and steam (-5.5%) obtain higher share, but reduce in magnitude.

Total CO_2 emissions decrease in 2030 by 8% compared to the baseline. In the demand-side this reduction is 9% and in the supply-side it is 7%. In 2030, the system emits 305 Mtn of CO_2 less than the baseline, of which 185 come from the demand-side and 119 from the supply-side. We remind that these results correspond to the case in which the supply-side does not benefit technological progress additional to that of the baseline.

The demand-side story also induces a decrease of total system cost: -3% in 2030 from baseline. In total, the system saves 540 billion ECU'90 in 2030, which mainly concern the purchase of energy and related equipment (85%). The savings in fuel costs represent about 10% of total savings. The transport sectors get most of the benefits (75%) followed by households (17%). Some small cost savings also occur in power and steam generation, through indirect system effects.

The indirect effects on power and steam generation mainly come from the reduction of electricity demand. The reduction affects, almost at a uniform manner, all technologies (compared to baseline).

The effects of the demand-side story in the case of high price baseline are similar in magnitude and nature.

The following table summarises the penetration of new technologies in the demand-side.

Table 11-5: Penetration of new technologies in the demand-side.

Technology penetration in 2030 (in cumulative investment)			
	Type of Technology	Baseline Low prices	Demand side story
Industry	improved	29.3%	30.0%
	advanced	33.4%	38.3%
Residential - Tertiary	improved	42.6%	43.7%
	advanced	40.8%	51.5%
Transports	improved	44.1%	31.7%
	advanced	49.9%	58.1%
All demand sectors	improved	39.4%	35.9%
	advanced	41.7%	49.8%

This table shows that whereas in the baseline the percent penetration of new technologies was 80% in 2030, in the demand-side story this percentage becomes 85%. In addition, the composition in terms of technology generation changes as well, allowing for higher penetration of advanced technology generations. The following figures summarise the findings from demand-side stories with low prices (compared to low price baseline scenario).

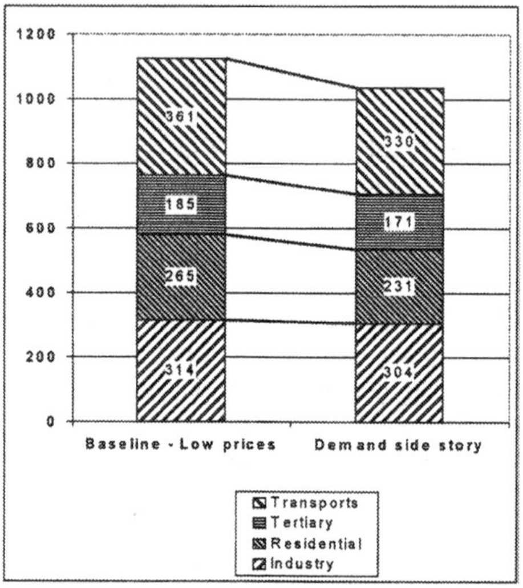

Figure 11-13: EU-14, final energy demand by energy form and sector in 2030 (scenarios with low prices in the baseline).

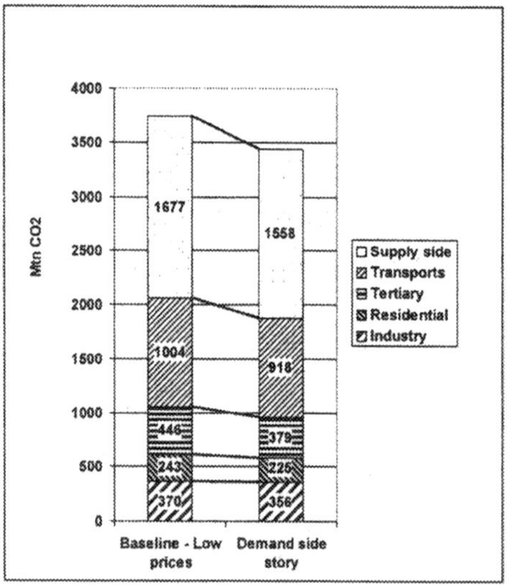

Figure 11-14: EU-14, CO_2 emissions in 2030 (scenarios with low prices in the baseline).

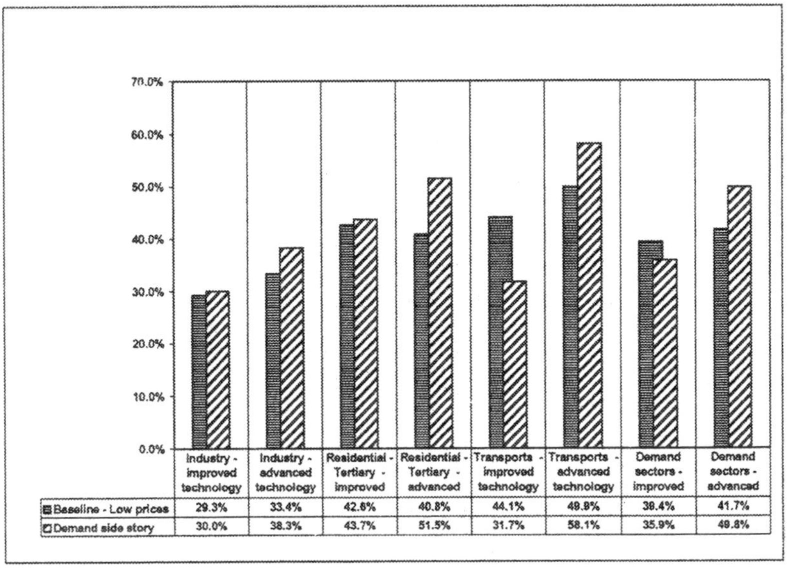

Figure 11-15: EU-14, penetration of new technologies in demand sectors in 2030 (scenarios with low prices in the baseline).

The following figures summarise the findings from demand-side stories with high prices compared to high price baseline scenario.

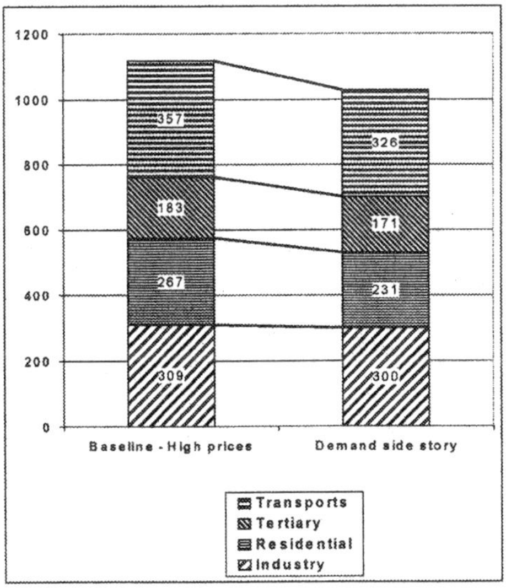

Figure 11-16: EU-14, final energy demand by energy form and sector in 2030 (scenarios with high prices in the baseline).

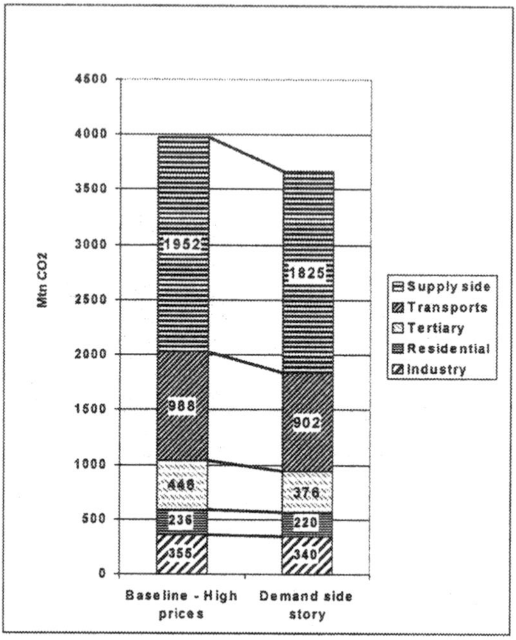

Figure 11-17: EU-14, CO_2 emissions in 2030 (scenarios with high prices in the baseline).

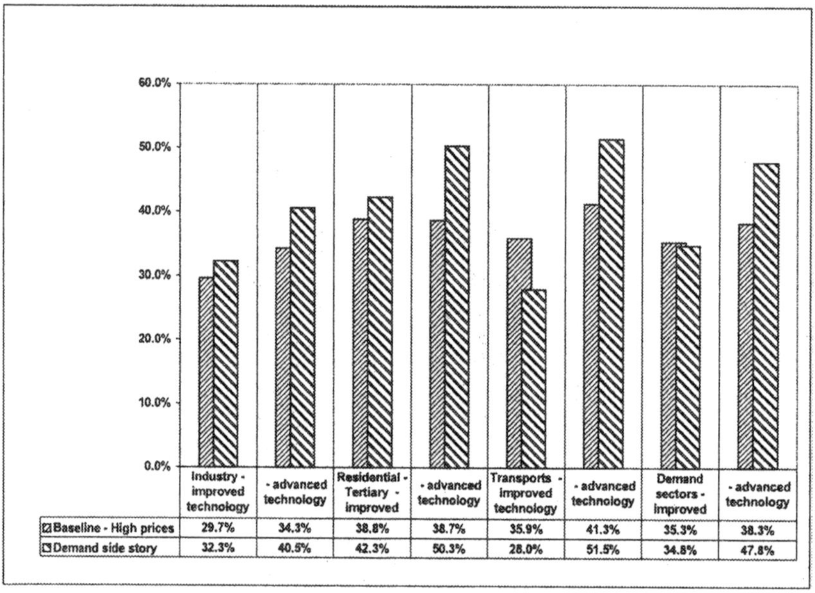

Figure 11-18: EU-14, penetration of new technologies in demand sectors in 2030 (scenarios with high prices in the baseline).

11.5.2 Effects from Combining Demand and Supply Progress

The model runs that analyse the combined effects from technology progress in the demand and the supply sectors are based on the assumptions presented above. For example, in a nuclear story with demand-side effects, progress is assumed to occur both for the nuclear power plants and the end-use technologies. The progress in the latter induces lower demand for electricity and steam (however higher market shares). The progress in the nuclear technology provides higher opportunities for penetration for this technology, but in a smaller overall market size. An interesting result is whether the size effect alters the conclusions regarding the supply-side stories, through differentiated economies of scale, since the model introduces different possibilities for economies of scale by type of supply technology.

The following table summarises the CO_2 emission reduction effects in 2030 on power and steam generation from the combined demand and supply progress and compares with the same effect coming from the supply-side progress.

Table 11-6: CO_2 emission reduction effects in 2030 for power and steam generation sector.

Mtons of CO_2 avoided	Nuclear Story	Clean Coal Story	Gas Story	Fuel Cells	Renewables	Pessimistic
Combined demand and supply progress	-86	+355	-215	-272	-273	+171
Supply side progress only	-94	+388	-229	-352	-272	+184

The table shows that the economies of scale mechanism play an important role mainly in those scenarios that involve decentralisation of generation, like for example the fuel cells. The additionality of the effects can be generally assumed, except the fuel cell and gas related stories.

The combined effects from progress in the demand and the supply sides are spectacular in terms of reducing CO_2 emissions and cost savings.

The fuel cell story combined with progress in the demand-side leads to stabilisation of CO_2 emissions in total (at the level of 2000) and in the power and steam generation sector. The decrease of emissions from the baseline reaches 15%

in 2030. The total system cost saving amount to more than 600 billion ECU per year (in 2030) of which 85% come from the demand-side and 15% from the supply-side. This represents in total –3.5% in 2030.

The gas story (with demand-side progress) also provides significant benefits. CO_2 emissions in 2030 decrease by 14% from baseline.

The renewables story combined with demand-side progress also lead to stabilisation of CO_2 emissions at the level of emissions in 2000. The cost savings are of the order of 550 billion ECU in 2030.

The nuclear story (with demand-side progress) leads to a drop of CO_2 emissions in 2030 (-10%), but this scenario does not involve stabilisation of emissions.

The combination of the clean coal story with the demand-side progress leads to a small increase of CO_2 emissions (1.5% more in 2030 compared to the baseline). The effects from the demand-side almost cancel the adverse effects on CO_2 emissions from the supply-side. Similar is the result for the pessimistic supply-side story, leading however to a decrease of emissions by 4% in 2030.

The following figures summarise the findings from combined demand and supply-side effects compared to low price baseline scenario.

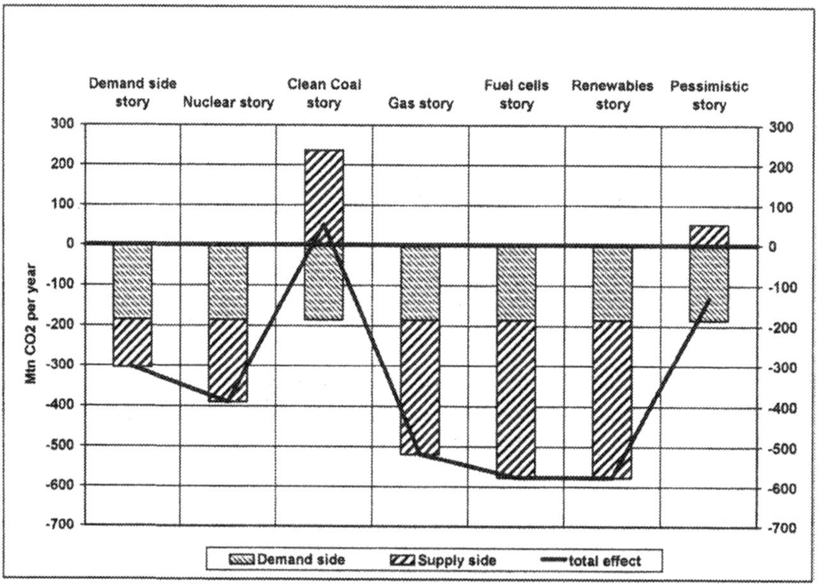

Figure 11-19: EU-14, change of CO_2 emissions in 2030 (scenario with low prices in the baseline).

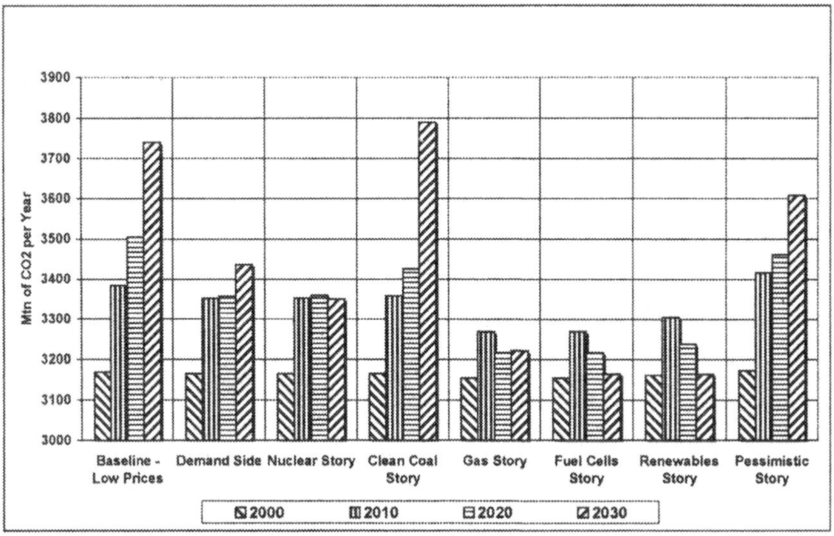

Figure 11-20: CO_2 emissions of EU-14, (scenarios with low prices combining demand and supply side effects).

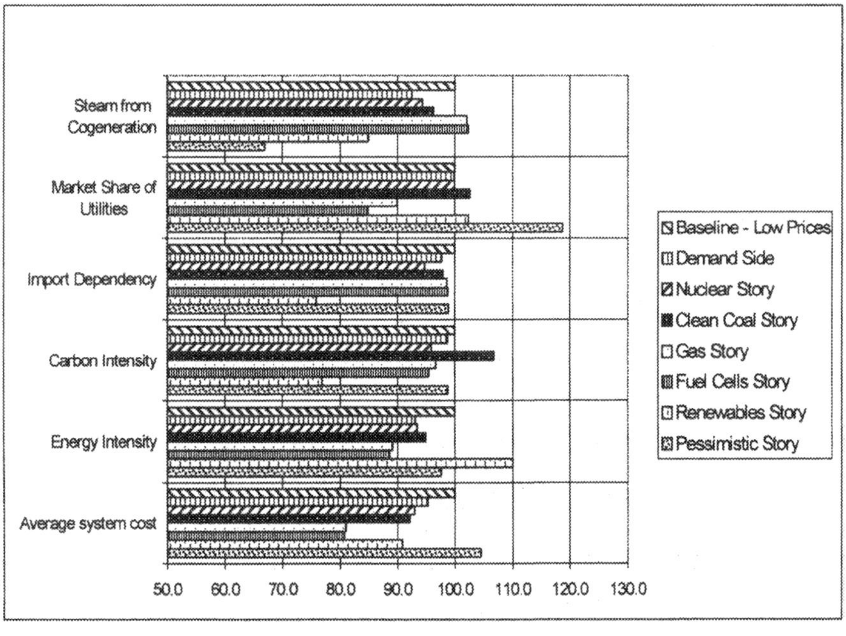

Figure 11-21: EU-14 Indicators (index=100 in baseline in 2030) under low price baseline scenario.

The following figures summarise the findings from combined demand and supply-side effects compared to high price baseline scenario. They show even more pronounced effects on CO_2 emissions from combining demand-side and supply-side technological progress.

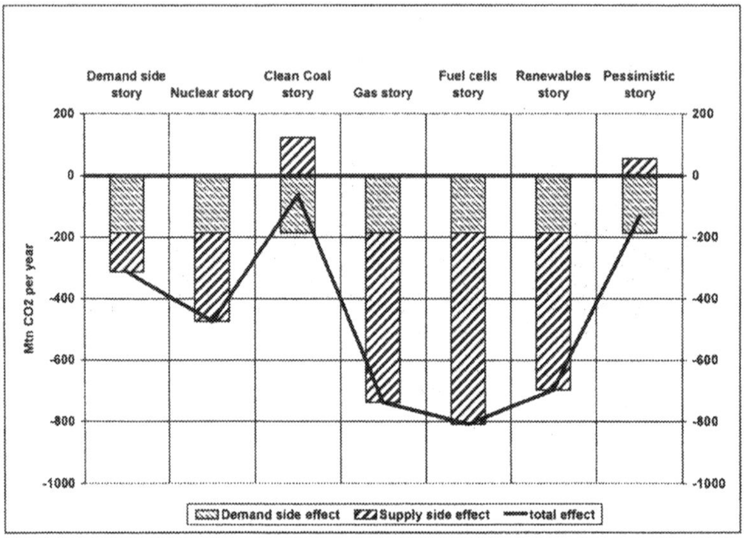

Figure 11-22: EU-14, change of CO_2 emissions in 2030 (scenario with high prices in the baseline).

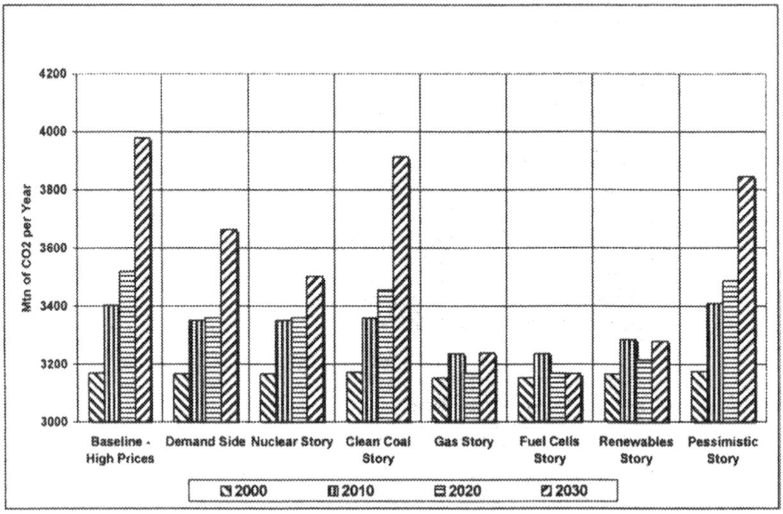

Figure 11-23: CO_2 emissions of EU-14 (scenario with high prices combining demand and supply side effects)

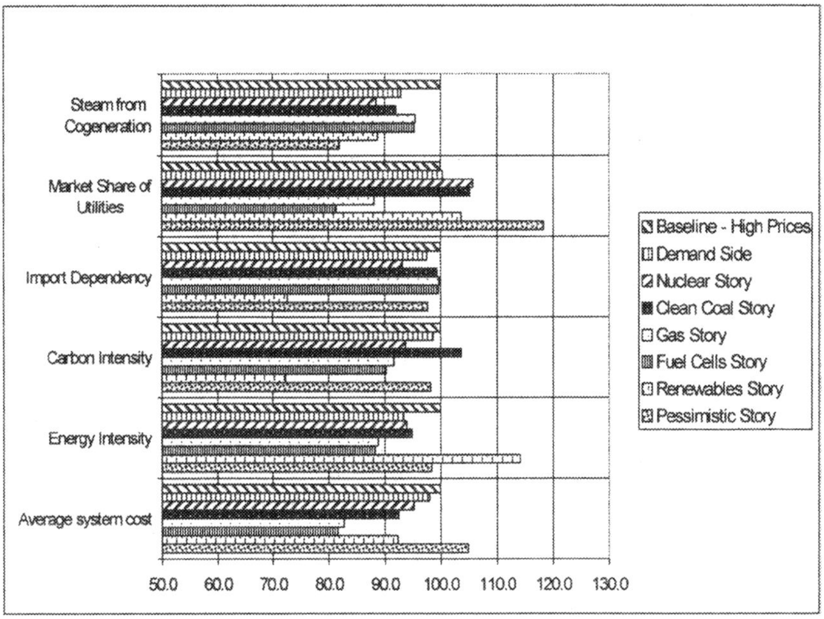

Figure 11-24: EU-14 Indicators (index=100 in baseline in 2030) under high price baseline scenario.

In both baseline cases, the results from combining demand-side with supply-side technological progress can be summarised as follows:

- the gas and fuel cell stories have the best performance in terms system cost (-20% from baseline); this is also facilitated from abundance of natural gas resources
- the same cases (gas-related) lead to the highest energy efficiency gains at the level of the overall energy system (-12% from baseline)
- the renewable story lead to highest performance for carbon intensity (-25% from baseline) and import dependence (-26% from baseline)
- maximum results in favour of power production decentralisation and steam cogenerated are obtained in the gas-related stories
- the demand-side progress alone does improve energy intensity but acts very little on carbon intensity
- the nuclear progress has little interaction with demand-side, while the coal progress offsets gains obtained from demand-side

11.6 Concluding Remarks

The analysis through the idea of technology stories proved analytically successful. Conclusions are drawn regarding the analysis of differential effects on energy and emissions from the development of technology progress in power and steam generation techniques and in the end-use sectors.

The case of the European Union energy system seems to be very sensible in the long run, as after 2020 the massive decommissioning of old plants opens up a large decision area regarding technology choice in power and steam generation. The differential technology progress may directly influence the strategic choice for the EU leading to highly varying profiles of the future power and steam system.

The stories revealed the strategic importance of natural gas and the related technologies in considerating the emission reduction, cogeneration development and generation decentralisation objectives. The clean coal technological development, event though attractive for exports of power equipment, may have adverse effects for the emissions within the EU. The development of nuclear energy seem to be limited in scope and gains, as restricted from load constraints and the lack of interaction with decentralised steam generation. The incremental improvement of renewable technologies may be significant in size to change the profile of the system in terms of emissions and costs, yet improving import dependency and carbon intensity.

The importance of technology progress in the end-use sectors is without any doubt. It leads to significant CO_2 emissions both directly in the demand sectors and indirectly in energy conversion sectors. It also induces considerable system cost savings and has interesting distributional effects as resulting from high savings for the households and transports.

Three of the stories that combine supply-side and demand-side technology progress lead to stabilisation of CO_2 emissions in the EU, namely the gas, the fuel cell and the renewables stories.

The following table summarises the results of the combined demand-side and supply-side stories.

Table 11-7: Summary results for the combined demand-side and sypply-side stories.

	Baseline Low prices	Demand side story	Nuclear story	Clean Coal story	Gas story	Fuel cells story	Renewables story	Pessimistic story
Import dependency	100.0	97.7	94.7	97.9	98.6	98.8	89.0	98.9
Energy intensity	100.0	93.0	93.4	94.9	89.1	88.6	93.8	97.6
Carbon intensity	100.0	91.9	89.6	101.3	86.1	84.6	84.6	96.4
System cost	100.0	97.0	96.9	96.9	96.8	96.8	96.9	97.1
Avoided CO_2 emissions in 2030 (Mtn)		-304	-390	50	-519	-577	-577	-133
Avoided cumulative CO_2 emissions 2010-2030 (Mtn)		-3149	-3551	-660	-6057	-6335	-5946	-937

Further analysis is needed to further check the above results and confirm their accuracy. It would also be important to combine the analysis with a macro-economic equilibrium model to analyse the significance of the results for the economy and the welfare.

12 Issues for EU R&D Strategy[45]

Technological innovations often require public support during the early stages of their development. Since public support can only be limited it must of necessity be selective and the setting of priorities of EU R&D policy becomes an important issue.

A number of difficulties arise during the process of formulating a European energy R&D strategy. First, such a strategy must take into account the multiple objectives of EU member states. Thus, the allocation of R&D funds, apart from fostering the transition towards a more sustainable energy system, should also be compatible with other main policy objectives such as ensuring secure energy supplies, improving the competitiveness and efficiency of the European industry and guaranteeing the technological leadership of EU countries within this field.

Second, market and international developments as well as non energy EU policy objectives must be taken into account.

- The creation of a single energy market is an important issue within the Union's political agenda. The gradual creation of European energy markets will have positive effects on the transparency of prices, on competition for the provision of energy services and, consequently, on the efficiency with which these services are produced and supplied.
- The transition towards a single market is being accompanied in the EU by a massive change in the market structures: national companies are being privatised and broken up and, in many cases, national markets are being simultaneously open to domestic and foreign competition. Some of the energy companies are being unbundled and broken up prior their privatisation while some others retain their size and expand outside their national market. However, even these larger companies are likely to face much more

---

[45] The principal author of this chapter is Antonio Soria of IPTS.

competition in a larger European market. Many new technological developments have reduced the optimal size of plants, at least in the case of power generation, thus encouraging the creation of smaller and more diverse players. In considering the R&D priorities in the EU the prospects for more competitive energy markets need to be taken into account. It is important to recall that, at least in the EU, the degree of centralisation of power generation declines rapidly between 1995 and 2030 almost under all circumstances and irrespectively of the technological scenario.

- EU technology policy has to be designed within the framework of the subsidiary principle that holds for the European-wide co-ordinated programmes: actions should be taken at a European level only when the instrumentation of such a co-ordinated policy is more efficient than separate implementation of national plans. There are many indications that energy R&D is a field where significant gains may be obtained as a result of a co-ordinated, EU-wide R&D action plan. For example, the presence of important economies of scale in executing R&D actions favours R&D on a EU scale, because of the complementarities and savings induced by conducting research in trans-European pools of R&D corporations: larger, co-ordinated groups avoid the possibility of effort duplication and allow for a more exhaustive exploration of potential lines of research. In addition, subcritical research teams that are maintained by inertial reasons may regain efficiency if stimulated by the exchange of ideas within a broader research consortium. Besides this, there are research fields whose costs are simply too high for most Member States and are feasible only through concerted and co-operative research. The European Commission plays a very significant role in such efforts. Economies of scope are also found in extending the overall R&D framework to include energy technology programmes: the possibility of application of innovations originated within the energy technology cluster may diffuse towards other industrial sectors more efficiently if the R&D programmes are co-ordinated at the same European level.

12.1 Medium Term Issues: Faster Adoption of Improved Technologies

One of the most important conclusions from the first part of this volume was that the global primary energy system is likely to remain dominated by fossil fuels. The role of new technologies in the medium term is quite limited. However, as was seen in the second part of this volume a large number of measures can be undertaken to limit, and in the case of Europe and other advanced regions, reduce the amount of CO_2 emissions. Indeed, as can be seen from Annex I, the number measures discussed in Part II is small when compared to the number of possible ways to improve efficiency through the faster penetration of technology in the various sectors.

Within the power generation sector, it is important to note the fundamental changes that the system is undergoing. The very sharp increase in gas use has been accompanied by the slowdown in coal's penetration and the reversal in the outlook of nuclear power.

Clean Fossil Fuels

It is possible that there will never be a revival of nuclear or a massive introduction of a "new" carbon free fuel. In view of this, it seems that a significant effort should be put into the development of technologies for *the clean use of fossil fuels, with particular emphasis on coal and liquid-fuels* for combustion engines, as well as for its rational use. The abundance of coal, compared to natural gas and oil, should be taken into account in selecting clean fossil fuel technologies for support. The importance of developing and putting into the market such technologies is of course linked to the increasing awareness of the potential risks associated to the global warming.

In deciding on the relative support between clean coal (and other clean fossil fuel technologies) and renewables, it is very important to take into account the possible higher costs of renewables and the possible higher environmental damage that even clean fossil fuel technologies may produce. *The analysis of energy/economy/energy interactions* should therefore be pursued and improved, as well as the analysis coming from the meteorology *and geology* that aim at refining the damage functions (including dynamic effects) of carbon emission and its effects onto the atmosphere.

Energy Efficiency Measures

Given the relative inelasticity of energy demand, energy efficiency measures can contribute to mitigate the pressure on natural resources and the environment (as far as they are available). Subsidies to incorporate energy saving devices and mandatory rules imposing an adequate degree of energy efficiency in new (and old) constructions are appropriate tools for what concerns the domestic and buildings sector.

Energy efficiency measures may be especially appropriate in the transportation sector, which is one of the fastest growing energy consuming sectors and on of the least responsive to market-based tools.

Joint Implementation Schemes

Last, but not least, it should be remembered that, for these measures concerning clean coal (and, on a broader basis, clean fossil fuel) technologies to be fully operative, they have to be implemented world-wide. In particular, they could be more effective in those parts of the globe experiencing the fastest transition from primitive and poor energy use towards advanced energy systems. A formidable challenge for the advanced countries is to find imaginative and effective schemes for technology transfer that simultaneously preserve the interests of technology-owners and produce incentives in technology demanders to go for the best available technologies. The means to achieve this scope may be addressed along many policy lines, implemented via separate, but (hopefully) co-ordinated programmes. These lines may involve from the set-up of efficient know-how diffusion mechanisms to the implementation of adequate channels of technology transfer. The flow of know-how, bringing to the emerging economies exactly what they are looking for, should be combined with increasing awareness on the necessity to satisfy the fast growing demand with the environmental-friendly technologies. As mentioned in Part II, the scope for joint implementation may be limited in the period to 2010 but it may be an extremely promising method in the longer term.

12.2 Long Term Issues: Need for More R&D

Coming back to the future world energy mix sketched in Chapter 2, there is wide agreement on the unsustainability of such a coal-dominated panorama from an environmental point of view. Bearing this in mind, and assuming for granted (with different probabilities depending on the scenario considered) that a sort of coal-era is about to start, a basic objective of the EU energy technology policy should be the minimisation of the length of this coal age, fostering the emergence of a really sustainable backstop cluster of technologies. The efforts in R&D should be directed, under this point of view, towards renewable energy sources and its integration within the energy system.

Renewable Energy Technologies and their Integration

It is evident that, in the longer term, only renewable energy sources are sustainable. It is therefore a priority for policy making to foster technological advancements that could put these technologies in a position closer to the commercialisation phase. The support given in the past to solar technologies, wind power and other RE technological schemes should therefore be continued, taking into account the relative successes and failures that have been encountered in the past. The technologies that are not yet close to enter the market should be the target of basic research efforts: this is the case of solar technologies (low temperature solar for domestic use, photovoltaics, high temperature solar thermal for power production). Other RE schemes, exhibiting less technological difficulties, are finding obstacles due logistic problems, lack of financial support, structural inertia and/or information opacity. This may be the case for wind power and several biomass exploitation schemes. These technologies require a comparatively higher effort on dissemination under appropriate finance planning.

One of the main difficulties hindering the penetration of renewables relates to its intrinsic intermittence. Overcoming this difficulty would certainly represent a significant step towards their entering in the market as real competitors of the already-established technologies. One of the factors that make fossil fuels valuable as energy carriers is precisely their time-flexibility for use. New renewable energy candidates have to prove to be at least as flexible as standard peaking schemes (gas turbines and hydro).

Particular emphasis should be given to RE-hybrid schemes, since they could represent the natural transition towards a fully decarbonised energy system. The experimental implementation of such hybrid schemes (wind-gas, solar thermal-gas) could start to be applied to isolated systems and then be progressively incorporated to the main market via decentralised schemes.

Technological R&D in RE is, therefore, a critical issue for those RE schemes that, being technology-intensive, did not achieve yet competitive unit costs or have an intrinsic lack of operativeness due to intermittence.

Energy Storage Devices

It is clear from the above considerations that an efficient R&D program on long-term energy options should not underestimate the research line devoted to energy storage and batteries. The availability of cheap, efficient and environmental friendly electric batteries may indeed contribute not only to the development of decentralised, renewable-based electric systems, but also (and possibly with more importance) to the emergence of the electric car. Under technological formats similar to the ones available today, batteries could increase the degree of technological diversity of the transport sector and could help prepare the way for more advanced and sustainable transportation schemes.

High efficiency devices, such as fuel cells, are the natural candidates to integrate RE-obtained energy carriers onto the electrical system. Although fuel cell technology has reached a significant degree of development, with remarkable cost declines in the recent past, they are still well above the competitive threshold. This is also due to the fact that the technology directly competing with fuel cells within the natural gas fuel line has (i.e. the gas turbine in combined cycle) has become extremely competitive.

The topic of energy-storage goes even beyond these issues. Indeed, the development of efficient, safe and versatile energy carriers would make it possible to overcome the intrinsic limitations of use of many specialised energy technologies, able to operate using particular fuels of energy forms. The most appealing case, that puts again into evidence the inflexibility of the energy system, is the technological impermeability between the electric and transport sectors. These sectors are leading the bulk of the expansion of the world primary energy demand, and jointly contributing for more than 50% of the global carbon emissions. The possible candidates to play a significant role as inter-technology energy carrier are well known: hydrogen, methanol, and, of course, battery-stored electricity. Developing these technologies is a challenge in which the R&D effort will have a significant role, and for which public R&D programmes have a room to complement and reinforce the private efforts, that would be concentrated on research domains closer to the market place.

Nuclear Energy: the Fission and Fusion Programmes

Nuclear energy supplies today a non negligible share of the world primary energy, which is considerably higher in some countries of the OECD. Many developing countries whose demand for electricity increases rapidly are considering a progressive nuclearization of their electrical sectors. This is the case even for countries such China and India, with significant coal reserves. On the contrary, there are advanced countries that decidedly endeavoured a dismantling policy (Canada, Sweden) or at least nuclear moratoria (USA, Germany, and many others). The outcome of the process is not yet clear. In any case, nuclear power is a carbon-free option that may or may not expand according first to the social willingness to accept the associated disadvantages and second to the degree with which the global warming threat will emerge in the forthcoming decades.

There are a number of nuclear-based advanced technology initiatives that offer a huge potential, and could indeed become the ultimate backstop technology, but whose degree of advancement is still far from the commercialisation phase, namely nuclear fusion. Basic research is still required to determine the feasibility today's design. The private sector still feels this alternative so remote, that only public support could contribute to the exploration of these new technology lines.

Future Energy Technology Markets and EU Competitiveness

It has already been mentioned in Chapter 3 that the electricity demand and primary energy consumption in Europe is likely to grow very modestly in the forthcoming decades. If carbon emissions need to be reduced, this reduction should take place by improving the energy conversion and the efficiency in end-use. Energy technologies could be oriented along the gas line, including not only advanced improvements in traditional equipment (gas turbines, combined cycles), but also innovative exploitation schemes (fuel cells, etc.), as well as renewables (wind, solar, biomass), and energy efficiency and energy saving measures.

The market for fossil fuel technologies in many parts of the developing world is expanding at a very rapid pace, but priorities are somehow different. Countries with large fossil fuel resources and with rapidly growing electricity demand are likely to decidedly go for fossil-fuel-based electricity in the next decades. The typical cases of such a situation are China and India. These countries, possessing huge coal reserves, are endeavouring a massive construction of coal-based power plants to fulfil the electricity demand accompanying a very rapid economic development. Only in China, more than 200 GW of new coal-fired capacity is projected up to 2010. This would imply the installation of 10-20 new 1000 MW power plants per year. The estimated costs of this capacity expansion have been estimated around US$ 200 billion.

The electrification pattern is, therefore, moving towards a reduction of the capital/variable cost ratio in the developed world (with the exception of renewables), whereas this ratio is likely to remain constant or even grow for those developing countries having cheap fossil fuels. Having passed already through this stage of development of the electric industry, these trends open a very good opportunity for European capital equipment suppliers, which have exhausted the expansion possibilities of their domestic market. Possible future bindings concerning environmental restrictions could be easy to fulfil by these countries if they select advanced technologies. The European energy technologies can offer to the developing countries the additional advantage of having been designed already including many environment-friendly issues: the reliability is high, the conversion efficiency is higher than the average performance, acid emissions are lower. Two main drawbacks may weaken the EU position on this issue: first, the competition of local manufacturers, and second, the competition of other advanced countries. For what concerns the former, the difference of the capital costs may be very large: Chinese domestic coal-firing technology could cost 50% less than technology imported from the industrialised countries, and, in order to limit the balance of payments deficit, Chinese authorities may be willing to limit the share of coal-based power generation equipment imports. In addition, less advanced technology would require less skilled staff and lower operation and maintenance costs. The possibility of losing the market for advanced coal technologies because of the US competitiveness should not be disregarded, even considering the leading role of EU manufacturers in such technologies. The main reason is that coal has a brighter future as a power-oriented fuel in the US than in the EU (at least in the short term). This creates a dynamic domestic market in the US that is missing in the EU, motivated not only by the higher environmental consciousness of the EU public opinion, but also because of the cheaper domestic coal price in the US.

In order to overcome these difficulties, efforts should be made by the EU manufacturers to maintain an adequate degree of competitiveness. To achieve this, imaginative joint implementation schemes should be set up in order to produce incentives for the developing economies to adopt the advanced technologies. The production of the capital equipment could be transferred to the concerned countries: benefits for the European companies should also arise because of the cheaper labour costs. An adequate approach to technology transfer, via the formation of risk-sharing joint enterprises is therefore mandatory to preserve the future of the European fossil-fuel power equipment industry.

For what concerns new technologies, like renewables, the European relative position varies depending on the type of technological branch is considered. During the late 1980s and early 1990s, Europe gained in competitive advantage with respect to the USA within the wind power generation field, not only in installed capacity terms, but also regarding equipment manufacturing capacity. Offshore wind plants have started to be considered in some parts of the EU as an alternative for the provision of power to remote coastal areas (mainly in Germany, Denmark, UK and the Netherlands). On the contrary, the leading solar-based initiatives have been and remain outside Europe, even if remarkable efforts have been made by governments and industrialists to change this situation. Photovoltaic power has not yet reached the operatively level of wind turbines, in terms of equipment costs and the most salient solar thermal initiatives and demonstration plants are outside the EU (USA, Israel, Australia). The importance of creating, via legal instruments and other type of public intervention, a domestic market large enough to produce incentives for firms to innovate and compete is even more important for new technologies than for conventional power generation schemes. Indeed, at the very beginning of a technological outburst, the possible technology paths are still unclear, and the earnings of early adopters motivated by setting references, procedures and standards, as well as by generating skills and exportable know-how may be huge.

13 Carbon Removal, Fuel Cycle Shift and Efficiency Measures: a Sectoral View

The Intergovernmental Panel on Climate Change (IPCCa 1996) distinguishes between technological options and policy measures. The dividing line between the two is fuzzy because technological options stand behind most policy measures. (The only other major source of emission reductions is consumer behaviour). In this report, we focus on technological measures, but we discuss policies and consumer behaviour where we think it influences our analysis in an important way.

There is a close interplay between technological possibilities, policies, and consumer behavior in situations in which recipients of benefits are different from those who carry the costs. A prominent example of this is the construction of a residential facility with rental units. Here, the costs of effective insulation are borne by the owner, who has an incentive to keep construction costs low. The benefits in the form of reduced expenditures for room heating and cooling accrue to the tenants. Whenever market rates for rents do not properly reflect efficient insulation costs, the incentive for reaching an overall economic optimum by high levels of insulation is low. The policy response to this and similar problems is standard setting. In the U.S., it is estimated that the National Appliance and Energy Conservation Act (NAECA) which establishes residential appliance standards during the period 1995–2015 will yield an overall net benefit of US$60 billion (IPCC 1996b).

Another important distinction can be made between measures that are taken solely in response to the threat of global climate change and those that represent a simple continuation of historical trends in technological progress. The latter is often referred to as dynamics-as-usual (DAU). Distinguishing between these two kinds of measures is conceptually important but beyond the scope of this chapter. To illustrate the difference between the historical trend of carbon intensity and "conventional wisdom," we show, in Figure 13-1 the average global carbon intensity since 1850 and ranges CO_2 projections under different scenarios.

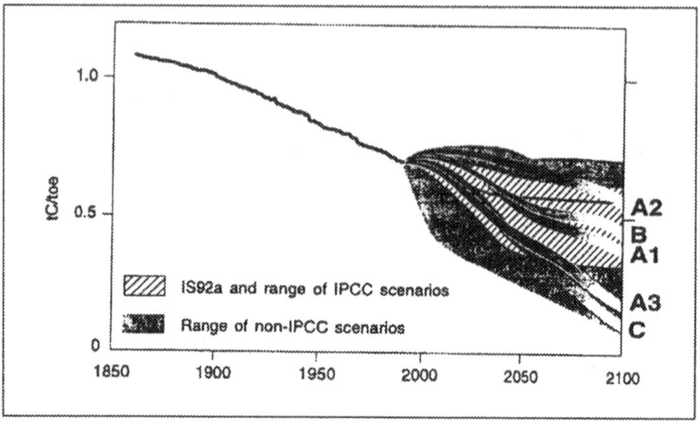

Source: WEC-IIASA 1995.

Figure 13-1: Decarbonization of global primary energy, historical development and ranges of contemporary scenarios.

According to the figure, a break in the historical trend has already occurred. The question is, how temporary is this? The answer is intimately connected to the choice of a reference scenario. Such a choice must be made to assess the costs and benefits of new technological options and is therefore a key to designing successful R&D strategies.

13.1 Carbon Removal and Sequestration

Since combustion of fossil fuels always emits CO_2, any measure aiming at a substantial reduction of carbon emissions from fossil fuel use must include the decarbonization of such fuels or of the flue gas. The removal of carbon in the course of technical processes is usually energy-intensive but leads to high rates of carbon emissions reduction. Alternatively, the sequestration of carbon from the atmosphere by biomass growing is independent of energy-related carbon emissions because it directly reduces atmospheric concentrations of carbon dioxide.

The eventual disposal of carbon removed can be thought of as a two-step process. The first step is the separation and recovery of CO_2; the second step is CO_2

storage. We shall look at the technical descriptions of these two steps separately and then look at total mitigation costs.

13.2 Separation and Recovery Processes

Several CO_2 separation and recovery processes are considered for the energy conversion sector. Most of them have already been applied in industrial processes of CO_2. Figure 13-2 summarises these options.

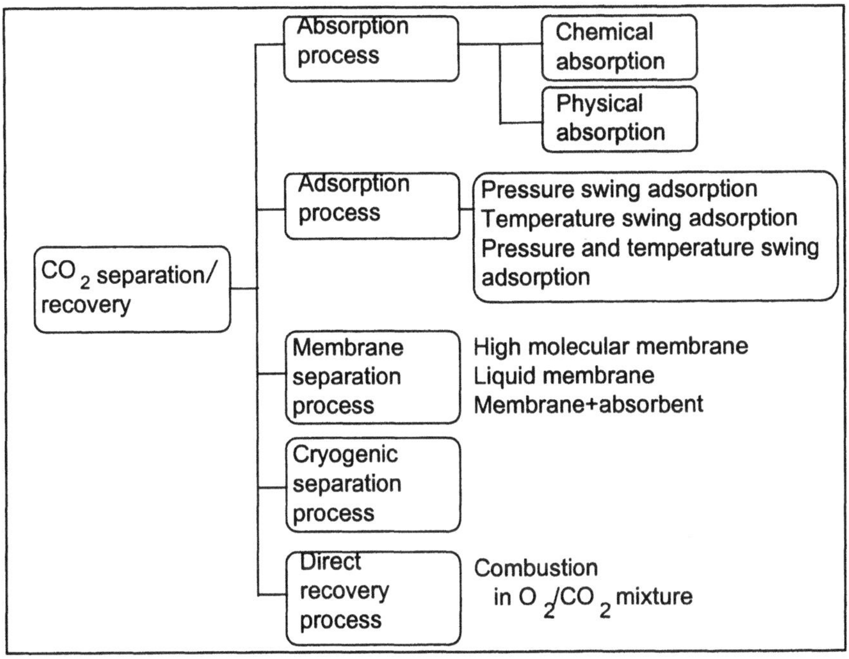

Figure 13-2: The main processes of CO_2 separation and recovery.

Chemical Absorption Process

Absorption of CO_2 is used in chemical plants for the commercial production of CO_2. Current demand for CO_2 is small in comparison with total emissions in the global power sector. As a countermeasure against global warming, chemical absorption would have to be introduced at a much larger scale. Generally, it is thought that a regenerable amine scrubber is the best candidate for CO_2 removal schemes. Many absorbents have been proposed for this process, but the monoethanol-amine (MEA) process is the most widely used, owing to its high absorption capacity. Recently, researchers have reported the development of new, high performance absorbents. Mimura et al. (1994) developed sterically hindered amines (SHAs) and tested them in a pilot plant. They found that the power reduction due to CO_2 absorption is about 8 percent when this process is used to reduce CO_2 emissions from 270 to 25 tons per hour at a power of 600 MW. In contrast, conventional absorbents reduce the output by 28 percent.

Other absorption processes in operation or in the planning stage include Statoil's Sleipner project (IEA, 1996) in which CO_2 is separated from natural gas. Here the main purpose is purification. In addition, there are a few research plants for CO_2 separation by the same process in Japan. Romania has also been conducting pilot plant tests and plans the construction of a demonstration plant.

Physical Absorption Process

Physical absorption processes are studied mainly in Europe and in the U.S.A. (Perry and Green 1983 and DOE 1993). The process uses either organic (e.g., dimethylether sulfolane) or inorganic (e.g., water) absorbents. Absorbents are regenerated by reducing pressure and heating absorbents in the regenerator. Research on this process is still theoretical and on a laboratory scale. This method is not expected to be superior to the chemical absorption process for today's atmospheric combustion processes. However, for pressurised combustion processes, this method has an advantage because it involves high-pressure gases. Also, physical absorption processes may turn out to be cheaper than the chemical absorption process in coal gasification plants.

Direct Carbon Recovery by Combustion in an O_2-CO_2 Mixture

In the Netherlands, one focus of research is CO_2 removal from ICGCC (Integrated Coal Gasification Combined Cycle) power plants. Using a technology that is still to be developed, gaseous fuel can be combusted in a mixture of oxygen and recycled CO_2. Such CO_2-saving gas turbines could lead to specific CO_2 emissions as low as 5g/kWh, but development of the turbines could be costly (Turkenburg 1995).

Physical Adsorption Process

The main advantage of physical adsorption when compared to chemical or physical absorption is its simple and energy-efficient operation and regeneration, which can be achieved through a temperature swing or pressure swing cycle. The primary adsorption material under consideration is zeolite. Adsorption can be divided into three principal methods shown in Table 13-1.

Table 13-1: Classification of physical adsorption methods.

	Adsorption	Desorption
PSA	High pressure Low temperature	Low pressure Low temperature
TSA	Middle pressure Low temperature	Middle pressure High temperature
PTSA	High pressure Low temperature	Low pressure High temperature

Research and development of adsorption processes has progressed furthest in Japan, where pilot plants have been built. Based on their research on pressure and temperature swing adsorption (PTSA), Ishibashi et al. (1995) reported that energy consumption of CO_2 separation was 46 percent of the electricity output generated. Although this value may seem very large, the authors estimate that it could drop to as low as 20 percent, mainly through increased performance and process scale.

According to the IEA (1994), the thermal efficiency of a pulverised fuel and flue gas desulfurization plant with pressure swing adsorption will be reduced from 40 percent to 28 percent, and the cost of CO_2 recovery will be on the order of 300$/tC.

However, according to the U.S. Department of Energy (DOE 1993), this technology is at present not likely to compete with other removal technologies, such as the chemical absorption process. It may be a long time before new, CO_2-specific adsorbent materials can be developed. According to DOE's assessment, this technology is likely to become economical in small applications first – if is to become competitive at all.

Cryogenic Separation

Cryogenic separation involves cooling the gases to very low temperature so that frozen CO_2 can be separated. The potential advantages of the process include the possibility of direct disposal of CO_2 ice (e.g., in the deep ocean) and the high purity of the separated gas which is close to 100 percent. The disadvantages include the high-energy inputs required for reaching cryogenic temperature. This process exists mainly as a theoretical idea.

Schüssler and Kümmel (1989) have estimated that starting from a coal-to-busbar efficiency of 38 percent, cryogenic CO_2 separation would decrease power plant efficiency to 26 percent if the CO_2 emissions were reduced from 0.26 to 0.04 kg/kWh with scrubbing. The major energy requirements are incurred during the compression stages so that improvement in this area would greatly reduce the overall power plant efficiency loss.

Membrane Separation Process

Under the membrane separation process, an appropriate membrane is used to separate the flue gases into CO_2-rich and lean gas streams (Blok et al. 1991). Membrane separation processes have been used in the field of CO_2 separation from EOR (enhanced oil recovery) gas with high pressure, but only theoretical and laboratory research has been conducted thus far on CO_2 separation from thermal power plants.

A number of process variations have been proposed, including liquid membranes (Saha and Chakma 1992), membrane gas absorption (Nishikawa et al. 1995, and Feron and Jansen 1995).Some of the methods exhibit high separation performance, but they lack durability. Existing commercial membranes exhibit low separation performance; high performance membranes are at present only studied in laboratories.

For a two-stage cascade, Golomb et al. (1989) have estimated that the overall efficiency of a coal-fired power plant would drop from 35 percent to 9-18 percent, if about 80 percent of the CO_2 were separated from the flue gases at a purity of 95 percent. Using the same assumptions, the researchers estimate the mitigation cost at 271$/tC recovered, about half of this sum having lowest cost estimates, but also a lower carbon removal rate and product gas purity.

According to the IEA (1994) the thermal efficiency of a pulverised fuel and flue gas desulfurization (PF+FGD) plant would be reduced from 40 percent to 31 percent, and the cost of CO_2 recovery would be 172$/tC.

The Carnol System

Researchers at the Brookhaven U.S. National Laboratory in the United States have developed the so-called Carnol system. (For a detailed description, see Steinberg 1996). This system combines CO_2 removal from coal-fired power plants with methanol generation. It thus eliminates the need for CO_2 disposal. Part of the CO_2 emitted is used for methanol production, and some of the carbon is separated in solid form during the process. As such, it can either be stored or sold. The proposed system also uses the waste heat from the methanol production to decrease the energy needs for CO_2 extraction from the stack gases. Natural gas is considered to be the most natural feedstock for methanol synthesis in the Carnol system, but biomass can be used if CO_2 emissions are to be reduced further.

The general Carnol system consists of several variants, some of which include non conventional methods. The future system costs after commercialisation can only be estimated. Under favorable conditions, the system could be economic even without credits for CO_2 reduction. More conservative assumptions by the same author lead to reduction costs of $175 per ton of carbon (Steinberg 1996).

13.2.1 CO_2 Disposal and Storage Processes

Carbon Sequestration

The potential of carbon conservation and sequestration over the time period 1995–2050 is estimated to be somewhere between 60 and 86 billion tons (IPCCb 1996). This corresponds to 10 to 14 times the global energy-related carbon emissions in the year 1990. Sequestering carbon in biomass immediately suggests the possibility of using this biomass as a primary energy source. Commenting on this possibility, Hall (1994) concludes that "displacing fossil fuel with biomass sustainably grown and converted into useful energy with modern conversion technologies would be more effective in decreasing atmospheric CO_2 than sequestering carbon in trees". Independent of this general conclusion, however, there are pilot projects in which forestation projects are launched for the specific purpose of offsetting carbon emissions. One of the best known examples is that of AES (Applied Energy Services) in the U.S. AES contributes two million dollars to a 10-year sustainable forest project in Guatemala to offset 378,000 tons of CO_2 emitted annually at a coal-fired power station in Connecticut. This project is an example of "Joint Implementation" with a specific cost of 0.10 $/tC to the power company.

Enhanced Oil Recovery

The most appealing means of disposing of CO_2 is to put the gas to productive use. For example, CO_2 can be injected into oil fields, thus improving the oil recovery rate. At present oil prices, only inexpensive CO_2 from natural sources is used for enhanced oil recovery in the U.S. CO_2 is transported over distances of several hundred kilometers at a typical cost of $8/tC (IPCC 1996a). Based on estimates of future enhanced oil recovery of original reserves plus undiscovered conventional oil by Masters et al., (1987, 1991), the global potential of using CO_2 for enhanced oil recovery is approximately 80 GtC, assuming a 50 percent probability of discovery. The biggest part of this potential is located in the Middle East, followed by North America and the Former Soviet Union. However, this is a theoretical maximum. In the Second Assessment Report of the IPCC (IPCCa 1996), a minimum potential of 20 GtC is estimated for this option.

Chemical Feedstock and Basic Materials

CO_2 is widely used as a feedstock by the chemical industry and has also some other limited uses as a basic material. In the chemical industry, most CO_2 is used in the synthesis of urea. At present, this process uses little more than 10MtC per year. Other chemical uses include the production of polycarbonates (Kemi 1987) and alcohol. After consumption and use, many of the chemicals degrade and release CO_2. Existing processes are therefore not suitable as major carbon disposal options, and in order to dispose of high levels of CO_2, it would be necessary to convert the CO_2 to other materials.

One variant Steinberg's Carnol process (described in the previous subsection) involves as a key design element the production and storage of solid-state carbon. It should therefore also be considered as a possible storage option.

Carbon Storage in Underground Reservoirs

Certain kinds of underground structures, such as depleted oil and natural gas reservoirs, aquifers, natural salt domes, and excavated rock caverns can serve as CO_2 reservoirs (Blok et al. 1991). Of these, depleted oil and natural gas reservoirs and aquifers are the most important options. The IPCC (1996a) estimates the storage potential of depleted oil and gas fields to be in the range of 130 and 500GtC. Storage potential in saline aquifers could reach more than 90GtC.

Most of the studies involve only theoretical research, e.g., the estimation of storage capacities. However, the Norwegian company Statoil oversees a project at the Sleipner West field that involves the separation of CO_2 from natural gas containing 7.9 percent CO_2. The recovered CO_2 is injected into the Utsira Salina Deep Aquifer. The main purpose of the separation is to increase the purity of the natural gas. Statoil plans to inject 1 million tons of CO_2 each year. Esso and Pertamina have also agreed to inject 6 billion tons of CO_2 into the aquifer in the Natuna Sea in Indonesia.

Carbon Injection into the Shallow Ocean

Haugan (1992) proposed a disposal method, which relies on density flow. This means that seawater containing dissolved CO_2, because of its greater density, settles down to the bottom of ocean. The seabed around Norway slopes gently downward. As CO_2 is injected into the shallow ocean, it is absorbed by the seawater and it will flow downward into the deep ocean, much as a ball rolls down a slope.

Herzog and Adams (1995) proposed an idea of the droplet plume. Under this method, injected CO_2 will initially rise, owing to its greater buoyancy. After the CO_2 is absorbed, however, the CO_2-rich seawater with its higher density will move like a plume towards the seabed, enabling deepwater storage of the CO_2.

Carbon Injection into the Deep Ocean

Deep-ocean storage of CO_2 was first proposed by Marchetti (1976). His original proposal was to generate a "gigamixer" by injecting CO_2 into sinking thermohaline currents that eventually reach the deep ocean bottom. This proposal combines the advantages of injection into shallow parts of the ocean with the advantage of the huge storage capacity of the deep ocean. The concrete proposal involved the Gibraltar subduction undercurrent, which would provide storage capacity on the order of 10 billion tons of CO_2 per year. Other sinking thermohaline currents include the subduction currents of Bab-al-Mandab in the Red Sea, the Wedell Sea, and the North Atlantic.

In the absence of sinking currents, injection of CO_2 into the deep ocean requires large-scale facilities, such as tankers or floating platforms. This option is therefore more expensive than storage in the shallow ocean, but it offers an enormous increase in storage potential.

If injected at depths of about 3000 meters, CO_2 exists as a negatively buoyant liquid, which tends to sink to the sea bottom. Therefore, CO_2 would form a negatively buoyant plume in the open ocean (Nordhaus 1975).

Another disposal method is solidification of CO_2. The resulting "dry ice" has a specific gravity of about 1.5. It is important in this scheme that the solid ice "torpedoes" are large enough to reach depths of at least 3 km so that the CO_2 can continue to sink spontaneously even in liquid form. Based on preliminary field tests, cubes of 3 to 4 meters would sink below a depth of 3 km before 50 percent of the CO_2 is dissolved, thus yielding excellent sequestration potential.

The natural net flow of carbon from the surface to deep ocean waters is approximately two billion tons per year. Once higher transport rates are available, the amount of CO_2 injected directly into the deep ocean can be greatly increased. The total storage potential is estimated to be enormous. IEA (1994) estimates that the potential for CO_2 stored as dissolved carbonate ion (CO_3^{2-}) is 1440 GtC and an even more if the chemical reaction leading to the formation of calcium carbonate ($CaCO_3$) sediment is taken into account.

Among the major outstanding uncertainties are the possible ecological effects of higher concentrations of CO_2 in the oceans and their effects on local chemistry in the vicinity of storage sites. According to the IEA (1994), the cost of basic disposal will be 4.1 \$/t-CO_2.

A potential problem with CO_2 storage in the deep ocean – as well as with its storage in the shallow ocean – is the formation of solid clathrates. Upon release, they form sinking plumes and eventually settle on the ocean floor (Saji et al. 1992; Nishikawa et al. 1992; and Austivik and Loken 1992). A major uncertainty in this approach concerns the possibility that hydrates will dissolve in seawater as they are transported within a plume or as they rest on the seabed. Experimental research is underway to study this problem and the dissolution of liquefied CO_2 in general, simulating CO_2 concentrations in the ocean. So far, these are only theoretical studies, and empirical research is still required.

13.2.2 Costs and Potentials of Carbon Sequestration, Removal, and Storage

To give a rough approximation of carbon reduction costs in the power sector, we reproduce, in Figure 13-3 the trend line derived by EC (1995), which is based on the relationship of carbon emissions to electricity generation costs.

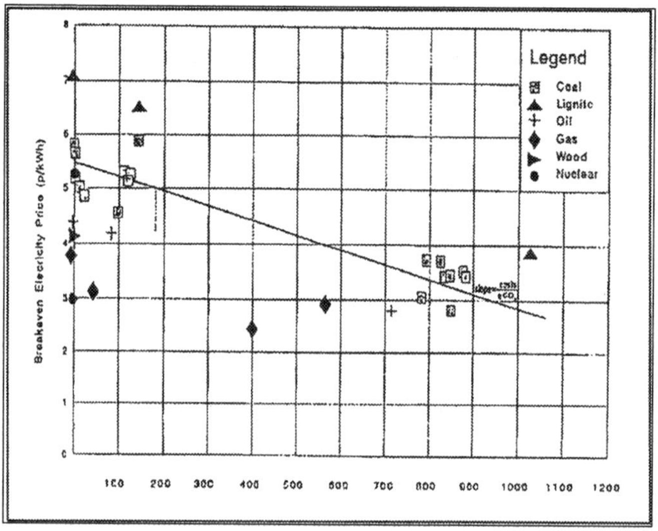

Source: EC 1995.

Figure 13-3: Electricity costs and carbon emissions.

The trend line in Figure 13-3 corresponds to reduction costs of US $150 per ton of carbon. Note that the trend line just approximates the relation between electricity costs and carbon emissions. The lowest cost in the figure has lower specific emissions than some higher cost options. This means that the trend line is only indicative and that concrete reduction costs depend on the particular situation, including the reference technology that applies in a given case.

For the chemical absorption method, Blok et al. (1989, 1991) have estimated the process costs and energy requirements in detail. The authors concluded that the efficiency of coal-fired power plants decreases from 41 percent to 29.7 percent, if

CO_2 emissions are reduced from 230 g-C/kWh in the unabated case to 30 g-C/kWh with separation. The electricity costs would increase by about 80 percent, resulting in CO_2 mitigation costs of about $140/t-C removed. According to the IEA (1994), the thermal efficiency will be reduced from 40 percent to 29 percent, and the cost will increase by 80 percent. These assumptions result in CO_2 mitigation costs of about $130/t-C removed in the case of 90 percent CO_2 removal from flue gas of the pulverised coal-fired plant with flue gas desulfurization (PF+FGD). Both the reduction of the thermal efficiency and the costs of two sources are identical. This means that using a conventional absorbent will significantly increase primary energy demands for coal-fired power.

Table 13-2: Summary of costs and potentials of removal and storages options.

Technology	Specific abatement costs		Potential	Reference
	US$/t C		Gt C	
Removal & sequestration	PF+FGD	IGCC		
Chemical absorption	128	319	–	IEA, 1994a
Physical absorption	n.a.	84	–	IEA, 1994a
Physical adsorption	308	751	–	IEA, 1994a
Cryogenic separation	n.a.	84	–	IEA, 1994a
Membrane separation	172	458	–	Golomb et al., 1989
Carnol system	0–175		–	Steinberg, 1996
O_2/CO_2 combustion	130–175		–	Blok et al., 1995
Forestation	6–30		–	IPCC, 1996b
Disposal & storage				
Enhanced oil recovery	0		>20	IPCC, 1996b
Underground reservoirs	11		130–500	IPCC, 1996a
Aquifers	7–30		90–2500	IEA, 1994b
Ocean disposal	4		1200>	IEA, 1994b
Transportation				
Pipeline, 1000km	32			IPCC, 1996a

13.3 Other Reduction Options

In this section, we summarise other options to reduce carbon emission. After the classification in this and the previous sections, the rest of this chapter's sections are devoted to discussing all the reduction options for each economic sector.

13.3.1 Demand Side Measures

Carbon emission reduction measures on the demand-side aim to reduce the specific energy demand of a given energy service. As such, they are not the direct targets of R&D strategies, but they are nevertheless included here for the sake of completeness. In the sequel, they are discussed in less detail than technological options.

13.3.2 Dematerialization and Recycling

The reduction in material intensity, particularly in the industrial sector, is the result of sufficiently high levels of per capita income and is generally observed in post-industrial economies. This "dematerialization" (Herman et al. 1989) goes along with a decrease in overall energy intensity of the economies. Dematerialization is thus largely an "autonomous" phenomenon, but it can also serve as a guiding concept for policy making. Recycling is an example because it amounts to the dematerialization of primary inputs. The reuse of scrap metal, paper, glass, and plastics, as well as the composting of organic matter will not only reduce carbon emissions, but will also address the general problem of waste disposal.

Okken and Gielen (1994) estimate that 86 percent of energy inputs necessary for the production of primary aluminum can be saved by using recycled aluminum instead of bauxite. At present, recycling rates in the EU are low. However, products with a long lifespan currently comprise a large share of European consumption (Okken and Gielen 1994).

13.3.3 Efficiency Improvements and Technological Change

Decarbonization through efficiency improvements is a fundamental option for achieving environmentally compatible energy development. Efficiency improvements reduce most of the adverse environmental impacts of energy, including greenhouse gas emissions, while also leading to lower primary energy inputs and therefore to lower fuel costs.

In the context of climate change, the appropriate measures of efficiency are primary energy and energy services. Energy services, in general, are not measured in energy units, but rather in other units such as, for example, person-kilometers.

Moreover, in some cases, useful energy for a given energy service can be"borrowed" from the environment (as with heat that is transferred by heat pumps for the purpose of room heating), and final energy is used only for heat transfer. Useful-energy efficiency can therefore not be expressed meaningfully in percentages. The question therefore arises as to how much energy is needed for a particular energy service. No straightforward answers are possible, but the IPCC (1996a) summarizes studies that use the concept of exergy to estimate primary-to-service (or exergy) efficiencies with the result that, typically, these are on the order of a few percent. Granted, the lower limits of energy needs derived in these studies are theoretical, but the underlying concept is of paramount importance in all long-term studies of the subject.

Technological change in past centuries has improved the efficiency of the factor inputs and substituted new or more abundant raw material resources for dwindling supplies, particularly labor and energy. The growing need to protect the global environment should lead to increasing efforts in the field of technological progress over and above historical trends.

To illustrate the costs and potential of efficiency improvement measures, we show here, in an example of efficiency improvement measures supply curves for total primary energy, derived by de Beer et al. (1995) for the Netherlands.

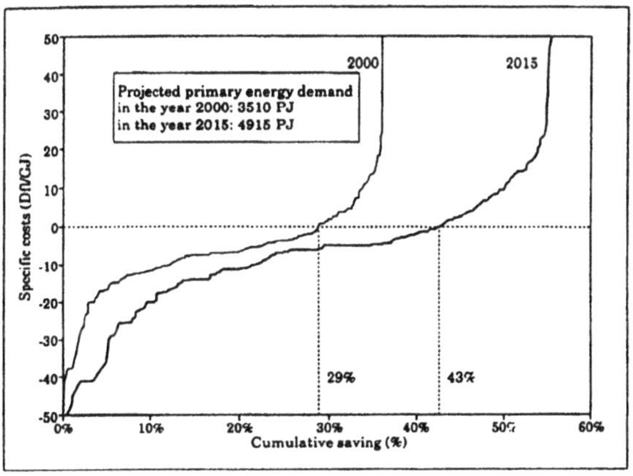

Source: de Beer et al. 1995.

Figure 13-4: Supply curves of energy efficiency improvement measures for the periods 1990–2000 and 1990–2050. A discount rate of 5 percent is used.

According to Figure 13-4, 29 percent of the Netherlands' primary energy demand between 1990 and 2000 can be saved at zero or negative costs. For the period 1990–2050, the value is 43 percent. These figures are the result of a bottom-up study (ICARUS) which, by definition, does not include "take-back" effects as discussed below where the transportation sector is analyzed.

13.3.4 Fuel Mix Changes

The most straightforward way to reduce carbon emissions is simply to use fuels that either do not emit carbon at all (e.g., nuclear, solar) or that are carbon-neutral (e.g., sustainable use of biomass). As far as the practical aspects of carbon emissions are concerned, switching from fuels with relatively high specific carbon emissions, such as coal, to fuels with lower specific emissions, such as natural gas, is an important option.

Hall (1994) quantified alternative ways of using biomass for CO_2 mitigation. His findings are summarized here in Figure 13-5.

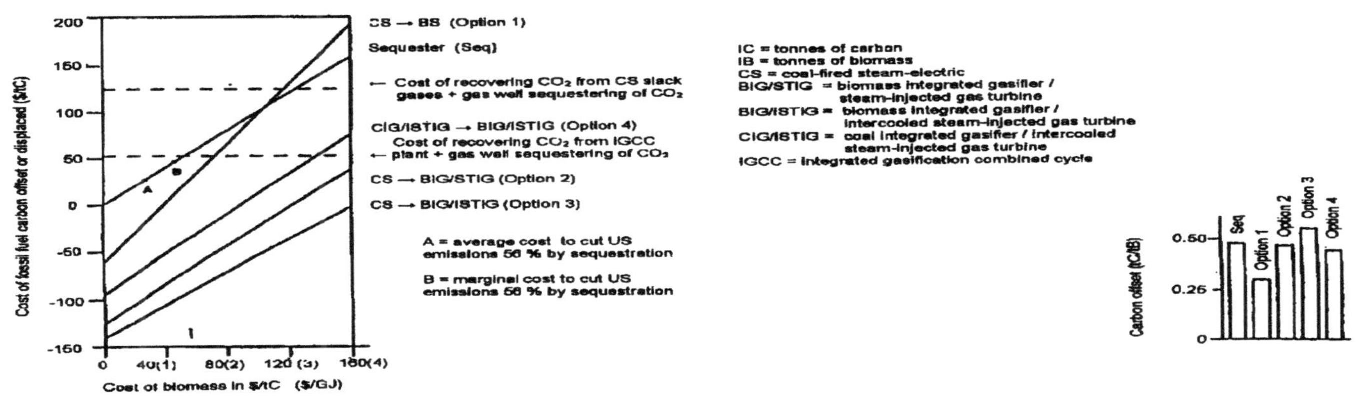

Alternative ways to use biomass to cope with greenhouse warming. Sequestering carbon in forests is compared to alternative strategies for substituting biomass for coal in power generation.

The bar chart on the right shows the CO_2 emissions reduction (in tonnes C/dry tonne of produced biomass) for sequestering and for four alternative options in which biomass is substituted for fossil fuels. The graph on the left side shows the cost of fossil CO_2 emissions offset or displaced (in $/tonne C) versus the cost of biomass for these alternative strategies. On the lines labeled "sequester", A = the average cost and B = the marginal cost of offsetting US CO_2 emissions 56 % through tree-planting on 139 million hectares, at an average total forest ecosystem sequestering rate of 5.3 tonnes C/ha/yr. The cost of fossil fuel displaced is $(C_b - C_f)/E_f$, where C_b (C_f) is the unit cost of the energy output from the biomass (fossil fuel) system, and E_f is the CO_2 emission rate of the fossil fuel displaced (in tonnes of C per unit of energy output). The carbon dioxide emissions avoided are assumed to be 23 kg C/GJ of input coal (HHV basis) and 0.76 kg C/liter of gasoline displaced

Option 1: Substitute biomass-fired steam-electric power for coal-fired steam-electric power.
Option 2: Substitute biomass-integrated gasifier/STIG power for coal-fired steam-electric power.
Option 3: Substitute biomass-integrated gasifier/ISTIG power for coal-fired steam-electric power.
Option 4: Substitute biomass-integrated gasifier/ISTIG power for coal-integrated gasifier/ISTIG power.

(Source: Hall 1994).

Figure 13-5: Alternative ways of using biomass for CO_2 mitigation: sequestering carbon in forests versus substitution of coal with biomass for electricity.

Approximately 40 percent of energy-related CO_2 was emitted by the energy conversion sector (IPCC 1996b). Theoretically, if all final energy were carbon-free, the entire carbon emissions problem could be solved in the energy sector alone. It is therefore clear that many measures and policy options exist for this sector to achieve deep reductions of total energy-related carbon emissions.

13.3.5 Removal and Sequestration

Figure 13-6 summarises findings of the Joule II Programme study on R&D in Clean Coal Technology.

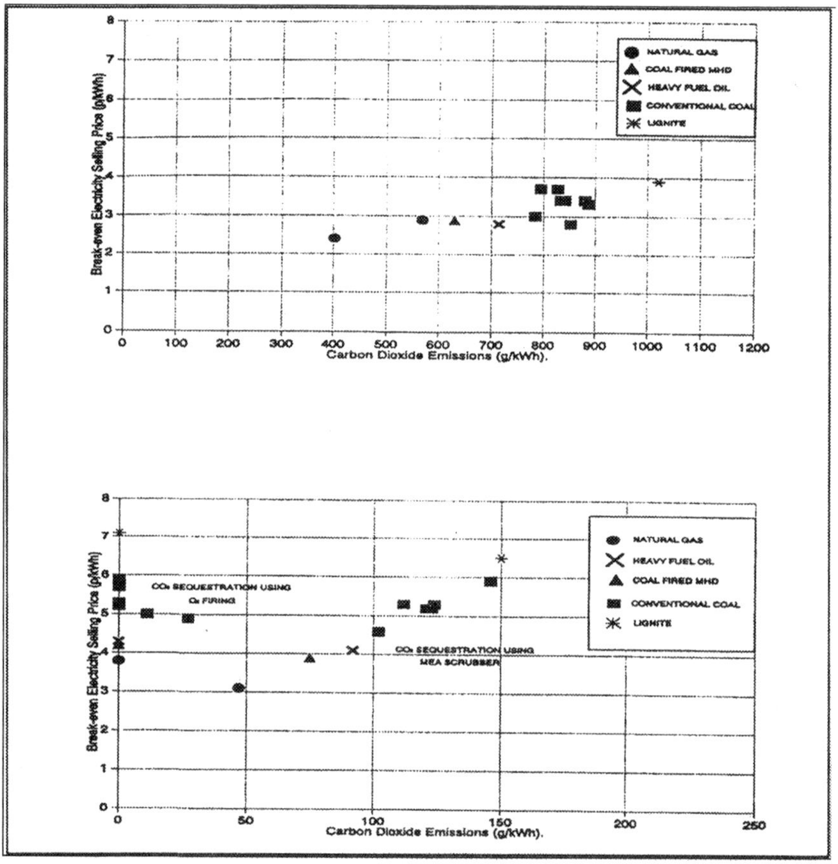

Source: EC 1995.

Figure 13-6: Break-even electricity selling prices for base-case technologies (upper part) and for technologies involving CO_2 sequestration (lower part).

The figure displays several interesting features. First, it indicates that CO_2 sequestration becomes competitive even for lignite power plants at approximately 7 UK pence (approximately 11 US cents) per kWh. The other noteworthy point is that there are areas in the two parts of Figure 13-6 that show a positive correlation between costs and carbon emissions. One can see that natural gas tends to be an economical and relatively clean source of power production. At electricity prices of 4 UK pence (6.5 US cents) per kWh, it is even profitable to sequester all CO_2 produced by a natural gas-fired power plant. However, this is only a static cost comparison; a scenario analysis would be necessary to determine the potential of such a cheap power generation option over time.

13.3.6 Energy Conversion Efficiency

Energy efficiency improvements have already been achieved for almost all types of energy conversion facilities. For example, dramatic improvements in fuel conversion efficiencies have been achieved during this century in electricity generation. The efficiency of electricity generation at the turn of the century was about 5 percent, whereas today the average efficiency in OECD countries is about 36 percent. This amounts to approximately a sevenfold efficiency increase in less than one hundred years. Today, the best combined-cycle natural gas-fired power plants can achieve more than 50 percent efficiency (Nakicenovic et al. 1993).

Figure 13-7 illustrates the historical improvements in electric conversion efficiency in the U.S the Former Soviet Union, Western and Eastern Europe.

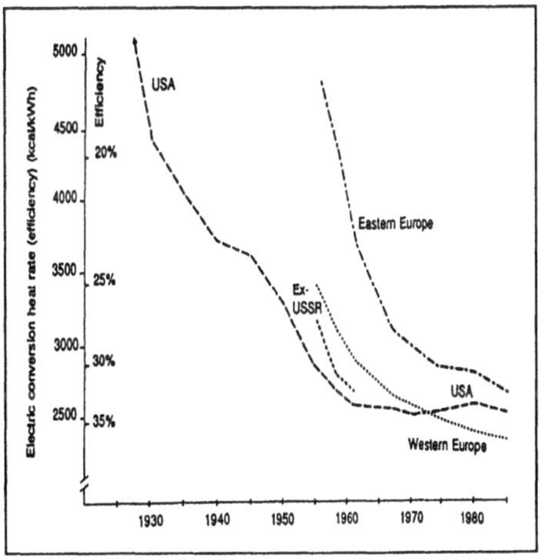

Source: Nakicenovic et al. (1993).

Figure 13-7: Improvements in electric conversion efficiency in the U.S., the former Soviet Union, Western and Eastern Europe.

The IPCC (1996b) estimates that more efficient combustion of all fossil fuel could reduce 30 percent of this sector's GHG emissions at zero or negative costs by 2020 and even by 60 percent in the longer run. For example, the average efficiency of a new coal-fired power plant today is about 34 percent but can be expected to reach 50 to 60 percent in the medium- to-long run (IPCCb 1996).

The cogeneration of electricity and district heat can substantially increase the overall efficiency of energy conversion. Cogeneration has efficiencies in the range of 60-80 percent. Often, introduction of cogeneration is hindered by the lack of sufficient demand for low-temperature heat.

Today, carbon emissions result primarily from the production of power. In future, the centralized production of hydrogen or methanol could become a major contributor to carbon emissions in the energy sector. This way, some emissions would be transferred from the transportation to the energy sector. In this case, it is therefore important to consider life-cycle GHG emissions of alternative automotive fuels. See Table 13-3 for a summary of recent estimates.

Table 13-3:Life cycle GHG emissions and costs.

Fuel	Greenhouse gas emissions in g/km CO_2 equivalent average driving cycle; car consuming 7 litres petrol/100 km				Pre-tax costs, based on Renault Clio 1.4 litre; 13,800 km per year; 10 year life. 10% d.r.			
	Vehicle manufacture	Fuel Supply	Operation	Total (not final)	Vehicle cost ($)	Fuel cost ($/litre petrol equiv.)	Total levelised Cost ($/km)	Cost relative to petrol (US¢/km)
Reformulated Petrol	25–27	17–48	180–193	222–268				
Petrol	25–27	15–63	182–207	222–297	15,168	0.26	0.29	0.00
Diesel	27–29	7–35	139–202	173–266	15,168–17,443	0.26–0.26	0.29–0.33	-0.35–3.64
LPG	26–28	7–20	147–155	180–203	16,083–15,384	0.19–0.26	0.29–0.30	-0.55–1.02
CNG	29–31	5–68	130–154	164–253	16,083–15,600	0.18–0.24	0.29–0.30	-0.28–0.99
Methanol from NG	25–27	76	149	250–252	16,128–15,168	0.25–0.35	0.28–0.31	-0.72–1.45
Methanol from Wood	25–27	0–44	15–16	40–87	16,128–15,168	0.68–0.82	0.31–0.34	2.30–4.79
Ethanol from Sugar Beet	25–27	50–220	15–16	90–263	16,128–15,168	0.94–1.03	0.34–0.36	4.61–6.74
Ethanol from Wood	25–27	0–41	15–16	40–84	16,128–15,168	0.68–0.82	0.32–0.34	2.79–5.27
Liquid Hydrogen	26–28	0–48	3–12	29–88	19,968–18,048	0.38–1.44	0.33–0.43	4.10–13.97
EV using electricity generated from:	Emissions based on urban driving cycle							
European average	44–48	170	0	114–118	24,768–20,928	0.48–0.96	0.36–0.44	6.81–14.74
Coal	44–48	300–310	0	344–358				
Oil	44–48	253	0	297–301				
Gas	44–48	130–140	0	174–188				
Nuclear	44–48	15	0	59–63				
Hydro	44–48	0	0	44–48				

Source: IEA 1994.

13.3.7 Fuel Mix Changes

Renewable use of biomass and non emitting renewable sources, such as solar and wind energy, can reduce significantly carbon emissions of the power sector. Today, these technologies are not competitive except in some niches. However, significant cost reductions are expected in future due to "technological learning".

Table 13-4: Electricity costs of biomass, solar photovoltaic, and wind, US cents/kWh.

Biomass[(1)]	4.2-9.9
Solar[(2)]	9.5-13.8
Wind[(3)]	2.0-5.0

[(1)]The high end of the range corresponds to conventional technology; the low end to future advanced technology. Source: IEA 1994.

[(2)]Annualized capital costs assuming a 5 percent discount rate and 30 years of service life, an availability factor of 25%, fixed O/M costs of $100 per year; range based on a statistical analysis of future capital costs of solar PV in industrialized countries. Source: Strubegger and Reitgruber 1995.

[(3)]Future costs, expected for the time 2020-2030.

Source: IEA 1994.

Nuclear energy is another source of carbon-free energy. To date, it has been introduced for reasons other than GHG emission reduction. In many parts of the world, particularly in Europe, nuclear energy faces the well-known problem of public acceptance. As a consequence, only a small part of the discussion about carbon mitigation concerns nuclear power, and carbon reduction costs of nuclear energy are hardly a topic. Long-term R&D strategies should therefore directly address the problem of public acceptance by developing schemes such as the Inherently Safe Reactor.

Carbon abatement costs of fuel production from biomass are summarized inTable 13-5. The lowest figure in the table is for methanol from wood. Still, the estimated cost ratio between wood-based methanol and gasoline is more than 3, equivalent to carbon abatement costs of almost $700 per ton of carbon. Values on the higher end of the scale lead to abatement costs of more than $1,800/tC.

Table 13-5: Cost ranges of biofuels for transportation.

	Ethanol from wheat	Ethanol from beet	Rape methyl ester	Methanol from wood	Electricity from wood
Net energy yield (GJ/ha)	58.4	200.6	48.9	158.1	209.9
Product energy credit / net input energy	4.74	12.21	4.98	13.33	22.85
% gasoline or diesel life-cycle use	52%	38%	43%	22%	8%
Gasoline or diesel substitution (l/ha)	1,695	3,961	885	4,908	-
Net CO_2 abatement (kg/ha)	4,072	13,628	3,722	10,661	9,480
% of gasoline or diesel life-cycle CO_2 emissions	40%	28%	31%	18%	9%
Cost of gasoline or diesel replaced (c/l), EU feedstock prices – world feedstock prices	89–103 54–62	92–100 64-72	98–136 50–59	33–39 33–39(US)	6.4–9.5 (c/kWh)
Cost ratio to gasoline or diesel (EU)	5.1–5.9	5.3–5.7	5.6–7.8	3.4–4.2	1.3–1.9
Net cost (subsidy) required (annual, $/ha)	1,207-1,444	2,960–3,284	710–1,052	1,068–1,436	142–447
Cost of CO_2 abated ($/t)	424–507	374–415	304–450	186–250	27–86
Net energy yield if 20% of OECD Europe cropland set aside and used in 2050 (EJ)	1.4	4.8	1.2	3.8	5.0

Source: IEA 1994.

As for electricity, significant price reductions are expected from technological learning. The Second Assessment Report by the IPCC (1996a) estimates that "advanced biofuels," such as ethanol derived from wood via enzymatic hydrolysis, might be competitive with gasoline at oil prices over approximately $25/bbl. This calculation assumes an annual yield of 150 dry tons per hectare, a value thought to be realistic by Hall (IEA 1994) in the year 2050.

13.4 The Industrial Sector

13.4.1 Dematerialization

At present, about 40 percent of steel, aluminum, and lead are produced from scrap worldwide. The scrap input for copper is estimated to be 15 percent or higher, whereas the global recycling rate for paper is around one-third, a figure which is similar to the present recycling rate of glass bottles in OECD countries.

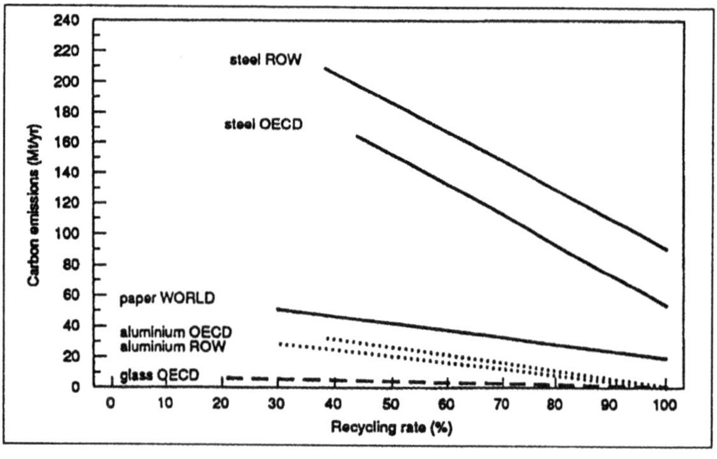

Figure 13-8: Carbon reduction potential of industry as a function of the recycling rate for selected materials.

In the future, recycling could significantly reduce specific resource requirements and resulting carbon emissions as illustrated in Figure 13-8. The largest carbon reduction potential would be in enhanced recycling of iron and steel products in conjunction with energy-efficient electric arc furnaces. By increasing the recycling rate to some 70 percent, the global steel industry could reduce its annual carbon emissions by about 110 Mt C, or approximately 24 percent. Increasing the recycling rate of aluminum, paper and glass to 70 percent each could reduce carbon emissions by some additional 50 Mt. Globally, an aggressive recycling strategy, e.g., raising the recycling rate for the major energy and carbon-intensive products as high as 70 percent, could reduce emissions by at least about 160 Mt C, or 8 percent of current total industrial emissions.

13.4.2 Energy Efficiency Improvement and Process Changes

For the OECD region as a whole, continuous casting yields the highest energy savings potential per ton of steel produced, followed by dry quenching of coke, hot direct rolling and sensible heat recovery at basic oxygen furnaces. Part of these measures have already been implemented and have contributed to the energy efficiency improvements in the steel industries of OECD countries over the last two decades.

Ranked by their cost effectiveness per ton of carbon reduced, the measures analyzed could reduce carbon emissions in the OECD steel industry by some 24 Mt C or 14 percent, including 12 percent carbon emission reductions at negative or zero cost per ton of carbon reduced. These negative costs are derived from a (hypothetical) comparison of an average OECD steel plant with the individual efficiency measures outlined above. For instance, applying top-gas recovery turbines to blast furnaces throughout the EU would cost about $60 million (1990 dollars) annually, while the resulting energy savings are estimated to correspond to $220 million (Springmann, 1991). This would produce annual carbon emission reductions of 0.6 Mt (hence a negative reduction cost per ton of carbon reduced).

Figure 13-9 compares various steel production processes in terms of their technologies offer possibilities for carbon emission reductions in steel manufacturing. Among these are (1) direct smelting (e.g., pre-reduction with a coal-derived reduction gas; followed by a second stage in which sponge iron is melted down either with coal, the Corex process, or electricity, the Plasmamelt process); (2) the use of electric-arc furnaces in combination with direct reduction of iron ore (using, for example, the HYL III process, which is particularly interesting from a CO_2 perspective, since it includes CO_2 scrubber for hydrogen and CO_2 recycling from the reduction gas); or (3) the direct reduction with hydrogen followed by an electric arc furnace, which can be a steel production technology with practically no carbon emissions, provided both hydrogen and electricity are produced carbon-free.

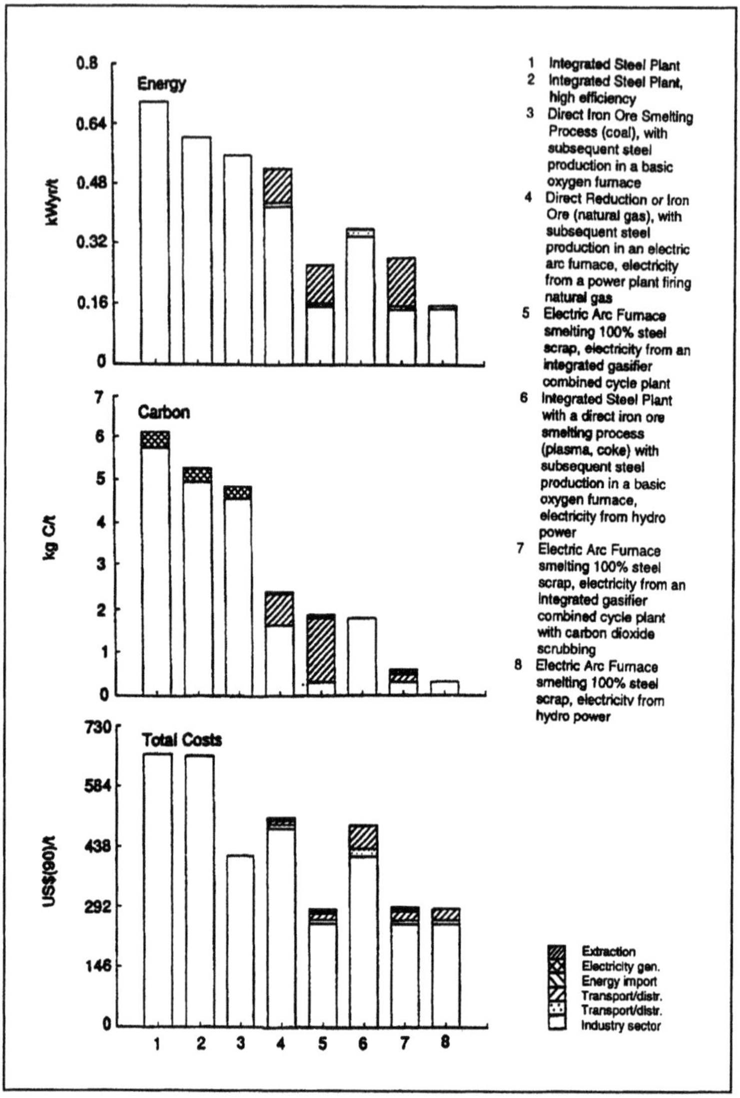

Figure 13-9: Steel production chains: energy consumption, carbon emissions, and costs of various production process routes.

Process technologies to reduce carbon emissions also exist in many other industrial sectors. Using the energy efficiency of the best plants as a guide, specific energy consumption for aluminum production in the EU could be reduced to some 47 GJ/ton (that is, by 36 percent) and for ammonia synthesis to some 28

GJ/ton (22 percent), compared to the industry average (Worrell et al. 1992). In the paper industry, a process change from thermal to mechanical dewatering in pulp mills would yield an estimated energy reduction of 0.07 kWyr (15 percent) per ton of paper. In the cement manufacturing industry, replacement of the wet by a dry process, in conjunction with precalcination and computer process control, could reduce energy requirements from 5.5 to less than 3 GJ (from 0.17 to 0.09 kWyr, respectively) per ton of concrete, although this process change is very capital-intensive (see Figure 13-10). As the CO_2 concentration in flue gases of cement plants reaches 30 percent, an alternative strategy might involve carbon scrubbing. Assuming a cost of \$150 per ton of carbon removed, this would correspond to an additional \$50 per ton of cement produced, nearly doubling current average cement production costs.

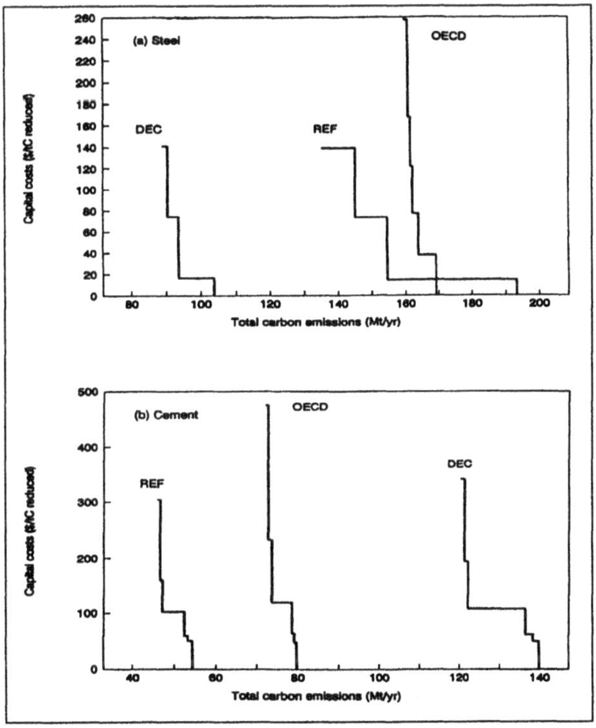

Figure 13-10: Capital costs (per ton of carbon reduced) for (a) steel and (b) cement process technology improvements for major world regions versus emission reduction potential.

In addition to process technology changes, a number of more generic technology options exist to further improve industrial energy efficiency, including waste heat recovery, cogeneration, optimized control and improvements of combustion, as well as efficiency improvements of electricity use. Giovannini and Pain (1990) estimate savings potentials from waste heat recovery of between 36 and 67 percent for the Netherlands, Germany and Japan, e.g., through heat exchangers or industrial cogeneration, or through heat upgrading with heat pumps. For industries with high demands for both steam and electricity, industrial cogeneration and heat cascading (e.g., via top and bottom-cycle cogeneration) appear promising. The OTA (1991) estimates that industrial cogeneration could reduce industrial energy demand in the U.S. by as much as 140 GWyr by the year 2015. For low-temperature heat applications, improved boiler efficiency (e.g., fluidized-bed boilers) and fuel substitution are possible CO_2 reduction options.

Improvements in electricity use via improved industrial drives or lighting could offer significant reduction potentials. Fisher (1990) estimates a global potential for industrial electricity reduction of 2–40 percent through replacement of industrial drives by the most efficient models, as well as the use of variable-speed drives. An additional reduction of 60–80 percent might be realized through installation of the most efficient lighting systems. EPRI (1990) has estimated a maximum technical potential for U.S. industrial electricity reduction of close to 40 percent (compared to a "base case" in the year 2000). The electricity reduction potential in industrialized countries for high-efficiency adjustable speed drives is estimated to be around 30 percent (IEA 1991). The NAS (1991) estimates that electricity saving of 200 TWh from improved electric drives would cost an average of about 2.6 U.S. cents/kWh in the U.S. (costs assume immediate replacement of existing units). Assuming an average carbon intensity of U.S. electricity production of 0.18 kg C/kWh (Nishimura, 1991), this translates to specific carbon reduction costs of $144 per ton of carbon.

An example of a full fuel-cycle analysis with IIASA's CO_2DB data base is presented in Figure 13-11, comparing the combination of different motor drives with different means of electricity generation. It shows that as far as technology is concerned, the overwhelming emission reduction potential is on the power generation side.

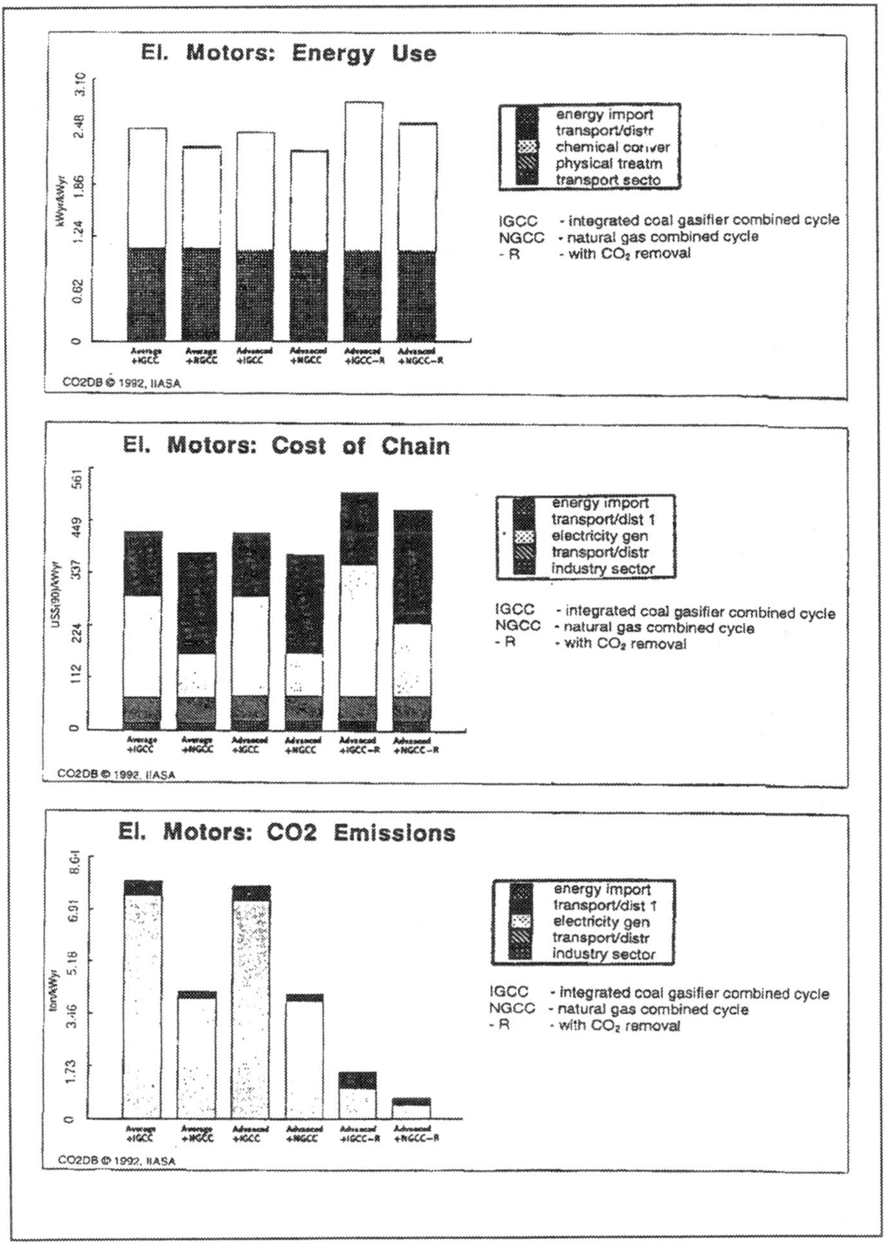

Source: Messner and Nakicenovic 1992.

Figure 13-11: Energy requirements, costs, and CO_2 emissions for six energy chains ending with industrial motor drives.

13.4.3 Fuel Mix Changes

The trend toward increased recycling and remelting of metal scrap requires electric arc furnaces leading to the increasing electrification and lower carbon emissions. The cost of this measure and its carbon emission reduction potential is described above.

13.4.4 Combined Measures in the Steel Industry

The recycling of scrap, in combination with measures in the electricity generating sector, (described in the section on carbon emission reductions in the energy sector), currently appears to be the least expensive process change measure by which industry can reduce its carbon emissions (see Figure 13-12). Compared with the production economics of the average steel plant, electric arc furnaces could achieve both carbon emission reductions and improved production economics, which would imply negative carbon reduction costs of $250–540 per ton of carbon. Direct smelting (e.g., via the Corex or Plasmamelt processes) could achieve carbon reductions at an additional cost of $10–70 per ton of steel, which translates into $140–370 per ton of carbon reduced. The process having the highest carbon reduction potential also has the highest costs (in excess of $900 per ton of carbon reduced). This is the zero-carbon steel production via direct reduction with hydrogen in combination with an electric arc furnace. Because of their high capital requirements, process technology changes on the positive side of steel industry's carbon reduction curve generally represent high-cost carbon reduction strategies.

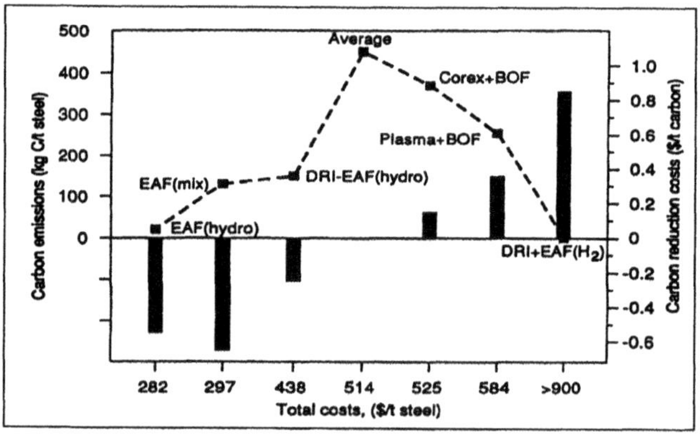

Figure 13-12: Summary of process technologies with carbon reductions in steel manufacture, carbon emissions and costs (line) versus specific carbon reduction costs (bars).

13.5 The Transportation Sector

13.5.1 Efficiency Improvements

The average fuel consumption of U.S. automobiles fell from 18.1 liters per 100 km in 1973 to 10.8 liters per 100 km in 1988, representing an average annual decrease of 3.4 percent. During the same 15-year period, average fuel consumption of new vehicles dropped from 16.5 to 8.2 liters per 100 km in the U.S. (4.7 percent per year) and from 8.4 to 6.8 liters per 100 km in Italy (1.4 percent per year) (Nakicenovic et al. 1993). In other European countries, such as Germany and France, there was no decrease of average fuel consumption between 1970 and 1990 (Michaelis 1994). In the OECD as a whole, total fuel consumption in the transport sector rose by some 470 GWyr, or at a rate of 2.5 percent per year (Nakicenovic et al. 1993). In general, the development of total demand for final energy in the transportation sector increases faster than the development of specific fuel consumption might suggest. Part of the reason for this divergence is the consumers' reaction to price reductions of automobile transportation. Consumers often respond to more economical automobiles by driving greater distances and by sharing a car less frequently. This phenomenon is referred to as "take back."

Today, the most fuel-efficient car available on the market is about twice as fuel Subaru Justy, or GM Chevette (diesel) have an average fuel consumption of less than 5 liters per 100 vehicle kilometer (47 mpg). Experimental vehicles combining all technical and design features can achieve specific fuel consumption of less than 3 liters per 100 vehicle kilometers, as exemplified by the Volkswagen E-80, the Toyota AXV, and the Renault Vesta2. This represents an improvement over current fleet averages of a factor of three to four.

The costs of achieving fuel efficiency improvements are difficult to determine. U.S. cost estimates for additional purchase costs to the consumer range from $60–130 per car (EPA 1990) to $500–750 per car (OTA 1991) for a vehicle about twice as efficient as the current U.S. average (55 versus 27 mpg, or 4.3 and 8.7 liters per 100 kilometers, respectively). The resulting fuel savings, based on an annual average of 10,000 miles driven, are about 180 gallons, or $230 per year at current U.S. fuel prices. Assuming an annuity factor of 20 percent, fuel savings imply a break-even point of additional costs on the order of $900 for an energy-efficient vehicle. From such a perspective, the fuel efficiency improvements in passenger cars constitute an attractive cost-saving option to reduce CO_2 emissions. As general experience suggest, however, automobile performance in terms of fuel efficiency and carbon intensity is only one of many criteria influencing consumer behavior. This compounds the "take-back" effect in the sense that it provides one more reason to explain why price effects can be expected to have only limited influence on energy savings by cars.

13.5.2 Demand Side Measures

On the demand-side, the Mitigation Panel of the American National Academy of Sciences (NAS 1991) estimates that a doubling of gasoline prices would induce consumers to prefer cars that are between 12 and 15 percent more energy efficient. The commercial transportation sector is expected to be more sensitive to such price signals. Still, even there, long-run elasticities are far away from unity. Kaya et al. (1991) estimate that the energy consumption of trucks in Japan could be expected to drop by only about 13 percent if fuel prices were to be increased by 50 percent.

Thus, it appears that price signals – especially in a politically acceptable range – will not lead to dramatic reductions in transportation fuel demands. This points to the need to investigate regulatory approaches, such as mandatory efficiency standards and consumer information programs, to more fully realize the CO_2 reduction potential of fuel efficiency improvements.

Behavior not only influences what transport modes are chosen, but also how they are used (usage efficiency). Usage efficiency comprises many components, ranging from traffic flow (congestion), driving modes and styles and, most importantly, to load factors. Automobile occupancy rates for U.S. commuters are estimated not to exceed 1.15 persons per car (NAS 1991). The average in the city of Denver is 1.2, and even in densely populated Tokyo, car occupancy is low, at not more than 1.4 persons/car (Newman and Kenworthy 1989). In cities such as London, Copenhagen, or Sydney, car occupancy is about 40 percent higher (above 1.7 persons per car), which means less energy use and emissions per passenger kilometer driven. But public transportation modes, which have to provide sufficient capacity even during rush hours, also illustrate the low usage efficiency of transport systems.

Usage efficiency is perhaps the least understood factor having an impact on the carbon efficiency of transportation systems. Improving usage efficiency not only depends on changes in social behavior and trip organization (such as carpooling or car sharing), but also on public policy incentives, such as the provision of special driving lanes or toll reductions for carpools (as practiced in some U.S. cities), or parking fees or city entrance fees (introduced in some European and Asian cities).

One way of looking at behavioral aspects in the transportation sector is to analyze modal split changes. The modal split of long-distance passenger transportation systems in the Former Soviet Union and China. Figure 13-13 shows that the modal split in these countries fits well to a model of logistic substitution. Good fits like these suggest significant system inertia. In this case, it reflects the limited substitutability between different transport modes as a consequence of different performance and quality requirements, relative economics, accessibility to infrastructures, and consumer preferences. In short, policies at aiming at influencing modals are likely to encounter strong "system opposition."

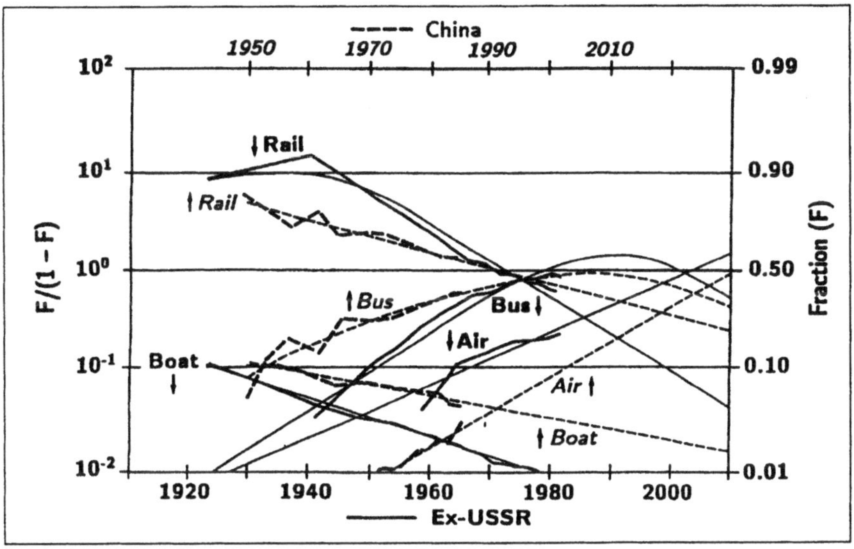

Figure 13-13: Modal split between long-distance passenger transportation systems in the former Soviet Union and China (shifted time axes), in fractional share of passenger kilometers.

13.5.3 Fuel Switching

A summary of fuel switching options for reducing carbon emissions by passenger cars is shown in Figure 13-14. The figure shows that the use of diesel and methanol fuels could reduce carbon emissions by about 20 percent, compressed natural gas (CNG) vehicles up to 30 percent. The specific carbon reduction costs of diesel and methanol are slightly positive, whereas for CNG vehicles – owing to fuel efficiency improvements and generally lower fuel costs – overall vehicle operating costs are lower, hence implying savings (negative carbon reduction costs). The data on ethanol admixture or ethanol-fueled cars are based on the Brazilian example. Such measures would entail additional costs, but these appear to be modest (5 percent) and are subject to the inevitable uncertainty range of such cost estimates.

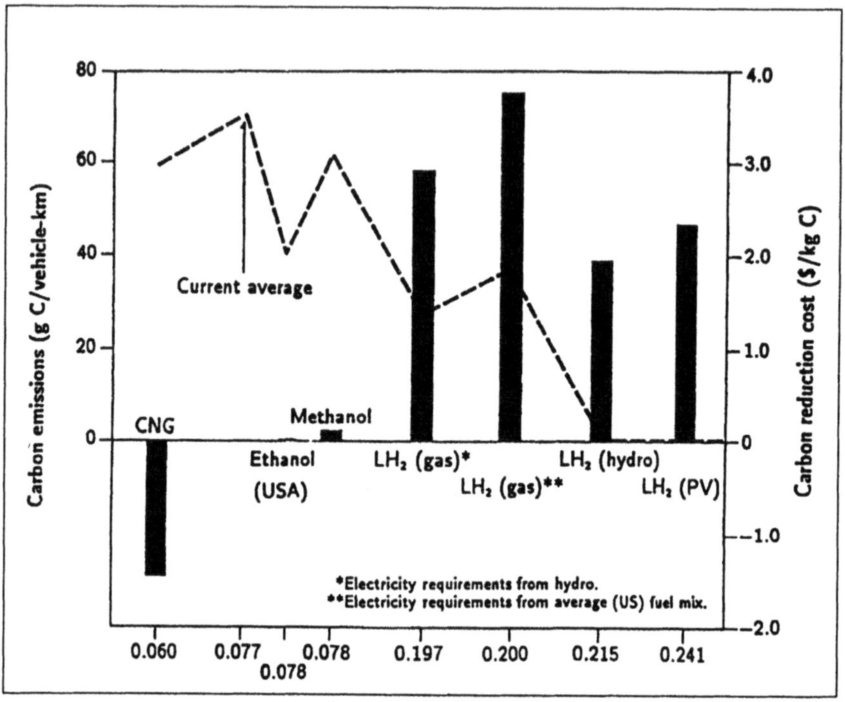

Figure 13-14: Carbon emissions (line) versus carbon reduction costs (bars) of various technological changes to passenger cars.

More revolutionary technological change would be represented by the massive introduction of a new generation of electric vehicles which, owing to the high end-use efficiency of electric cars, would yield lower overall carbon emissions even if electricity were to be produced in oil-fired power plants. Another alternative is hydrogen fuel cell-powered cars. However, the hydrogen options have specific carbon reduction costs in excess of $2,000 per ton of carbon, which makes them high-cost carbon reduction strategies indeed.

Figure 13-15 summarizes the technological options available in the passenger transportation sector with respect to energy and carbon intensity. Only hydrogen-fueled vehicles would result in zero carbon emissions, provided fossil fuels are not used in the production of hydrogen. Figure 13-15 also indicates the relative scope of efficiency improvements (see the example of a steam locomotive in 1855, 1955, and a modern, coal-based, electric intercity train) and fuel substitution (coal-versus gas-electric intercity train, or gasoline versus CNG car). The relative rankings of transport modes can, however, vary per unit of energy service delivered (e.g., passenger-km) according to differences in load factors. The latter is highest for aircraft (50 to 60 percent) and lowest for private cars and public mass transit systems (as low as 20 to 25 percent). There are, however, large regional variations.

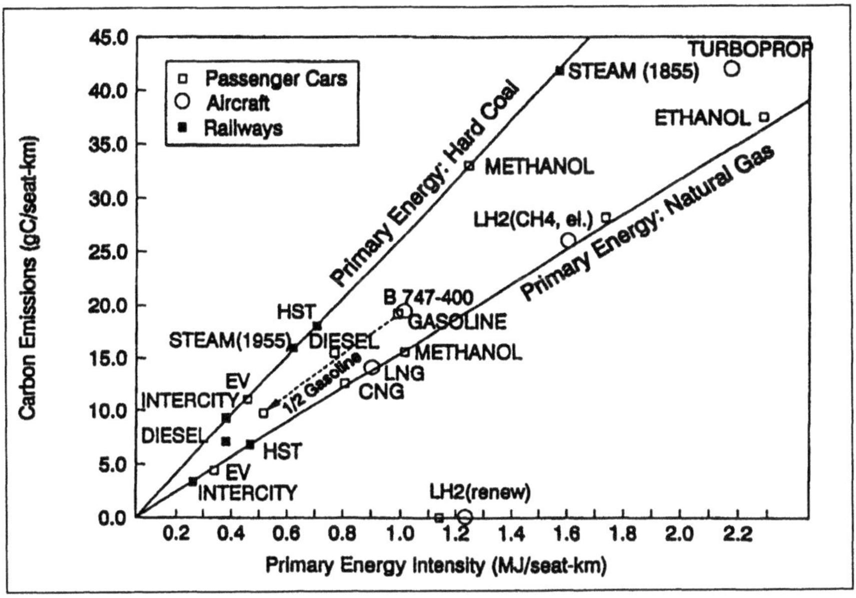

Source: Schafer, 1992.

Figure 13-15: Energy and carbon intensity of various passenger transportation technology chains.

Yanagisawa (1995) has compared the carbon emissions of a number of alternative automotive fuels. His results are summarized in Table 13-6. The table shows that CNG cars have the lowest emissions among all near-term options.

Table 13-6: Fuel-cycle analysis of automotive fuels.

Fuel Type	From Energy Station To Driving (Kg-CO_2/100km)	From Mining to the Energy Station (Kg-CO_2/100km)	Total Emissions (Kg-O_2/100km)
Gasoline	17.1	5.33 (5.47kg gasoline)	22.4
Methanol (N.Gas)	16.3	7.48 (11.97kg MeOH)	23.8
EV (coal)	26.3	1.50 (11.43kg coal)	27.8
CNG	14.5	3.12 (5.25kg NG)	17.7
H_2 (electrolysis,coal)	71.9	3.65 (27.9kg coal)	75.6
H_2 (N. Gas)	29.8	6.59 (10.84kg NG)	36.4
Methanol + EV[1]	15.8	0.81 (6.18kg coal)	16.6

[1]Electricity is assumed to be generated by a coal-fired plant. CO_2 discharged from the plant is converted into methanol.

Source: Yanagisawa 1995.

13.6 The Residential & Commercial Sector

Recent literature in the field suggests that the relative energy savings potential of the residential and commercial sector is larger than the potential in either the industrial or transportation sectors. Moreover, according to several results, a significant proportion of the energy and emissions savings can be achieved economically, i.e., by saving money at the same time. A typical size of this "no-regrets" potential to reduce carbon emissions in this sector is 50 percent. However, the resulting dollar savings and emissions reductions usually describe a theoretical potential: it should not be assumed that merely improving the information flow to consumers would result in a quick realization of this emission reduction potential. The market price of goods and services alone is an unreliable indicator of consumer preference in the residential and commercial sector.

One way to measure the degree of incompleteness of the market-price criterion is to calculate the discount rates that are implicit in purchasing decisions in the residential sector. Train (1985) found the following ranges: improvements of thermal integrity of buildings: 10–32 percent; space heating and fuel type: 4.4–36 percent; air conditioning: 3.2–29 percent; refrigerators: 39–100 percent; other home appliances: 18–67 percent. In a way, these high discount rates explain why the usual profitability criteria must be applied with caution in this sector. The question remains as to why the discount rates implied by consumer behavior are so much higher than market discount rates. Discussing this question goes beyond the scope of this report. Readers interested in this topic are referred to Manne and Schrattenholzer (1993). In any case, increasing the assumed discount rate in a study of conservation potentials very quickly reduces the size of the calculated savings.

As to the practical savings potential, the IPCC (1996b) estimates that energy efficiency measures for appliances, lighting, and office equipment with paybacks to the consumer of five years or less have the technical potential to reduce carbon emissions by approximately 20 percent in 2010, 25 percent in 2020, and 40 percent in 2050, all relative to a baseline that already includes some energy efficiency improvements. According to the same report (IPCC 1996b), improvements in the building envelope have the technical potential to reduce carbon emissions by 10 percent in 2010, 10 percent in 2020, and 15 percent in 2050, again assuming a payback time of up to five years.

13.6.1 Efficiency Improvements

There are numerous examples of improved efficiency of residential appliances. Compared to the 1986 stock, the 1991 stock of refrigerators, freezers, or room air conditioners in the U.S. consumes about 20 percent less, and 1991 models sold consume an average between 30 to 40 percent less than the average 1986 vintages (Levine et al. 1992). With the best available and most advanced technology, further significant improvements in energy efficiency are possible. In the U.S., where refrigerators are the second largest consumer of residential electricity, the average 1991 stock consumed about 1200 kWh/yr, down from 1450 kWh in 1986, and the best available model sold consumed 710 kWh. Further reductions to 200–500 kWh are considered within reach during the 1990s (Levine et al. 1992). According to Meier (1991), more than 25 percent of the current U.S. refrigerator electricity demand could be reduced by cost-efficient measures. Meier's calculations assume an annual discount rate of 10 percent, which looks low in light of the above discussion. Using a rate of 30 percent instead results in positive costs of all saving measures considered in this case.

An example analyzing costs and emissions of different refrigerator types, based on data of IIASA's CO_2DB database, is shown in Figure 13-16.

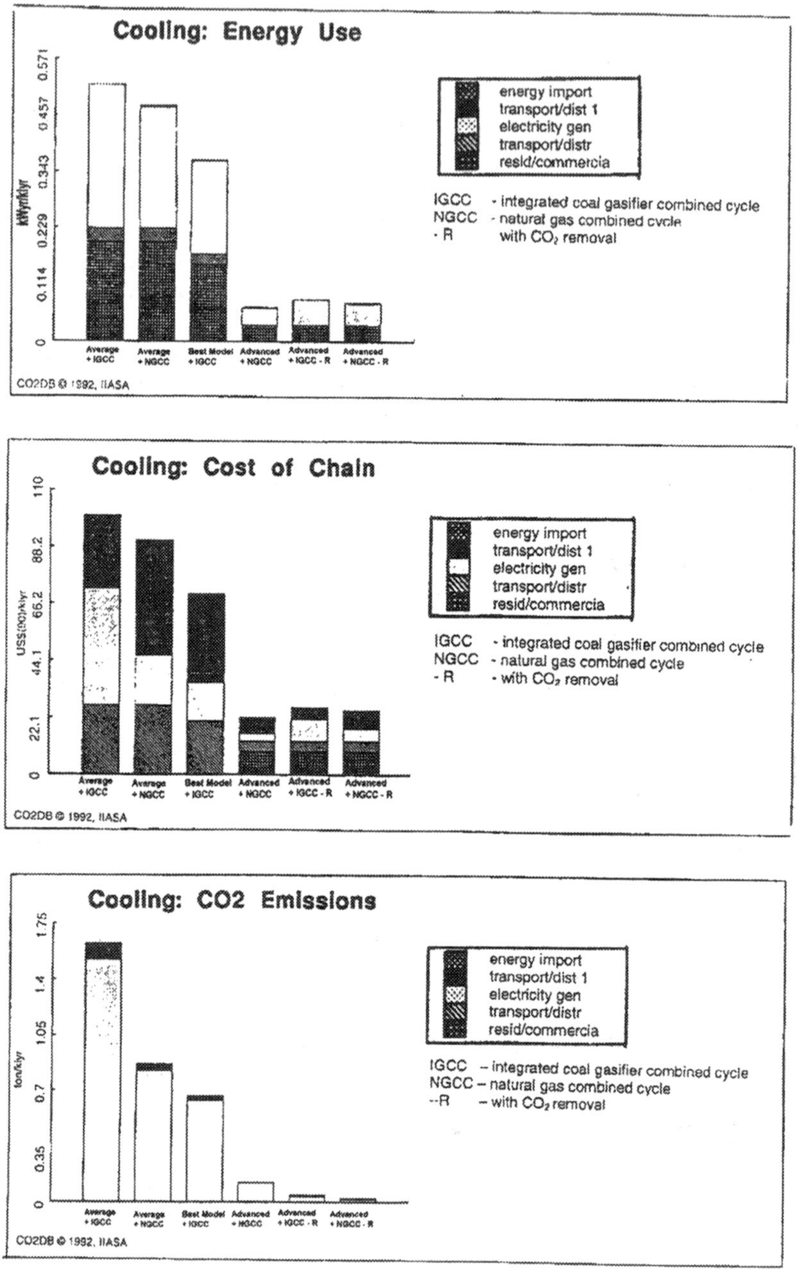

Source: Messner and Nakicenovic 1992.

Figure 13-16: Energy requirements, costs, and CO_2 emissions for six energy chains ending in refrigeration.

The figure shows that an advanced refrigerator model provides cheaper services than even the best model of today, even if its electricity comes from a source that includes carbon removal.

The message of this subsection is that successful R&D in energy-efficient household appliances is only the prerequisite for reducing final energy consumption and carbon emissions. To actually realize existing potentials, policy must help improve consumers' access to information and/or involve instruments such as standard setting, taxes, and appliance energy labeling.

13.6.2 Fuel Mix Changes

The table presents the calculated CO_2 emissions and CO_2 reduction costs for residential space heating. The data represent a residence with 150 square meters of living space at three levels of building insulation standards. The CO_2 reduction costs are calculated as the differences between the reference case (gas heating and 1990 insulation standards) and each alternative considering the levelized annual heating costs. In addition to the emissions related to residential heating systems, the CO_2 emissions account for all emissions along the energy delivery chain.

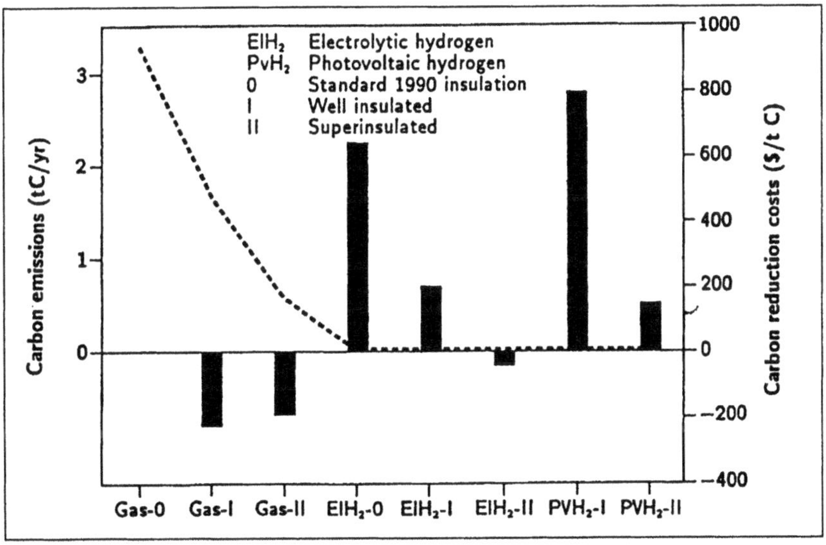

Figure 13-17: Carbon emissions (line) versus CO_2 reduction cost (bars) for residential heating. The abbreviations represent the three insulation categories in ascending order of technological sophistication (0, I, II): natural gas; a hydrogen delivery chain based on hydro-electricity and electrolysis (ElH2); and photovoltaic electricity and electrolytic hydrogen (PVH2).

Figure 13-17 conveys that improved building insulation is the single most important factor for immediate CO_2 emission reductions. This, and high-efficiency heating technology, even for natural gas-fueled systems, could lead to substantial emission reductions. There, the costs of CO_2 reduction are negative, i.e., lowering CO_2 emissions reduces heating costs at the same time.

Likewise, the combination of hydroelectricity and electrolytic hydrogen could generate net savings if used in superinsulated residences (ElH2-II). Photovoltaic hydrogen in conjunction with superinsulation (PVH2-II) achieves zero carbon emissions for $140 per ton.

Commercial heating follows the trend of residential heating. Here, the cogeneration of electricity and space/process heat production is an essential prerequisite. Improved insulation and phosphoric-acid fuel cells (PAFC) would result in CO_2 reduction costs of approximately $25–80 per ton of carbon.

14 APPENDIX I: Detailed Macroeconomic and Sectoral Projections

Table 14-1: Macroeconomic assumptions of baseline scenario for EU-15 (all numbers are annualized growth rates except if otherwise indicated).

	Observations		Forecasts										
	1985/90	1990/95	1995/00	2000/05	2005/10	2010/15	2015/20	2020/25	2025/30	1995	2030		
			Macroeconomic aggregates										
GDP Growth	3.22%	1.43%	2.57%	2.45%	2.35%	1.90%	1.72%	1.67%	1.67%				
Priv. Consumption	3.57%	1.23%	2.27%	2.31%	2.24%	1.82%	1.73%	1.69%	1.69%				
Consumer Price Index	4.79%	3.27%	2.26%	2.13%	2.01%	1.88%	1.76%	1.63%	1.50%				
Population Total	0.33%	1.30%	0.36%	0.28%	0.18%	0.10%	0.05%	0.03%	0.01%				
			Sectoral value added							***Share in value added***			
Manufacturing	3.20%	0.06%	2.22%	2.11%	1.92%	1.52%	1.41%	1.35%	1.33%	**28.17%**	**24.94%**		
- Intensive	3.57%	0.98%	2.20%	1.96%	1.76%	1.43%	1.27%	1.21%	1.14%				
- Metals	1.84%	-0.31%	0.92%	0.60%	0.39%	0.26%	0.14%	0.08%	0.05%	0.91%	0.51%	3.2%	2.0%
Iron and Steel	1.10%	-0.14%	0.67%	0.35%	0.09%	-0.05%	-0.18%	-0.24%	-0.28%				
Non-ferrous	3.12%	-0.58%	1.32%	1.00%	0.84%	0.72%	0.60%	0.53%	0.49%				
- Chemicals	4.15%	1.93%	2.09%	2.26%	1.98%	1.65%	1.49%	1.41%	1.33%	2.32%	2.09%	8.3%	8.4%
- Paper	3.62%	0.99%	2.48%	2.10%	1.99%	1.53%	1.35%	1.28%	1.20%	1.93%	1.71%	6.8%	6.9%
- Building Materials	3.91%	0.19%	2.89%	2.06%	1.84%	1.53%	1.33%	1.27%	1.23%	1.19%	1.07%	4.2%	4.3%
- Other Industries	2.76%	-0.10%	2.23%	2.15%	1.97%	1.55%	1.45%	1.38%	1.38%				
- Food	1.63%	0.80%	2.41%	2.07%	1.77%	1.48%	1.31%	1.23%	1.16%	3.19%	2.77%	11.3%	11.1%
- Textiles	0.32%	-1.38%	1.00%	1.02%	0.93%	0.75%	0.73%	0.72%	0.74%	1.53%	1.01%	5.4%	4.0%
- Engineering	2.75%	-0.04%	2.75%	2.82%	2.65%	2.04%	1.95%	1.86%	1.86%	9.31%	10.08%	33.1%	40.4%
- Others	3.44%	0.68%	1.47%	1.31%	0.85%	0.69%	0.61%	0.46%	0.40%	2.10%	1.38%	7.5%	5.5%
- Construction	3.92%	-0.62%	1.86%	1.61%	1.46%	1.09%	0.98%	0.91%	0.88%	5.68%	4.32%	20.2%	17.3%
Services	3.67%	1.99%	2.87%	2.71%	2.63%	2.12%	1.90%	1.85%	1.85%	**64.18%**	**69.42%**	100.0%	100.0%
- services	4.79%	2.76%	3.36%	3.20%	3.12%	2.49%	2.20%	2.14%	2.15%				
- non market	1.49%	0.96%	2.06%	1.98%	1.86%	1.52%	1.37%	1.34%	1.30%				
- trade	3.40%	1.04%	2.33%	2.00%	1.83%	1.50%	1.38%	1.35%	1.31%				
Agriculture	1.20%	0.78%	1.17%	1.23%	1.13%	0.94%	0.87%	0.80%	0.76%	**2.92%**	**2.02%**		
Energy Sector	0.74%	2.29%	1.39%	1.42%	1.40%	1.26%	1.14%	1.13%	1.12%	**4.73%**	**3.62%**		

Table 14-2 Macroeconomic assumptions of baseline scenario for Austria (all numbers are annualized growth rates except if otherwise indicated).

	Observations		Forecasts								
	1985/90	1990/95	1995/00	2000/05	2005/10	2010/15	2015/20	2020/25	2025/30	1995	2030
			Macroeconomic aggregates								
GDP Growth	3.22%	2.28%	2.22%	2.19%	2.14%	1.74%	1.59%	1.51%	1.49%		
Priv. Consumption	3.09%	1.51%	1.72%	1.94%	1.92%	1.56%	1.59%	1.51%	1.49%		
Consumer Price Index	2.18%	3.25%	1.98%	1.90%	1.82%	1.74%	1.66%	1.58%	1.50%		
GDP Deflator	2.89%	3.22%	1.76%	1.72%	1.67%	1.63%	1.59%	1.54%	1.50%		
Exchange Rate ($)	-11.28%	-2.35%	0.88%	0.00%	0.00%	0.00%	0.00%	0.00%	0.00%		
Population Total	0.42%	0.64%	0.23%	0.20%	0.15%	0.12%	0.10%	0.05%	0.01%		
Lending Rate (level)	-	-	-	-	-	-	-	-	-		
			Sectoral value added							***Share in value added***	
Manufacturing	3.12%	1.36%	2.36%	2.33%	2.20%	1.75%	1.58%	1.47%	1.41%	**34.75%**	**35.14%**
- Intensive	5.22%	1.39%	2.18%	2.10%	1.90%	1.67%	1.47%	1.33%	1.26%		
- Metals	5.94%	-1.08%	0.27%	0.80%	1.00%	1.10%	0.90%	0.70%	0.60%	1.51%	1.04%
Iron and Steel	6.41%	-0.71%	-0.16%	0.36%	0.55%	0.64%	0.43%	0.22%	0.11%		
Non-ferrous	5.52%	-1.44%	0.67%	1.20%	1.39%	1.49%	1.28%	1.07%	0.96%		
- Chemicals	6.44%	1.11%	1.37%	1.80%	1.80%	1.65%	1.60%	1.50%	1.50%	2.36%	2.18%
- Paper	6.48%	2.00%	3.18%	2.70%	2.20%	1.80%	1.60%	1.50%	1.40%	2.09%	2.24%
- Building Materials	2.10%	3.12%	3.36%	2.60%	2.20%	1.85%	1.50%	1.30%	1.20%	2.08%	2.20%
- Other Industries	2.39%	2.64%	2.41%	2.40%	2.29%	1.78%	1.62%	1.51%	1.45%		
- Food	2.71%	1.69%	2.26%	2.00%	1.60%	1.30%	1.30%	1.20%	1.00%	3.67%	3.29%
- Textiles	-2.01%	-4.29%	0.46%	0.10%	0.00%	-0.50%	-1.00%	0.00%	0.00%	1.22%	0.61%
- Engineering	2.43%	1.77%	1.71%	2.00%	2.20%	2.00%	1.80%	1.60%	1.60%	10.81%	10.83%
- Others	4.75%	3.33%	1.89%	1.70%	1.50%	1.30%	1.20%	1.00%	1.00%	2.19%	1.87%
- Construction	2.69%	5.34%	3.69%	3.40%	3.00%	2.00%	1.80%	1.70%	1.60%	8.82%	10.88%
Services	3.49%	2.46%	2.29%	2.21%	2.21%	1.80%	1.65%	1.59%	1.59%	**57.65%**	**58.94%**
- services	4.44%	3.07%	2.96%	2.95%	2.90%	2.28%	2.03%	1.98%	1.99%	28.55%	35.08%
- non market	1.11%	2.12%	1.80%	1.40%	1.40%	1.15%	1.10%	1.00%	0.95%	15.03%	12.30%
- trade	4.44%	1.63%	1.42%	1.42%	1.40%	1.25%	1.20%	1.10%	1.10%	14.07%	11.56%
Agriculture	1.70%	-1.23%	0.74%	1.20%	1.00%	0.90%	0.80%	0.70%	0.60%	**2.92%**	**2.07%**
Energy Sector	2.43%	2.17%	1.20%	1.35%	1.40%	1.30%	1.30%	1.20%	1.10%	**4.69%**	**3.84%**

Table14 –3: Macroeconomic assumptions of baseline scenario for Belgium (all numbers are annualized growth rates except if otherwise indicated).

	Observations		Forecasts								
	1985/90	1990/95	1995/00	2000/05	2005/10	2010/15	2015/20	2020/25	2025/30	1995	2030
			Macroeconomic aggregates								
GDP Growth	2.85%	1.51%	2.44%	2.29%	2.22%	1.82%	1.67%	1.59%	1.57%		
Priv. Consumption	2.79%	1.41%	1.58%	1.89%	1.87%	1.54%	1.67%	1.59%	1.57%		
Consumer Price Index	2.11%	2.47%	2.02%	1.93%	1.85%	1.76%	1.67%	1.59%	1.50%		
GDP Deflator	3.11%	2.81%	1.86%	1.80%	1.74%	1.68%	1.62%	1.56%	1.50%		
Exchange Rate ($)	-10.87%	-2.44%	0.90%	0.00%	0.00%	0.00%	0.00%	0.00%	0.00%		
Population Total	0.22%	0.35%	0.32%	0.22%	0.15%	0.12%	0.10%	0.05%	0.01%		
Lending Rate (level)	12.54%	7.50%	6.00%	5.00%	4.50%	4.30%	4.30%	4.30%	4.30%		
			Sectoral value added							***Share in value added***	
Manufacturing	3.24%	0.62%	3.61%	2.38%	1.98%	1.37%	1.20%	1.04%	0.95%	**26.66%**	**25.27%**
- Intensive	4.55%	1.45%	2.91%	1.95%	1.74%	1.31%	1.17%	1.04%	0.91%		
- Metals	1.57%	-1.60%	2.26%	1.00%	0.80%	0.40%	0.20%	0.10%	0.10%	1.90%	1.23%
Iron and Steel	-0.25%	-1.50%	1.68%	0.41%	0.20%	-0.22%	-0.44%	-0.56%	-0.58%		
Non-ferrous	4.78%	-1.76%	3.12%	1.82%	1.59%	1.15%	0.92%	0.80%	0.78%		
- Chemicals	4.35%	3.30%	2.57%	2.00%	1.90%	1.40%	1.30%	1.20%	1.00%	3.29%	2.95%
- Paper	7.39%	1.60%	4.06%	2.50%	2.30%	1.90%	1.70%	1.50%	1.30%	1.47%	1.59%
- Building Materials	7.98%	1.70%	3.36%	2.50%	1.90%	1.50%	1.30%	1.10%	1.00%	1.28%	1.22%
- Other Industries	2.17%	0.28%	3.90%	2.56%	2.07%	1.39%	1.22%	1.05%	0.97%		
- Food	1.35%	1.04%	2.94%	2.30%	1.90%	1.50%	1.30%	1.10%	1.00%	3.13%	2.89%
- Textiles	5.03%	0.58%	1.34%	1.00%	0.60%	0.30%	0.20%	0.10%	0.00%	1.58%	0.96%
- Engineering	-0.96%	-0.90%	5.40%	3.50%	3.20%	2.00%	1.80%	1.60%	1.50%	6.40%	8.31%
- Others	4.38%	0.03%	1.08%	0.50%	0.24%	0.20%	0.10%	0.00%	0.00%	1.98%	1.12%
- Construction	5.79%	1.30%	4.29%	2.50%	1.50%	1.00%	0.80%	0.60%	0.50%	5.63%	4.99%
Services	3.00%	1.81%	2.07%	2.35%	2.42%	2.08%	1.91%	1.85%	1.86%	**66.96%**	**70.10%**
- services	4.78%	1.68%	2.68%	2.85%	3.12%	2.63%	2.35%	2.28%	2.27%	37.23%	46.56%
- non market	0.49%	1.90%	0.73%	1.20%	1.10%	1.00%	1.00%	0.90%	0.90%	13.38%	9.58%
- trade	1.40%	2.05%	1.74%	2.00%	1.60%	1.35%	1.30%	1.20%	1.20%	16.34%	13.95%
Agriculture	0.08%	3.00%	0.22%	0.60%	0.40%	0.30%	0.20%	0.10%	0.10%	**2.11%**	**1.18%**
Energy Sector	2.21%	1.90%	1.69%	1.60%	1.40%	1.30%	1.20%	1.10%	1.00%	**4.27%**	**3.45%**

Table 14-4: Macroeconomic assumptions of baseline scenario for Denmark (all numbers are annualized growth rates except if otherwise indicated).

	Observations		Forecasts								
	1985/90	1990/95	1995/00	2000/05	2005/10	2010/15	2015/20	2020/25	2025/30	1995	2030
			Macroeconomic aggregates								
GDP Growth	1.82%	1.68%	2.81%	2.19%	2.09%	1.61%	1.44%	1.38%	1.16%		
Priv. Consumption	0.25%	2.87%	2.62%	2.12%	2.03%	1.56%	1.44%	1.38%	1.16%		
Consumer Price Index	3.92%	1.97%	2.38%	2.23%	2.09%	1.94%	1.79%	1.65%	1.50%		
GDP Deflator	3.91%	1.74%	2.24%	2.12%	1.99%	1.87%	1.75%	1.62%	1.50%		
Exchange Rate ($)	-10.20%	-1.94%	0.52%	0.00%	0.00%	0.00%	0.00%	0.00%	0.00%		
Population Total	0.11%	0.33%	0.32%	0.15%	0.10%	0.07%	0.04%	0.01%	0.00%		
Lending Rate (level)	14.70%	8.30%	6.30%	6.00%	5.00%	4.80%	4.60%	4.50%	4.50%		
			Sectoral value added							***Share in value added***	
Manufacturing	-0.19%	0.71%	2.62%	2.21%	1.93%	1.46%	1.26%	1.12%	0.98%	**22.93%**	**21.72%**
- Intensive	-0.21%	2.12%	2.76%	1.88%	1.53%	1.36%	1.25%	1.09%	0.96%		
- Metals	2.37%	0.22%	0.80%	0.70%	0.60%	0.50%	0.40%	0.30%	0.18%	0.12%	0.07%
Iron and Steel	9.59%	-3.57%	-0.06%	-0.20%	-0.34%	-0.49%	-0.64%	-0.79%	-0.98%		
Non-ferrous	0.76%	1.20%	0.99%	0.89%	0.79%	0.68%	0.58%	0.48%	0.36%		
- Chemicals	2.66%	5.09%	1.60%	1.30%	1.10%	1.00%	0.90%	0.85%	0.80%	2.02%	1.57%
- Paper	-1.77%	0.25%	3.32%	2.20%	2.00%	1.80%	1.70%	1.40%	1.20%	1.86%	1.94%
- Building Materials	-1.76%	0.42%	4.34%	2.50%	1.50%	1.20%	1.00%	0.90%	0.80%	0.89%	0.87%
- Other Industries	-0.33%	0.35%	2.59%	2.30%	2.04%	1.49%	1.26%	1.13%	0.98%		
- Food	1.89%	1.45%	2.90%	2.00%	1.80%	1.50%	1.30%	1.10%	1.00%	3.72%	3.52%
- Textiles	-7.18%	-2.83%	0.40%	0.00%	-0.50%	-1.00%	-0.50%	0.00%	0.00%	0.68%	0.33%
- Engineering	-1.03%	0.05%	3.68%	2.70%	2.40%	1.50%	1.20%	1.10%	0.90%	6.28%	6.52%
- Others	-0.83%	4.45%	2.52%	2.00%	2.00%	1.50%	1.20%	1.10%	0.90%	2.43%	2.26%
- Construction	0.63%	-1.28%	1.22%	2.40%	2.00%	1.70%	1.50%	1.30%	1.20%	4.94%	4.62%
Services	2.31%	1.82%	3.01%	2.28%	2.23%	1.70%	1.54%	1.49%	1.25%	**69.33%**	**72.17%**
- services	3.78%	2.18%	3.74%	2.75%	2.70%	2.05%	1.85%	1.82%	1.55%	33.70%	40.53%
- non market	0.85%	1.38%	2.38%	1.90%	1.80%	1.25%	1.15%	1.10%	0.80%	22.13%	19.76%
- trade	1.45%	1.63%	2.18%	1.60%	1.60%	1.40%	1.20%	1.10%	1.00%	13.49%	11.87%
Agriculture	1.96%	1.06%	1.34%	1.20%	1.00%	0.90%	0.80%	0.70%	0.50%	**4.14%**	**3.05%**
Energy Sector	11.51%	7.01%	1.80%	1.50%	1.40%	1.30%	1.20%	1.15%	1.10%	**3.60%**	**3.07%**

Table 14-5: Macroeconomic assumptions of baseline scenario for Finland (all numbers are annualized growth rates except if otherwise indicated).

	Observations		Forecasts								
	1985/90	1990/95	1995/00	2000/05	2005/10	2010/15	2015/20	2020/25	2025/30	1995	2030
			Macroeconomic aggregates								
GDP Growth	3.39%	-0.69%	3.17%	2.26%	2.13%	1.64%	1.42%	1.35%	1.17%		
Priv. Consumption	3.42%	-0.73%	3.14%	2.25%	2.12%	1.64%	1.42%	1.35%	1.17%		
Consumer Price Index	4.96%	2.13%	1.72%	1.68%	1.65%	1.61%	1.57%	1.54%	1.50%		
GDP Deflator	5.64%	2.16%	1.90%	1.83%	1.77%	1.70%	1.63%	1.57%	1.50%		
Exchange Rate ($)	-9.21%	2.74%	0.56%	0.00%	0.00%	0.00%	0.00%	0.00%	0.00%		
Population Total	0.34%	0.51%	0.45%	0.34%	0.28%	0.20%	0.12%	0.08%	0.05%		
Lending Rate (level)	10.41%	8.80%	6.10%	6.00%	5.00%	4.80%	4.60%	4.50%	4.50%		
			Sectoral value added							***Share in value added***	
Manufacturing	3.53%	-3.03%	3.66%	2.85%	2.52%	1.95%	1.59%	1.45%	1.28%	**29.37%**	**32.65%**
- Intensive	3.68%	1.72%	4.54%	4.04%	3.39%	2.48%	1.81%	1.70%	1.43%		
- Metals	6.55%	0.01%	0.80%	0.00%	0.00%	-0.50%	-1.00%	-1.50%	-2.50%	0.39%	0.16%
Iron and Steel	5.53%	0.67%	0.23%	-0.58%	-0.60%	-1.11%	-1.63%	-2.15%	-3.16%		
Non-ferrous	8.45%	-1.17%	1.85%	0.99%	0.94%	0.39%	-0.15%	-0.69%	-1.73%		
- Chemicals	3.60%	2.30%	2.20%	2.20%	1.80%	1.50%	1.20%	1.00%	0.90%	1.72%	1.54%
- Paper	3.34%	2.96%	5.69%	5.00%	4.10%	2.90%	2.10%	2.00%	1.70%	5.80%	9.57%
- Building Materials	4.41%	-4.77%	2.80%	1.50%	1.20%	1.00%	0.70%	0.50%	0.00%	0.91%	0.70%
- Other Industries	3.02%	-2.97%	3.27%	2.29%	2.08%	1.66%	1.47%	1.31%	1.19%		
- Food	2.15%	1.02%	2.92%	1.80%	1.50%	1.40%	1.30%	1.30%	1.00%	2.97%	2.70%
- Textiles	-7.37%	-4.45%	0.98%	0.00%	0.00%	-0.50%	-1.00%	-1.00%	-1.00%	0.74%	0.34%
- Engineering	2.77%	0.11%	4.45%	3.00%	2.70%	2.20%	2.00%	1.90%	1.80%	6.82%	8.67%
- Others	3.57%	1.19%	2.52%	2.50%	2.20%	1.60%	1.50%	1.20%	1.00%	3.68%	3.57%
- Construction	4.68%	-8.51%	2.82%	1.80%	1.70%	1.30%	1.00%	0.70%	0.60%	6.34%	5.41%
Services	3.91%	-0.39%	3.14%	2.09%	2.06%	1.57%	1.40%	1.39%	1.19%	**61.56%**	**60.67%**
- services	4.99%	1.17%	4.30%	2.68%	2.65%	2.00%	1.95%	1.90%	1.62%	32.76%	39.74%
- non market	2.17%	-0.46%	1.44%	1.05%	0.90%	0.70%	0.10%	0.10%	0.00%	20.30%	13.12%
- trade	4.40%	-5.20%	2.50%	1.90%	1.90%	1.40%	1.35%	1.25%	1.10%	8.50%	7.81%
Agriculture	0.23%	-0.16%	1.56%	1.20%	0.90%	0.70%	0.60%	0.40%	0.20%	**6.55%**	**4.51%**
Energy Sector	2.46%	3.14%	2.12%	1.60%	1.70%	1.50%	1.20%	1.00%	1.00%	**2.52%**	**2.18%**

Table 14-6: Macroeconomic assumptions of baseline scenario for France (all numbers are annualized growth rates except if otherwise indicated).

	Observations		Forecasts								
	1985/90	1990/95	1995/00	2000/05	2005/10	2010/15	2015/20	2020/25	2025/30	1995	2030
				Macroeconomic aggregates							
GDP Growth	2.99%	0.94%	2.39%	2.30%	2.20%	1.85%	1.67%	1.58%	1.58%		
Priv. Consumption	2.87%	1.23%	2.24%	2.23%	2.13%	1.80%	1.67%	1.58%	1.58%		
Consumer Price Index	3.09%	2.19%	1.72%	1.68%	1.65%	1.61%	1.57%	1.54%	1.50%		
GDP Deflator	3.42%	2.17%	1.62%	1.60%	1.58%	1.56%	1.54%	1.52%	1.50%		
Exchange Rate ($)	-9.53%	-1.71%	0.31%	0.00%	0.00%	0.00%	0.00%	0.00%	0.00%		
Population Total	0.52%	0.53%	0.50%	0.50%	0.40%	0.30%	0.18%	0.12%	0.08%		
Lending Rate (level)	17.33%	7.60%	6.00%	5.90%	5.80%	5.70%	5.60%	5.55%	5.50%		
				Sectoral value added						***Share in value added***	
Manufacturing	2.64%	-0.81%	1.44%	1.92%	1.89%	1.59%	1.54%	1.43%	1.39%	**24.23%**	**21.55%**
- Intensive	3.34%	0.21%	1.87%	1.18%	1.14%	1.16%	1.11%	1.05%	0.97%		
- Metals	3.69%	-1.06%	0.46%	0.10%	0.00%	0.00%	0.00%	0.00%	0.00%	0.83%	0.44%
Iron and Steel	1.78%	-1.44%	0.32%	-0.04%	-0.14%	-0.14%	-0.14%	-0.14%	-0.14%		
Non-ferrous	5.72%	-0.71%	0.59%	0.22%	0.12%	0.12%	0.12%	0.12%	0.12%		
- Chemicals	4.32%	1.46%	2.10%	2.30%	2.00%	1.70%	1.60%	1.50%	1.35%	2.08%	1.98%
- Paper	2.93%	0.07%	1.80%	0.52%	0.80%	1.30%	1.20%	1.10%	1.00%	1.61%	1.21%
- Building Materials	1.86%	-0.99%	2.70%	0.60%	0.50%	0.40%	0.40%	0.40%	0.40%	0.93%	0.62%
- Other Industries	2.46%	-1.10%	1.32%	2.14%	2.10%	1.70%	1.65%	1.52%	1.50%		
- Food	1.00%	0.40%	2.07%	2.12%	2.00%	1.80%	1.55%	1.40%	1.35%	2.78%	2.61%
- Textiles	-1.93%	-3.09%	-0.95%	-1.78%	-1.88%	-2.50%	-2.50%	-2.50%	-2.50%	1.14%	0.28%
- Engineering	2.97%	-0.51%	1.75%	3.00%	2.90%	2.25%	2.20%	2.00%	1.95%	8.57%	9.67%
- Others	2.82%	-0.13%	1.18%	1.50%	1.40%	1.30%	1.20%	1.00%	0.80%	1.85%	1.44%
- Construction	3.61%	-2.86%	0.60%	1.50%	1.40%	1.15%	1.05%	1.00%	1.00%	4.43%	3.31%
Services	3.39%	1.64%	2.86%	2.52%	2.37%	1.99%	1.76%	1.68%	1.68%	**68.44%**	**72.85%**
- services	4.58%	1.89%	3.10%	3.04%	2.85%	2.40%	2.13%	2.05%	2.05%	38.50%	46.88%
- non market	1.54%	2.59%	2.65%	2.20%	2.00%	1.55%	1.30%	1.17%	1.13%	17.59%	16.27%
- trade	2.61%	-0.31%	2.40%	1.25%	1.20%	1.05%	0.95%	0.90%	0.90%	12.36%	9.69%
Agriculture	1.75%	-0.05%	0.88%	1.00%	0.90%	0.77%	0.70%	0.60%	0.57%	**3.37%**	**2.26%**
Energy Sector	0.01%	1.29%	1.16%	1.60%	1.72%	1.72%	1.52%	1.22%	1.20%	**3.96%**	**3.34%**

Table 14-7: Macroeconomic assumptions of baseline scenario for Germany (all numbers are annualized growth rates except if otherwise indicated).

	Observations		Forecasts								
	1985/90	1990/95	1995/00	2000/05	2005/10	2010/15	2015/20	2020/25	2025/30	1995	2030
			Macroeconomic aggregates								
GDP Growth	3.23%	1.84%	2.52%	2.47%	2.30%	1.81%	1.66%	1.64%	1.62%		
Priv. Consumption	3.90%	1.58%	2.20%	2.31%	2.17%	1.71%	1.66%	1.64%	1.62%		
Consumer Price Index	1.02%	1.91%	1.90%	1.83%	1.77%	1.70%	1.63%	1.57%	1.50%		
GDP Deflator	2.43%	3.52%	1.72%	1.68%	1.65%	1.61%	1.57%	1.54%	1.50%		
Exchange Rate ($)	1.45%	-4.58%	0.88%	0.00%	0.00%	0.00%	0.00%	0.00%	0.00%		
Population Total	0.72%	5.43%	0.61%	0.35%	0.15%	0.02%	0.00%	-0.01%	-0.02%		
Lending Rate (level)	12.11%	4.56%	3.72%	4.30%	4.30%	4.30%	4.30%	4.30%	4.30%		
			Sectoral value added							***Share in value added***	
Manufacturing	2.36%	-0.50%	2.02%	1.96%	1.69%	1.44%	1.36%	1.30%	1.29%	**31.38%**	**27.14%**
- Intensive	2.46%	0.50%	2.11%	1.88%	1.60%	1.26%	1.17%	1.11%	1.05%		
- Metals	1.32%	-0.23%	1.32%	1.00%	0.35%	0.25%	0.15%	0.10%	0.10%	0.98%	0.58%
Iron and Steel	1.28%	0.22%	1.17%	0.85%	0.20%	0.10%	0.00%	-0.05%	-0.05%		
Non-ferrous	1.52%	-2.44%	2.11%	1.76%	1.08%	0.95%	0.83%	0.75%	0.73%		
- Chemicals	2.54%	0.90%	2.52%	2.35%	2.00%	1.70%	1.65%	1.58%	1.50%	2.59%	2.50%
- Paper	2.42%	0.36%	1.84%	1.70%	1.60%	1.10%	1.05%	1.00%	0.90%	2.07%	1.63%
- Building Materials	3.37%	0.52%	2.31%	1.80%	1.60%	1.20%	0.90%	0.80%	0.75%	1.18%	0.94%
- Other Industries	2.25%	-0.77%	1.99%	1.98%	1.72%	1.49%	1.42%	1.35%	1.35%		
- Food	-0.10%	-0.41%	2.16%	2.00%	1.56%	1.25%	1.05%	1.00%	1.00%	2.81%	2.31%
- Textiles	-0.78%	-3.77%	-1.16%	-0.80%	-0.50%	0.00%	0.00%	0.00%	0.00%	0.85%	0.38%
- Engineering	3.11%	-1.23%	2.70%	2.58%	2.37%	2.00%	1.90%	1.80%	1.78%	13.47%	14.23%
- Others	3.51%	0.59%	1.68%	1.50%	0.24%	0.24%	0.25%	0.10%	0.08%	2.34%	1.43%
- Construction	1.46%	0.24%	0.60%	0.80%	0.70%	0.64%	0.55%	0.50%	0.50%	5.08%	3.14%
Services	4.12%	3.27%	2.88%	2.80%	2.65%	2.02%	1.83%	1.82%	1.79%	**63.61%**	**69.36%**
- services	5.33%	4.15%	3.22%	3.10%	3.00%	2.30%	2.07%	2.07%	2.04%	39.86%	47.93%
- non market	1.55%	1.72%	2.25%	2.20%	2.00%	1.55%	1.40%	1.35%	1.30%	13.19%	11.97%
- trade	3.67%	2.08%	2.36%	2.35%	2.00%	1.40%	1.30%	1.22%	1.15%	10.57%	9.47%
Agriculture	2.67%	0.06%	0.78%	1.20%	1.00%	0.60%	0.50%	0.40%	0.35%	**1.40%**	**0.89%**
Energy Sector	-0.56%	0.26%	1.08%	1.05%	1.20%	1.10%	1.00%	1.00%	1.00%	**3.61%**	**2.61%**

Table 14-8: Macroeconomic assumptions of baseline scenario for Greece (all numbers are annualized growth rates except if otherwise indicated).

	Observations		Forecasts								
	1985/90	1990/95	1995/00	2000/05	2005/10	2010/15	2015/20	2020/25	2025/30	1995	2030
			Macroeconomic aggregates								
GDP Growth	1.68%	1.40%	2.86%	3.41%	3.71%	3.41%	3.40%	3.36%	3.11%		
Priv. Consumption	2.27%	1.77%	2.14%	2.98%	3.29%	3.03%	3.40%	3.36%	3.11%		
Consumer Price Index	17.37%	13.93%	6.03%	5.27%	4.52%	3.76%	3.01%	2.25%	1.50%		
GDP Deflator	16.11%	13.28%	6.35%	5.54%	4.73%	3.92%	3.12%	2.31%	1.50%		
Exchange Rate ($)	2.77%	7.90%	1.39%	0.00%	0.00%	0.00%	0.00%	0.00%	0.00%		
Population Total	0.31%	0.76%	0.58%	0.35%	0.20%	0.05%	0.01%	0.00%	-0.01%		
Lending Rate (level)	20.50%	17.30%	7.50%	6.00%	5.00%	4.80%	4.60%	4.50%	4.50%		
			Sectoral value added							***Share in value added***	
Manufacturing	1.87%	-1.29%	1.77%	3.04%	3.21%	2.70%	2.64%	2.42%	2.29%	**22.07%**	**17.15%**
- Intensive	1.92%	0.17%	1.69%	2.66%	2.49%	2.14%	1.93%	1.62%	1.41%		
- Metals	0.56%	2.51%	-2.84%	0.20%	0.10%	0.10%	0.00%	0.00%	0.00%	0.86%	0.24%
Iron and Steel	-2.29%	4.38%	-4.25%	-1.37%	-1.60%	-1.75%	-2.04%	-2.28%	-2.57%		
Non-ferrous	1.51%	1.93%	-2.39%	0.65%	0.54%	0.53%	0.42%	0.42%	0.41%		
- Chemicals	3.20%	0.53%	1.70%	2.00%	1.80%	1.50%	1.30%	1.00%	0.70%	2.34%	1.22%
- Paper	0.05%	0.70%	1.50%	1.70%	1.50%	1.30%	1.10%	0.90%	0.70%	1.10%	0.54%
- Building Materials	2.22%	-2.54%	4.89%	5.70%	5.00%	4.00%	3.50%	2.90%	2.60%	1.11%	1.43%
- Other Industries	1.85%	-1.74%	1.80%	3.16%	3.43%	2.87%	2.84%	2.64%	2.52%		
- Food	0.12%	3.65%	2.90%	3.50%	4.30%	4.00%	3.90%	3.70%	3.50%	3.84%	4.34%
- Textiles	-0.61%	-5.60%	-3.50%	0.00%	1.50%	1.70%	2.20%	2.50%	2.70%	2.45%	1.10%
- Engineering	0.93%	-2.34%	1.74%	1.60%	1.50%	1.30%	1.20%	1.20%	1.10%	2.35%	1.21%
- Others	1.32%	-2.26%	-1.00%	1.50%	2.00%	2.00%	1.70%	1.50%	1.30%	2.01%	1.00%
- Construction	4.75%	-2.47%	3.84%	4.70%	4.20%	3.00%	2.90%	2.50%	2.30%	6.03%	6.08%
Services	2.67%	1.94%	4.21%	4.18%	4.30%	3.93%	3.88%	3.84%	3.49%	**59.04%**	**73.56%**
- services	2.99%	4.19%	7.18%	6.27%	5.48%	4.58%	4.45%	4.36%	3.87%	28.91%	53.66%
- non market	2.29%	-0.18%	-0.19%	-0.50%	1.00%	2.00%	2.00%	2.00%	2.00%	18.23%	8.76%
- trade	2.62%	0.34%	2.71%	3.30%	3.50%	3.20%	3.20%	3.10%	2.90%	11.90%	11.14%
Agriculture	-2.27%	2.96%	-1.20%	0.00%	1.20%	1.50%	1.70%	1.80%	1.90%	**15.66%**	**7.02%**
Energy Sector	3.48%	4.81%	3.00%	3.20%	2.70%	2.40%	1.90%	1.50%	1.30%	**3.24%**	**2.27%**

Table 14-9: Macroeconomic assumptions of baseline scenario for Ireland (all numbers are annualized growth rates except if otherwise indicated).

	Observations		Forecasts								
	1985/90	1990/95	1995/00	2000/05	2005/10	2010/15	2015/20	2020/25	2025/30	1995	2030
			Macroeconomic aggregates								
GDP Growth	5.57%	6.30%	5.63%	3.78%	2.67%	2.18%	1.97%	1.87%	1.81%		
Priv. Consumption	5.22%	2.84%	4.47%	3.40%	2.42%	1.98%	1.97%	1.87%	1.81%		
Consumer Price Index	3.28%	2.42%	2.32%	2.18%	2.05%	1.91%	1.77%	1.64%	1.50%		
GDP Deflator	2.80%	1.55%	2.36%	2.22%	2.07%	1.93%	1.79%	1.64%	1.50%		
Exchange Rate ($)	-8.55%	0.68%	-0.29%	0.00%	0.00%	0.00%	0.00%	0.00%	0.00%		
Population Total	-0.21%	0.41%	0.32%	0.40%	0.40%	0.30%	0.25%	0.20%	0.15%		
Lending Rate (level)	12.64%	8.30%	6.20%	6.00%	5.00%	4.80%	4.60%	4.50%	4.50%		
			Sectoral value added							***Share in value added***	
Manufacturing	6.51%	5.28%	5.17%	4.54%	3.05%	2.44%	2.15%	2.05%	1.99%	**33.04%**	**35.48%**
- Intensive	4.46%	7.83%	5.26%	4.55%	2.74%	2.20%	1.70%	1.55%	1.47%		
- Metals	3.78%	-2.87%	5.50%	2.50%	1.10%	0.60%	0.30%	0.10%	0.00%	0.29%	0.18%
Iron and Steel	9.12%	-3.03%	4.60%	1.59%	0.16%	-0.38%	-0.73%	-0.98%	-1.14%		
Non-ferrous	2.95%	-2.84%	5.66%	2.66%	1.25%	0.75%	0.45%	0.25%	0.15%		
- Chemicals	4.33%	11.41%	6.49%	5.50%	3.20%	2.50%	1.90%	1.77%	1.70%	6.21%	7.22%
- Paper	8.59%	7.62%	3.16%	2.70%	1.80%	1.70%	1.40%	1.00%	0.80%	1.25%	0.87%
- Building Materials	2.94%	-0.09%	1.60%	1.50%	1.10%	1.00%	0.80%	0.70%	0.60%	1.57%	0.85%
- Other Industries	7.26%	4.38%	5.13%	4.53%	3.16%	2.53%	2.32%	2.23%	2.17%		
- Food	5.14%	4.01%	2.98%	2.85%	2.20%	1.40%	1.10%	1.00%	0.80%	5.32%	3.68%
- Textiles	3.46%	-0.85%	1.76%	1.40%	1.20%	1.10%	1.00%	0.90%	0.80%	0.97%	0.55%
- Engineering	11.91%	5.85%	7.29%	6.00%	4.00%	3.20%	3.00%	2.85%	2.80%	11.42%	17.81%
- Others	4.91%	4.70%	2.52%	2.50%	1.80%	1.50%	1.40%	1.30%	1.20%	1.68%	1.15%
- Construction	3.11%	2.45%	3.34%	3.00%	2.00%	1.70%	1.30%	1.20%	1.00%	4.32%	3.17%
Services	6.68%	7.34%	6.38%	3.67%	2.63%	2.16%	1.97%	1.87%	1.80%	**57.91%**	**59.44%**
- services	8.61%	8.80%	7.45%	4.00%	2.90%	2.34%	2.20%	2.10%	2.02%	33.23%	38.53%
- non market	2.54%	6.88%	5.20%	3.24%	2.30%	1.85%	1.55%	1.40%	1.35%	15.38%	13.27%
- trade	8.24%	3.51%	4.30%	2.99%	2.00%	1.85%	1.65%	1.55%	1.50%	9.30%	7.64%
Agriculture	2.09%	1.80%	2.24%	1.20%	1.00%	0.90%	0.80%	0.70%	0.50%	**6.41%**	**3.47%**
Energy Sector	-10.56%	10.04%	2.04%	1.80%	1.50%	1.30%	1.10%	1.00%	1.00%	**2.64%**	**1.61%**

Table 14-10: Macroeconomic assumptions of baseline scenario for Italy (all numbers are annualized growth rates except if otherwise indicated).

	Observations		Forecasts								
	1985/90	1990/95	1995/00	2000/05	2005/10	2010/15	2015/20	2020/25	2025/30	1995	2030
			Macroeconomic aggregates								
GDP Growth	3.00%	1.33%	2.12%	2.11%	2.05%	1.64%	1.51%	1.47%	1.49%		
Priv. Consumption	3.61%	0.90%	1.42%	1.78%	1.76%	1.41%	1.52%	1.48%	1.49%		
Consumer Price Index	5.67%	5.11%	2.82%	2.60%	2.38%	2.16%	1.94%	1.72%	1.50%		
GDP Deflator	6.87%	4.99%	2.90%	2.66%	2.43%	2.20%	1.97%	1.73%	1.50%		
Exchange Rate ($)	-8.90%	6.35%	-1.07%	0.00%	0.00%	0.00%	0.00%	0.00%	0.00%		
Population Total	-0.14%	-0.22%	0.08%	0.10%	0.05%	0.00%	0.00%	-0.01%	-0.02%		
Lending Rate (level)	13.71%	11.90%	6.50%	5.00%	4.50%	4.30%	4.30%	4.30%	4.30%		
			Sectoral value added							***Share in value added***	
Manufacturing	3.51%	0.86%	2.12%	1.49%	1.41%	1.28%	1.21%	1.17%	1.15%	**27.96%**	**24.64%**
- Intensive	4.80%	1.04%	1.63%	1.65%	1.56%	1.41%	1.28%	1.24%	1.17%		
- Metals	-0.19%	0.69%	0.96%	0.50%	0.35%	0.25%	0.15%	0.10%	0.10%	0.93%	0.57%
Iron and Steel	-3.82%	1.43%	0.39%	-0.09%	-0.25%	-0.37%	-0.49%	-0.56%	-0.58%		
Non-ferrous	2.09%	0.30%	1.27%	0.81%	0.65%	0.55%	0.44%	0.39%	0.38%		
- Chemicals	6.89%	2.04%	1.00%	2.00%	1.90%	1.70%	1.50%	1.40%	1.30%	2.10%	1.94%
- Paper	4.61%	1.57%	1.82%	1.70%	1.60%	1.40%	1.30%	1.25%	1.20%	1.42%	1.28%
- Building Materials	5.94%	-0.48%	2.68%	1.80%	1.70%	1.60%	1.50%	1.50%	1.40%	1.53%	1.52%
- Other Industries	3.09%	0.81%	2.26%	1.45%	1.36%	1.25%	1.19%	1.15%	1.15%		
- Food	3.65%	1.27%	2.29%	1.80%	1.60%	1.40%	1.30%	1.20%	1.10%	2.34%	2.15%
- Textiles	2.51%	0.98%	2.14%	2.00%	1.70%	1.40%	1.30%	1.20%	1.20%	3.55%	3.30%
- Engineering	4.11%	0.43%	2.33%	1.90%	1.90%	1.80%	1.70%	1.70%	1.70%	7.91%	8.15%
- Others	2.44%	1.19%	1.08%	0.24%	0.20%	0.10%	0.10%	0.00%	0.00%	2.38%	1.40%
- Construction	2.13%	0.89%	2.69%	0.80%	0.70%	0.64%	0.60%	0.55%	0.50%	5.80%	4.33%
Services	2.94%	1.41%	2.26%	2.49%	2.43%	1.85%	1.69%	1.64%	1.67%	**63.22%**	**68.50%**
- services	3.66%	1.73%	2.71%	2.90%	2.88%	2.05%	1.88%	1.83%	1.90%	34.14%	41.02%
- non market	1.17%	1.10%	1.36%	1.95%	1.90%	1.50%	1.40%	1.35%	1.30%	13.79%	12.72%
- trade	3.12%	1.00%	2.05%	2.00%	1.80%	1.65%	1.45%	1.40%	1.35%	15.28%	14.76%
Agriculture	0.10%	2.55%	1.15%	1.20%	1.00%	0.80%	0.70%	0.65%	0.60%	**3.46%**	**2.53%**
Energy Sector	3.22%	2.07%	1.06%	1.30%	1.20%	1.20%	1.10%	1.10%	1.10%	**5.37%**	**4.33%**

Table 14-11: Macroeconomic assumptions of baseline scenario for Netherlands (all numbers are annualized growth rates except if otherwise indicated).

	Observations		Forecasts								
	1985/90	1990/95	1995/00	2000/05	2005/10	2010/15	2015/20	2020/25	2025/30	1995	2030
			Macroeconomic aggregates								
GDP Growth	3.06%	1.82%	2.86%	2.64%	2.54%	2.12%	1.86%	1.78%	1.75%		
Priv. Consumption	2.78%	2.18%	2.32%	2.39%	2.33%	1.94%	1.86%	1.78%	1.75%		
Consumer Price Index	0.74%	2.51%	1.98%	1.90%	1.82%	1.74%	1.66%	1.58%	1.50%		
GDP Deflator	0.86%	2.06%	1.86%	1.80%	1.74%	1.68%	1.62%	1.56%	1.50%		
Exchange Rate ($)	-11.33%	-0.15%	-1.44%	0.00%	0.00%	0.00%	0.00%	0.00%	0.00%		
Population Total	0.63%	0.77%	0.73%	0.45%	0.35%	0.30%	0.18%	0.12%	0.08%		
Lending Rate (level)	7.34%	6.90%	6.00%	6.00%	5.00%	4.80%	4.60%	4.50%	4.50%		
			Sectoral value added							***Share in value added***	
Manufacturing	3.08%	0.92%	2.37%	2.45%	2.30%	1.96%	1.84%	1.69%	1.59%	**23.28%**	**21.78%**
- Intensive	2.62%	1.14%	1.87%	2.12%	2.08%	1.78%	1.65%	1.49%	1.48%		
- Metals	-0.57%	1.63%	1.00%	0.60%	0.60%	0.50%	0.40%	0.30%	0.18%	0.72%	0.40%
Iron and Steel	-0.58%	2.21%	0.43%	0.01%	0.00%	-0.12%	-0.24%	-0.36%	-0.50%		
Non-ferrous	-0.55%	0.41%	2.22%	1.75%	1.69%	1.53%	1.38%	1.23%	1.07%		
- Chemicals	2.75%	1.03%	2.82%	2.80%	2.70%	2.30%	2.20%	2.00%	2.00%	2.83%	3.01%
- Paper	3.61%	0.67%	0.40%	1.50%	1.50%	1.20%	1.00%	0.80%	0.70%	2.13%	1.41%
- Building Materials	2.64%	2.19%	2.80%	2.30%	2.20%	1.90%	1.70%	1.60%	1.60%	0.98%	0.91%
- Other Industries	3.05%	0.84%	2.56%	2.57%	2.38%	2.03%	1.90%	1.76%	1.63%		
- Food	2.57%	1.67%	2.82%	2.20%	2.20%	1.85%	1.56%	1.50%	1.50%	3.24%	2.95%
- Textiles	0.96%	-3.37%	0.41%	0.50%	0.30%	0.20%	0.10%	0.00%	-0.50%	0.51%	0.25%
- Engineering	2.74%	0.32%	3.19%	3.50%	3.20%	2.80%	2.70%	2.50%	2.30%	6.03%	7.56%
- Others	5.34%	0.77%	2.52%	2.00%	2.00%	1.30%	1.00%	0.50%	0.50%	1.40%	1.06%
- Construction	3.41%	1.40%	1.90%	2.00%	1.70%	1.40%	1.30%	1.20%	1.00%	5.43%	4.24%
Services	3.42%	2.04%	3.09%	2.78%	2.72%	2.23%	1.90%	1.85%	1.82%	**65.86%**	**68.57%**
- services	3.99%	2.48%	3.40%	3.14%	3.10%	2.50%	2.08%	2.02%	1.98%	41.79%	47.57%
- non market	1.39%	1.11%	2.32%	2.00%	1.80%	1.50%	1.40%	1.30%	1.30%	10.59%	8.73%
- trade	3.48%	1.46%	2.73%	2.20%	2.10%	1.80%	1.60%	1.60%	1.60%	13.49%	12.27%
Agriculture	4.99%	2.63%	3.50%	3.00%	2.50%	2.20%	2.00%	1.80%	1.80%	**4.44%**	**4.72%**
Energy Sector	-0.90%	2.46%	1.80%	1.60%	1.50%	1.40%	1.30%	1.30%	1.30%	**6.42%**	**4.93%**

Table 14-12: Macroeconomic assumptions of baseline scenario for Portugal (all numbers are annualized growth rates except if otherwise indicated).

	Observations		Forecasts								
	1985/90	1990/95	1995/00	2000/05	2005/10	2010/15	2015/20	2020/25	2025/30	1995	2030
			Macroeconomic aggregates								
GDP Growth	8.04%	1.72%	3.12%	3.45%	3.62%	3.38%	3.10%	3.10%	2.79%		
Priv. Consumption	9.08%	1.75%	2.58%	3.15%	3.33%	3.12%	3.10%	3.10%	2.79%		
Consumer Price Index	11.36%	7.21%	2.88%	2.65%	2.42%	2.19%	1.96%	1.73%	1.50%		
GDP Deflator	14.18%	8.96%	3.14%	2.87%	2.59%	2.32%	2.05%	1.77%	1.50%		
Exchange Rate ($)	-3.49%	1.09%	0.32%	0.00%	0.00%	0.00%	0.00%	0.00%	0.00%		
Population Total	-0.20%	0.22%	0.20%	0.17%	0.10%	0.05%	0.01%	0.01%	0.00%		
Lending Rate (level)	20.75%	7.50%	4.10%	6.00%	5.00%	4.80%	4.60%	4.50%	4.50%		
			Sectoral value added							***Share in value added***	
Manufacturing	7.36%	-0.80%	1.39%	2.71%	3.08%	2.58%	2.35%	2.19%	2.02%	**28.04%**	**20.70%**
- Intensive	10.67%	-0.56%	1.46%	2.84%	2.54%	2.03%	1.52%	1.17%	1.10%		
- Metals	19.62%	0.16%	0.08%	0.30%	0.20%	0.00%	0.00%	-0.20%	-0.50%	0.93%	0.31%
Iron and Steel	28.22%	-2.39%	-0.63%	-0.44%	-0.57%	-0.80%	-0.83%	-1.06%	-1.40%		
Non-ferrous	15.29%	1.79%	0.47%	0.68%	0.58%	0.37%	0.36%	0.15%	-0.15%		
- Chemicals	11.40%	-5.29%	-0.30%	0.50%	0.45%	0.30%	0.20%	0.15%	0.12%	1.62%	0.57%
- Paper	8.82%	-0.62%	-1.00%	1.50%	1.25%	0.70%	0.50%	0.30%	0.20%	1.49%	0.58%
- Building Materials	8.06%	3.72%	4.68%	5.50%	4.50%	3.50%	2.50%	1.90%	1.80%	2.19%	2.39%
- Other Industries	6.04%	-0.87%	1.36%	2.67%	3.24%	2.73%	2.57%	2.45%	2.24%		
- Food	9.40%	2.13%	3.12%	3.50%	4.00%	3.50%	3.20%	3.10%	2.80%	6.66%	6.88%
- Textiles	1.19%	-2.79%	-1.17%	1.50%	2.50%	2.20%	2.00%	1.90%	1.80%	4.72%	2.64%
- Engineering	8.76%	-3.91%	-1.76%	-0.50%	1.00%	1.20%	1.40%	1.50%	1.50%	3.83%	1.56%
- Others	6.40%	-2.53%	-0.61%	1.00%	1.90%	2.00%	2.00%	2.00%	2.00%	1.59%	0.87%
- Construction	5.96%	0.56%	3.94%	4.50%	4.00%	2.80%	2.60%	2.30%	2.00%	5.00%	4.90%
Services	8.74%	3.14%	4.07%	3.94%	4.00%	3.82%	3.48%	3.51%	3.10%	**62.63%**	**73.72%**
- services	9.71%	4.37%	5.46%	5.31%	5.22%	4.88%	4.30%	4.29%	3.63%	30.40%	50.44%
- non market	8.52%	2.31%	2.65%	2.60%	2.40%	2.30%	2.20%	2.10%	2.00%	15.04%	11.07%
- trade	7.49%	1.83%	2.71%	2.20%	2.42%	2.20%	2.00%	2.00%	2.00%	17.19%	12.21%
Agriculture	3.15%	-0.54%	1.34%	1.70%	2.00%	1.80%	1.70%	1.70%	1.70%	**5.54%**	**3.30%**
Energy Sector	19.58%	2.91%	2.06%	2.00%	1.80%	1.70%	1.60%	1.50%	1.50%	**3.79%**	**2.28%**

Table 14-13: Macroeconomic assumptions of baseline scenario for Spain (all numbers are annualized growth rates except if otherwise indicated).

	Observations		Forecasts								
	1985/90	1990/95	1995/00	2000/05	2005/10	2010/15	2015/20	2020/25	2025/30	1995	2030
			Macroeconomic aggregates								
GDP Growth	8.04%	1.72%	3.12%	3.45%	3.62%	3.38%	3.10%	3.10%	2.79%		
Priv. Consumption	9.08%	1.75%	2.58%	3.15%	3.33%	3.12%	3.10%	3.10%	2.79%		
Consumer Price Index	11.36%	7.21%	2.88%	2.65%	2.42%	2.19%	1.96%	1.73%	1.50%		
GDP Deflator	14.18%	8.96%	3.14%	2.87%	2.59%	2.32%	2.05%	1.77%	1.50%		
Exchange Rate ($)	-3.49%	1.09%	0.32%	0.00%	0.00%	0.00%	0.00%	0.00%	0.00%		
Population Total	-0.20%	0.22%	0.20%	0.17%	0.10%	0.05%	0.01%	0.01%	0.00%		
Lending Rate (level)	20.75%	7.50%	4.10%	6.00%	5.00%	4.80%	4.60%	4.50%	4.50%		
			Sectoral value added							***Share in value added***	
Manufacturing	7.36%	-0.80%	1.39%	2.71%	3.08%	2.58%	2.35%	2.19%	2.02%	**28.04%**	**20.70%**
- Intensive	10.67%	-0.56%	1.46%	2.84%	2.54%	2.03%	1.52%	1.17%	1.10%		
- Metals	19.62%	0.16%	0.08%	0.30%	0.20%	0.00%	0.00%	-0.20%	-0.50%	0.93%	0.31%
Iron and Steel	28.22%	-2.39%	-0.63%	-0.44%	-0.57%	-0.80%	-0.83%	-1.06%	-1.40%		
Non-ferrous	15.29%	1.79%	0.47%	0.68%	0.58%	0.37%	0.36%	0.15%	-0.15%		
- Chemicals	11.40%	-5.29%	-0.30%	0.50%	0.45%	0.30%	0.20%	0.15%	0.12%	1.62%	0.57%
- Paper	8.82%	-0.62%	-1.00%	1.50%	1.25%	0.70%	0.50%	0.30%	0.20%	1.49%	0.58%
- Building Materials	8.06%	3.72%	4.68%	5.50%	4.50%	3.50%	2.50%	1.90%	1.80%	2.19%	2.39%
- Other Industries	6.04%	-0.87%	1.36%	2.67%	3.24%	2.73%	2.57%	2.45%	2.24%		
- Food	9.40%	2.13%	3.12%	3.50%	4.00%	3.50%	3.20%	3.10%	2.80%	6.66%	6.88%
- Textiles	1.19%	-2.79%	-1.17%	1.50%	2.50%	2.20%	2.00%	1.90%	1.80%	4.72%	2.64%
- Engineering	8.76%	-3.91%	-1.76%	-0.50%	1.00%	1.20%	1.40%	1.50%	1.50%	3.83%	1.56%
- Others	6.40%	-2.53%	-0.61%	1.00%	1.90%	2.00%	2.00%	2.00%	2.00%	1.59%	0.87%
- Construction	5.96%	0.56%	3.94%	4.50%	4.00%	2.80%	2.60%	2.30%	2.00%	5.00%	4.90%
Services	8.74%	3.14%	4.07%	3.94%	4.00%	3.82%	3.48%	3.51%	3.10%	**62.63%**	**73.72%**
- services	9.71%	4.37%	5.46%	5.31%	5.22%	4.88%	4.30%	4.29%	3.63%	30.40%	50.44%
- non market	8.52%	2.31%	2.65%	2.60%	2.40%	2.30%	2.20%	2.10%	2.00%	15.04%	11.07%
- trade	7.49%	1.83%	2.71%	2.20%	2.42%	2.20%	2.00%	2.00%	2.00%	17.19%	12.21%
Agriculture	3.15%	-0.54%	1.34%	1.70%	2.00%	1.80%	1.70%	1.70%	1.70%	**5.54%**	**3.30%**
Energy Sector	19.58%	2.91%	2.06%	2.00%	1.80%	1.70%	1.60%	1.50%	1.50%	**3.79%**	**2.28%**

Table 14-14: Macroeconomic assumptions of baseline scenario for Sweden (all numbers are annualized growth rates except if otherwise indicated).

	Observations		Forecasts								
	1985/90	1990/95	1995/00	2000/05	2005/10	2010/15	2015/20	2020/25	2025/30	1995	2030
			Macroeconomic aggregates								
GDP Growth	2.80%	0.92%	2.41%	2.30%	2.14%	1.62%	1.44%	1.42%	1.41%		
Priv. Consumption	2.39%	-0.97%	2.02%	2.12%	1.98%	1.50%	1.44%	1.42%	1.41%		
Consumer Price Index	6.21%	4.20%	2.28%	2.15%	2.02%	1.89%	1.76%	1.63%	1.50%		
GDP Deflator	6.99%	3.64%	2.30%	2.17%	2.03%	1.90%	1.77%	1.63%	1.50%		
Exchange Rate ($)	-7.21%	3.81%	-1.69%	0.00%	0.00%	0.00%	0.00%	0.00%	0.00%		
Population Total	0.57%	0.60%	0.58%	0.35%	0.20%	0.05%	0.01%	0.00%	0.00%		
Lending Rate (level)	16.89%	10.20%	6.50%	6.00%	5.00%	4.80%	4.60%	4.50%	4.50%		
			Sectoral value added							***Share in value added***	
Manufacturing	1.80%	0.70%	2.35%	2.37%	2.20%	1.58%	1.37%	1.39%	1.39%	**31.35%**	**31.20%**
- Intensive	1.90%	2.25%	3.22%	2.93%	2.70%	1.78%	1.53%	1.52%	1.45%		
- Metals	0.79%	2.45%	0.80%	0.50%	0.30%	0.15%	0.00%	-0.20%	-0.50%	1.50%	0.84%
Iron and Steel	1.00%	2.22%	0.23%	-0.09%	-0.30%	-0.47%	-0.64%	-0.86%	-1.18%		
Non-ferrous	0.07%	3.23%	2.63%	2.17%	1.84%	1.58%	1.33%	1.05%	0.67%		
- Chemicals	3.18%	5.42%	2.35%	2.00%	1.80%	1.30%	1.00%	0.90%	0.90%	2.61%	2.31%
- Paper	1.51%	1.84%	5.10%	4.50%	4.00%	2.50%	2.20%	2.20%	2.10%	3.56%	5.75%
- Building Materials	2.41%	-7.83%	0.22%	0.50%	0.30%	0.00%	-0.50%	-0.50%	-1.00%	0.45%	0.23%
- Other Industries	1.76%	0.19%	2.04%	2.16%	2.00%	1.50%	1.30%	1.34%	1.36%		
- Food	-0.96%	1.01%	1.50%	1.00%	0.80%	0.60%	0.40%	0.30%	0.20%	2.29%	1.55%
- Textiles	-3.44%	-4.91%	0.06%	0.50%	0.30%	0.00%	-0.50%	-0.50%	-1.00%	0.36%	0.18%
- Engineering	1.84%	1.70%	3.35%	3.50%	3.20%	2.30%	2.00%	2.00%	2.00%	11.45%	15.06%
- Others	1.31%	-1.21%	0.60%	1.00%	0.50%	0.50%	0.00%	0.00%	0.00%	2.17%	1.31%
- Construction	3.07%	-1.58%	0.50%	0.30%	0.20%	0.10%	0.10%	0.10%	0.10%	6.95%	3.97%
Services	3.43%	1.14%	2.56%	2.37%	2.20%	1.71%	1.54%	1.49%	1.47%	**63.03%**	**64.91%**
- services	3.85%	1.25%	2.96%	2.90%	2.70%	2.08%	1.85%	1.82%	1.80%	33.58%	39.60%
- non market	2.34%	1.56%	2.10%	1.70%	1.50%	1.10%	1.00%	0.90%	0.85%	19.77%	16.56%
- trade	4.24%	-0.05%	2.06%	1.80%	1.70%	1.40%	1.30%	1.20%	1.20%	9.69%	8.75%
Agriculture	1.72%	-0.62%	0.46%	0.50%	0.30%	0.20%	0.15%	0.10%	0.10%	**2.71%**	**1.58%**
Energy Sector	1.77%	0.09%	1.64%	1.50%	1.40%	1.00%	0.90%	0.80%	0.80%	**2.91%**	**2.31%**

Table 14-15: Macroeconomic assumptions of baseline scenario for United Kingdom (all numbers are annualized growth rates except if otherwise indicated).

	Observations		Forecasts								
	1985/90	1990/95	1995/00	2000/05	2005/10	2010/15	2015/20	2020/25	2025/30	1995	2030
			Macroeconomic aggregates								
GDP Growth	3.25%	1.27%	2.85%	2.59%	2.54%	1.95%	1.66%	1.61%	1.72%		
Priv. Consumption	4.37%	1.00%	3.04%	2.68%	2.61%	2.01%	1.66%	1.61%	1.72%		
Consumer Price Index	5.92%	3.24%	2.44%	2.28%	2.13%	1.97%	1.81%	1.66%	1.50%		
GDP Deflator	5.55%	3.71%	2.54%	2.37%	2.19%	2.02%	1.85%	1.67%	1.50%		
Exchange Rate ($)	-6.29%	2.40%	0.19%	0.00%	0.00%	0.00%	0.00%	0.00%	0.00%		
Population Total	0.28%	0.27%	0.14%	0.15%	0.10%	0.05%	0.01%	0.01%	0.00%		
Lending Rate (level)	12.33%	5.57%	4.77%	5.00%	4.50%	4.30%	4.30%	4.30%	4.30%		
			Sectoral value added							***Share in value added***	
Manufacturing	4.20%	0.41%	2.35%	2.10%	1.87%	1.03%	0.96%	0.97%	1.04%	27.79%	22.18%
- Intensive	4.55%	1.38%	2.06%	1.82%	1.63%	1.06%	0.81%	0.81%	0.78%		
- Metals	4.15%	-1.58%	0.45%	0.40%	0.75%	0.50%	0.15%	0.10%	0.10%	0.68%	0.37%
Iron and Steel	4.13%	-0.73%	0.17%	0.11%	0.46%	0.20%	-0.15%	-0.20%	-0.21%		
Non-ferrous	4.21%	-3.89%	1.28%	1.19%	1.52%	1.24%	0.86%	0.78%	0.76%		
- Chemicals	3.90%	2.87%	1.83%	1.80%	1.60%	1.10%	0.75%	0.75%	0.70%	2.23%	1.63%
- Paper	6.01%	1.52%	2.74%	2.20%	2.00%	1.20%	1.00%	1.00%	1.00%	2.34%	1.94%
- Building Materials	2.75%	-0.28%	2.05%	1.80%	1.20%	0.90%	0.80%	0.80%	0.75%	0.76%	0.55%
- Other Industries	2.36%	0.31%	2.43%	2.17%	1.94%	1.02%	1.00%	1.01%	1.11%		
- Food	1.48%	0.46%	2.44%	1.80%	1.30%	0.80%	0.70%	0.60%	0.50%	4.11%	2.94%
- Textiles	-0.72%	-2.29%	2.23%	1.50%	1.00%	0.60%	0.55%	0.50%	0.45%	1.11%	0.74%
- Engineering	0.00%	2.05%	3.30%	3.30%	3.10%	1.50%	1.50%	1.50%	1.70%	8.87%	9.31%
- Others	5.10%	0.21%	1.08%	1.00%	0.50%	0.24%	0.25%	0.30%	0.30%	1.74%	1.00%
- Construction	6.50%	-1.57%	1.50%	1.00%	0.80%	0.50%	0.45%	0.50%	0.50%	5.96%	3.70%
Services	4.01%	1.24%	3.22%	2.94%	2.93%	2.40%	1.98%	1.89%	2.00%	**63.85%**	**71.98%**
- services	6.09%	3.16%	3.87%	3.53%	3.50%	2.78%	2.22%	2.00%	2.19%	40.67%	52.35%
- non market	-0.36%	-4.13%	1.54%	1.50%	1.50%	1.50%	1.40%	1.70%	1.65%	11.53%	9.42%
- trade	4.25%	1.15%	2.50%	2.00%	1.90%	1.50%	1.40%	1.50%	1.40%	11.65%	10.20%
Agriculture	0.97%	1.31%	1.70%	1.20%	1.10%	0.70%	0.50%	0.40%	0.20%	**1.51%**	**0.96%**
Energy Sector	-1.33%	4.92%	1.60%	1.40%	1.30%	0.90%	0.80%	1.00%	1.00%	**6.86%**	**4.88%**

15 APPENDIX II: Brief Description of Models Used in the Study[46]

The POLES and PRIMES models have been developed within a series of collaborative research projects partially funded by the European Commission, Joule Programme of DGXII. Pierre Valette Head of Unit F1 of DGXII has supervised on behalf of the European Commission. In 1998 both of the models have been successfully peer-reviewed by external referees in the context of a special procedure organised by DG XII.

15.1 The PRIMES Model

15.1.1 Introduction

The history and the status of operationality of the PRIMES energy system model for the EU-15 countries in presented in this chapter.

From the very begnning, in 1993-1994, the PRIMES energy model was designed to focus on market-related mechanisms influencing the evolution of energy demand and supply and the context for technology penetration in the market. The PRIMES model also was designed to serve as an energy policy analysis tool including the relationships between energy policy and technology assessment.

Detailed technological models, often categorised in bottom-up approaches, have been among the first ever constructed energy system models. This was also the case of the models developed in the context of DG XII, as for example EFOM and MEDEE. A continuity from older technological (or engineering explicit) energy model has been set as one of the main requirements for PRIMES.

[46] Detailed manuals of the models contained in this Appendix can be found in the full report under the project JOS3-CT95-0008.

The older models have formulated a single, global mathematical programming (optimisation) problem that covers the energy system. Examples are EFOM, MARKAL, BESOM etc.. In policy analysis, these models have been criticised for the lack of explicit representation of markets, the absence of market-based policy instruments and the lack of realism in formulating demand and the "individual" behaviour of demanders or suppliers.

The efficiency gap problem illustrates the methodological failure of such bottom-up models, being unable to represent seemingly non cost-effective choices of consumers or producers. Micro-economic analyses suggested that the gap can be explained by specific conditions that prevail in the markets (distortions, barriers, etc.) and by the individual behaviour of economic agents (for instance, small consumers may use high subjective discount rates). In the context of technology assessment applications, it has been also stated that the lack of market mechanisms in the models may bring serious biases in the projections and the accurate estimation of future potential of technologies.

These critics, together with the need to represent the growing process of market liberalisation, motivated analysts to adopt market-oriented modelling approaches that involve explicitly market regimes and model separately the behaviour of economic agents. Such models can then incorporate detailed representations of policy instruments and structural options that may endogenously formulate the efficiency gap problem. These models are often called "new generation models" and currently prevail in policy analysis studies. Examples, which also include the modelling of energy-environment interactions, are IFFS, GEMS, GEMINI, ENPEP, NEMS (all in USA). PRIMES development for the European Union belongs to same family of models and should be characterised as a hybrid model combining engineering-orientation with economic market-driven representations.

These new generation models are often characterised as partial equilibrium models because they cover only the energy system and not the rest of the economy. They are also called generalised equilibrium models because they can formulate the behavioural conditions for the economic agents in a variety of mathematical formulations for the sub-models, and represent different market clearing regimes, reflected in the choice of algorithm for global model convergence (equilibrium). PRIMES can also be characterised as partial and generalised equilibrium model.

The development of PRIMES has required intensive research work. The modular design of the model, required first to develop the sub-models which cover demand and supply behaviour of the economic agents acting in the energy markets. The modules had to be designed simultaneously, in order to achieve consistency at the level of market integration. The cycle between construction of sub-models and integration has been repeated many times, as the overall model design proved to be more complex than initially planned. Hence, although several versions of the sub-models have been implemented (1995-1997), the integrated model ran as such for the first time in May 1997. We have named that model "version 1" of PRIMES. Version 2 of PRIMES has been released in June 1998.

The model version 1 has been used in 1997 in the evaluation of the set of policies and measures envisaged by the European Commission as the negotiation position of the EU for the Kyoto conference for climate change. The experience with the PRIMES use, proved to be very helpful for the modellers. Several model limitations have been depicted. They concerned the data and sectoral classifications, the lack of cost-supply responsiveness of the primary and secondary transformation sub-models, and the difficulty to control the interaction between centralised and independent power and steam producers.

These limitations gave rise to new important developments of the model, which has been profoundly redesigned in all the sub-models and the integration. The new version of the model (version 2 of PRIMES) adopted also a major innovation for the mathematical formulation and the integration, namely the non linear mixed complementarity (MCP) formulation. It is the first time that a market oriented, detailed energy model is formulated in MCP and solved through the GAMS/PATH algorithm. This choice considerably improved the completeness of the model design and the consistency of interactions between supply and demand. Also, it allowed a shift away from the linear programming limitations, permitting the incorporation of non linear mechanisms that relate to markets, resources and technological developments.

The second version of PRIMES is fully operational and calibrated on 1995 dataset for all European Union member states.

15.1.2 Scope and Objectives

PRIMES is a modeling system that simulates a market equilibrium solution for energy supply and demand in the European Union (EU) member states. The whole model is written in the GAMS algebraic specification language and involves Excel-based input and output interfaces. The model is applicable to any other, single, or multi- country by changing the data, the definition of sets (classification, technologies etc.) and eventually adapting the code.

In the current EU application of PRIMES, the energy market equilibrium is simultaneously achieved in the EU taken as a whole, while all country-specific energy systems are also balanced the model finds the equilibrium by finding the prices of each energy form such that the quantities producers find best to supply match the quantities consumers wish to use. The equilibrium is static (within each time period) but repeated in a time-forward path, under dynamic relationships.

The model design is explicit and detailed regarding the representation of energy and pollution abatement technologies. The system reflects considerations about market economics, industry structure, energy/environmental policies and regulation. These are conceived so as to influence market behaviour of energy system agents. The modular structure of PRIMES reflects a distribution of decision making among agents that decide individually about their supply, demand, combined supply and demand, and prices.

PRIMES is a general purpose model. It is conceived for forecasting, scenario construction and policy impact analysis. It covers a medium to long term horizon but it is not designed for the very long term. It is modular and allows either for a unified model use or for partial use of modules to support specific energy studies.

The model can support policy analysis in the following fields:

- environmental issues
- pricing policy, taxation, regulation and other policy instruments
- new technologies and renewable sources
- energy efficiency in the demand-side
- alternative fuels
- energy trade and EU energy provision
- conversion decentralisation, as well as
- traditional issues regarding electricity generation, gas distribution and refineries.

15.1.3 PRIMES Model Design

A fundamental assumption in PRIMES is that producers and consumers both respond to changes in price. The factors determining the demand for and the supply of each fuel are analysed and represented, so they form the demand and/or supply behaviour of the agents. Through an iterative process, the model determines the economic equilibrium for each fuel market. Price-driven equilibrium is considered in all energy and environment markets, including Europe-wide clearing of oil and gas markets, as well as Europe-wide networks, such as the Europe-wide power grid and natural gas network.

Due to the heterogeneity of the energy market no single methodology can adequately describe all demand, supply and conversion processes. On the other hand, the economic structure of the energy system itself facilitates its representation through largely separable individual units, each performing a number of individual functions.

Based on these principles, *PRIMES is organised around a modular design representing in a different manner fuel supply, energy conversion and end-use of demand sectors*. The individual modules vary in the depth of their structural representation. The modularity feature allows each sector to be represented in the way considered appropriate, highlighting the particular issues important for the sector, including the most expedient regional structure. The electricity module covers the whole Europe, while representing chronological load curves and dispatching at the national level. The natural gas market also expands over the whole Europe. However, coal supply, refineries and demand operate at the national level. Furthermore, the modularity allows any single sector or group of sectors to be run independently for stand alone analysis.

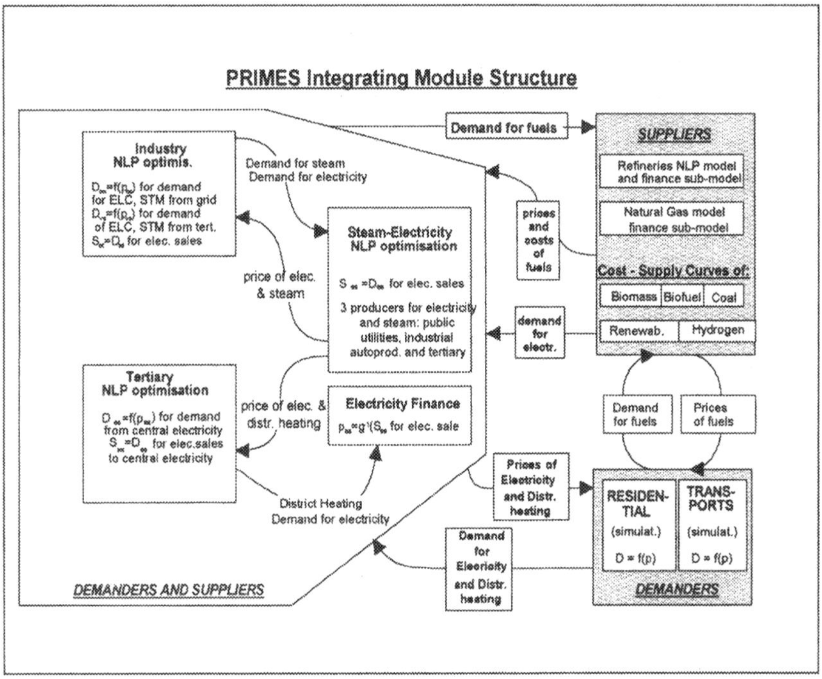

Figure 15-1: PRIMES integrating module structure.

The model is organised by energy production sub-system (oil products, natural gas, coal, electricity and heat production, others) for supply and by end-use sectors for demand (residential, commercial, transports, six industrial sectors). Some demanders may be also suppliers, as for example industrial cogenerators of electricity and steam.

The different modules interact via the exchange of fuel quantities and prices, leading to the global equilibrium of the system.

15.1.3.1 Model Formulation

At the global level, that is the market clearing level, the formulation of the model corresponds to a market equilibrium of the page:

- Demand = Function (Price)
- Supply = Demand
- Price = Inverse Function (Supply)

The behaviour in the supply-side, corresponding to cost minimisation, is formulated as a set of *non linear optimisation models* and the demand-side has the form of a system of (non linear) equations, hence the equilibrium model can be written:

Solve for		x,q,p,u	that satisfy:
Supply side:		Min $c \cdot x$	
	s.t.	$A \cdot x \leq b,$	$x \in X$
Demand-side:		$q = Q(p)$	
Cost Evaluation:	$u = f$ (c, x and other factors)		
Equilibrium Condition:	$p = u$ + taxes		

where x and q denote supply and demand quantities, while u and p stand for producer and consumer prices.

The supply-side may include more than one non linear mathematical programming problems corresponding to the behaviour of several supplying agents (for example, one for refineries, one for gas and one for electricity). In addition, the possibility that some suppliers of energy commodities may also be demanders for other energy commodities (for example, the electricity sector) is included in optimisation modules.

The design principles adopted for PRIMES follow the steps as below.

1. We define which are the agents, the commodities and the markets; we proceed to an activity analysis, i.e. define supplying and demanding activities of agents for each commodity.
2. We choose a mathematical formulation of each agent's behaviour and we specify the corresponding sub-model, while respecting activity analysis defined in the previous step; a sub-model may be a mathematical programming model (linear, non linear or mixedinteger) or a system of (non linear) equations; all sub-models consider commodity prices are given; the supply sub-models have to correspond to cost minimisation and should satisfy demand; the demand sub-models must relate demand quantities to commodity prices; a supply sub-model may include formulations of market allocation

among many suppliers, if necessary (for example between domestic production and imports).

3. We define a price setting mechanism for each commodity; this may be based on average cost or marginal cost pricing and may include any other external factors influencing prices (for example world-wide leading prices); it is interesting to note that this part of the model can reflect realistic situations concerning price-setting regimes (excluding artificial price-setting based on shadow prices); the price setting mechanisms should be related, but not exclusively, to the results of the supply sub-modules.

4. We define the equilibrium conditions for all commodities by equalising producer prices (from step 3) and consumer prices (used in demand sub-models in step 2); taxes and subsidies may be included in these equilibrium conditions, such as excise taxes, VAT, carbon-tax and so on (other types of regulations must be incorporated in the corresponding submodels at the agent level).

The model is formulated as a global non linear mixed complementarity problem (NMCP) that includes all the equations/inequalities of the different modules and replaces the objective functions of each module with the Karush-Kuhn-Tucker first order optimality conditions. The exact formulation of PRIMES in NMCP format is presented in a following chapter on the mathematical structure of the model integration.

Alternative model formulations (as followed for IFFS, NEMS and the previous version of PRIMES) adopted a Gauss-Seidel or Jacobi procedure. The solution algorithm for the overall model, in this case, would involve an iteration procedure. starting from an initial guess of the vector of commodity prices, demand quantities are computed from sub-models corresponding to demanders; mathematical programming models for supplying agents are then solved constrained to demand quantities computed before; based on the results of the supply sub-models, the cost evaluation equations compute produced prices, which augmented with taxes are used to evaluate a new approximation to consumer prices; these are compared to the ones used in the previous iteration and if they are found close enough the process is terminated, otherwise the process restarts by re-computing demand. Such a solution algorithm is not only time consuming, but also due to the non convexities and non linearity present in the sub-modules no guarantee of convergence exists.

15.1.3.2 Features of Sub Models

The supply modules simulate both the operation and the capacity expansion activities. The dynamic relationships involve stock-flow relations (for example capital accumulation), inertia in the penetration of new technologies, backward looking expectations (more formally, the model uses adaptive expectations) and consumer habits. Thus, the model integrates a static and a dynamic solution under myopic anticipation.

Also, the model fully integrates the national within the multinational energy system (for oil refinery, gas supply to Europe and generation and trade of electricity).

Demand is evaluated at a national level. Electricity dispatching and capacity expansion are determined at a national level, depending however on a complex market allocation mechanism, operating through the electricity grid, Europe-wide. The natural gas distribution market clears at a multinational level, even wider than the European Union. The refinery sector operates at a national level, but capacities, market shares and prices depend heavily on Europe-wide competition. Primary energy supply, for example coal and lignite supply curves, have, on the other hand, a national-specific character. Finally, energy savings, technology progress in power generation, abatement technologies, renewables and alternative fuels (biomass, methanol, hydrogen) are determined at each country-specific energy system.

Cost evaluation modules and price-setting mechanisms are at the core of the model. The former are attached to each energy supply module. The cost module considers total revenue requirements of the sector (based on total costs and other accounting costs) and allocates payments over the consumers, according to a general Ramsey pricing rules (parameters are user selected). The pricing parameters reflect alternative industrial economic circumstances and are linked to marginal and average values from the sector's optimisation. For example, these rules consider a peak-synchronisation characterisation of consumers or average cost rating of energy demand by consumers. The allocation of payments is further determined, by also considering eventual cross-subsidization policy or other distortions. In brief, the price-setting mechanism reflects the design considerations for the market clearing regimes. The value of parameters in these cost-pricing modules can be altered, in policy scenarios, to reflect structural change.

Prices of purchased fuels depend also on cost-supply curves that are exogenously specified, but operate within the equilibrium process. Such curves are used for all primary energy supply, including EU gas supply, coal, biomass and even renewable sources to reflect land availability constraints. They are also defined for imports.

15.1.3.3 Technology

As mentioned, PRIMES has been designed to support technology assessment at the energy system level. The dynamics, as simulated by the model, influence the penetration of new technologies.

Several parameters and formulations are built-in to represent non economic factors that affect the velocity of new technology penetration. For example, the modules include learning by doing curves, parameters that represent subjective perception of technology costs as seen by consumers, standards, etc. These can be used to represent market failures or inertia that may deprive the system from cost-effective technology solutions.

In addition, market related factors, as represented within the optimisation modules, can also explain the lack of decision for the most cost-effective solutions. These factors are related to the individual character of decision maker's optimality and this is represented in the model by design (different optimality conditions per module) and through the use of parameters, as for example by varying the discount rates with the consumer size.

Policy parameters can of course change the optimality conditions and influence technology choice and penetration. The model can in addition simulate accompanying policies that aim at structural improvements that may maximize the effects of policy measures. For example, true cost pricing, removal of barriers, new funding mechanisms etc. can be reflected to changes of parameters that will influence technology choices and penetration.

15.1.3.4 Environment

The mechanisms relating pollution with energy activities, also involving pollution abatement choices, are fully integrated into PRIMES. The optimisation modules simultaneously consider energy and environment costs. Constraints are built in to represent environmental regulation. The technology choice mechanisms also consider abatement equipment. Policy measures dedicated to pollution can affect optimality and can also be accompanied with policy aiming at structural change. Finally, a module computes dispersion and deposition of emitted pollutants.

The main policy instruments for the environment considered in PRIMES are:

- Regulation by sector (in the form of a constraint of emissions by sector);
- Regulation by country (in the form of a global constraint taking into account emissions from all modules);
- Taxation for the environment. This can be either exogenously given (in which case the emissions are not explicitly limited) or endogenously (as the shadow price of the constraint binding the emissions);
- Pollution permits. A separate market for pollution permits is implemented in the model. The different sectors can therefore trade (sell or buy) permits based on their initial endowment;
- Subsidisation of abatement costs for electricity and steam.

15.1.3.5 Policy Instruments

Special care has been devoted to the representation of various policy instruments in the model. For some policy instruments, it is straightforward to built scenario variants and evaluate implications. For other instruments, the analysis is more sophisticated and has to combine evaluations outside the model with results from model runs.

Economic and fiscal instruments constitute an obvious case of straightforward use of the model. Taxes, excise, VAT, carbon etc., are explicitly represented for all energy forms and uses. Fully detailed tax scenarios can be assessed, including differentiation of rates by sector, combination with subsidies and exemptions, harmonisation across member states, etc. The consequences of higher taxation for costs of derived energy forms (e.g. steam, electricity) is endogenously treated.

Other economic instruments, like the tradable emission rights (pollution permits) are also formulated in PRIMES. Other measures such as new funding mechanisms for energy technologies, information campaigns and measures aiming at removal of barriers, can be evaluated at the energy system level (regarding their total effects) through the built in mechanisms of PRIMES, like perceived costs, risk premium, etc.

Command and control regulation, that is the pursuit of objectives through administrative processes, can be analyzed through the use of constraints and binding within the optimization modules. The model can evaluate the effectiveness and compute proxies to the shadow cost of regulation. Emission norms, efficiency norms, regulations such as the "Non Fossil Fuel Obligation" can be represented and analyzed.

Voluntary agreements are one of the cases for which the model-based analysis must be combined with ad hoc evaluation. Voluntary agreements can be represented as constraints within the optimization modules. However, in reality, they are not necessarily imperative constraints, since deviations may be possible, although involving higher costs for the consumer. In such cases, voluntary agreements are possible when deviations are threatened by the risk of considerably higher costs. PRIMES has not such a mechanism built in. However, after evaluations outside the model, the analyst can formulate constraints, do sensitivity analysis with the model and compare shadow prices to known higher cost threats.

Demand-side Management and Integrated Resource Planning is also another example for which the model is not entirely sufficient. PRIMES, being explicit in technology representations, includes electricity consumption technologies and uses in all consumption modules. To each use, the model associates generic load patterns, the aggregation of which over the consumers' electricity uses gives the load shape faced by electricity generation. A DSM measure can be simulated by a change either in the shape or the area (efficiency) of a particular electricity use. This will alter the optimality conditions of electricity generation and will probably imply cost savings. Externally to the model, the analyst has to evaluate implementation costs of the measure and allocate the bearing of the costs between the consumer and the generator. In such a way, he can carry out cost-benefit analysis to evaluate DSM measures. The concept of Integrated Resource Planning seems now old fashioned within the on going liberalization of markets[47]. At the energy system level, PRIMES is a complete tool for IRP evaluations, but the model is totally inadequate if IRP is to be carried out at the generator level.

To study the general issue of internalization of externalities one has to use an accounting framework for externalities[48], consider internalization through economic instruments (thus compare them for effectiveness, as mentioned before) or define a regulation scheme that will oblige the actors to take into account external costs. Total cost pricing was recently brought up in the debate as a means to regulate decision making. Total cost (that is including external costs) can be imposed in all optimization modules of PRIMES. This will influence technology choice and pricing throughout the system.

15.1.4 The PRIMES Modules

15.1.4.1 Model Nomenclature

REGIONS	• 15 European Union countries
DEMAND	
Residential	• 4 energy uses, 4 types of households, 2 technology vintages, several types of electric appliances
Commercial	• 4 sectors, 7 energy uses, full technology vintages, thermal integrity, heat and cooling production, energy savings

[47] If the concept is limited to the obligation of generators to do IRP, instead of the society.
[48] For example, EXTERNE results.

Industry	• 9 industrial sectors, separately modelled: iron and steel, non ferrous, chemicals, building materials, paper and pulp, food drink tobacco, engineering, textiles, other industries other industries
	• Distinction between four types of sub-processes (as appropriate for each industrial sector): basic processing, scrap/recycled material processing, manufacturing, product finishing
	• Eleven different energy uses defined: electricity uses, electrolysis uses, electrical furnaces, electrical arc, furnaces, blast furnaces, kilns, steam uses, heat uses, motor energy uses and raw materials uses
	• Full technology vintages, heat production and recovery, energy savings and pollution abatement
Transports	• Transport by purpose: urban passenger transport, non urban passenger transport, freight transport
	• Transport modes: for urban passengers car, public transport, motorcycle, non urban passengers: car, bus, rail, air, navigation; for freight transport truck, rail, air, navigation.
	• 6 to 10 alternative technologies for each mode (car, bus, truck); more limited number of alternatives for rail, air and navigation
Electricity production	• 148 different plant types per country for the existing thermal plants; ·678 different plant types per country for the new thermal plants; 3 different plant types per country for the existing reservoir plants; 30 different plant types per country for the existing intermittent plants
	• Chronological load curves, interconnections, network representation; three typical companies per country;
	• Cogeneration of power and steam, district heating
Refineries	• 4 refineries with typical refinery structure
	• 6 typical refining units (cracking, reforming etc.)

Natural gas	• Regional supply detail (Europe, Russia, Middle Africa, North Sea etc.)
	• Transportation, distribution network
Fuel types	• 18 energy forms in total
Markets	• Country specific
	• European level markets: refining, gas, electricity exchanges
	• World oil market: exogenous

15.1.4.2 Time Horizon

PRIMES is a long-term model that is being set to consider the period 1990-2030, running by period of 5 years.

15.1.4.3 Output

Dynamic Annual Projections in Specific Units

- Full detailed EUROSTAT Energy Balance sheets per country and per year
- Energy demand at the above mentioned classification
- Energy costs, producer and consumer prices
- Power generation park, load curves, load factors, investment and marginal costs (central systems, combined heat-power, exchanges)
- Refining units, expansion, costs
- Natural gas transport and distribution: flows, capacities, costs
- Endogenous treatment of energy savings and new technologies
- Atmospheric emissions (CO_2, NO_x, SO_2, N_2O, CH_4, VOC, PM), abatement equipment and standards

15.1.4.4 Case Studies and Planning Applications

Scenario Building for the Medium Long Term

- Energy and environment technology assessment
- Energy system implications of policy instruments for the environment (taxation, abatement standards, pollution permits, ...)

- All issues of energy policy, investment plans and energy pricing policy
- European energy market integration, European networks
- Energy system implications and forecasting for the penetration of new energy technologies in energy savings, energy demand, power generation etc.
- Energy supply to Europe: dependency and vulnerability analysis for natural gas and oil.

15.1.4.5 Required Infrastructure

Hardware:	PC Pentium with Windows '95 or preferably Alpha Digital Equipment running NT-Windows with 128MB RAM or higher
Software:	GAMS Ver. 2.25 with PATH solver and Cplex (or OSL)
	MS EXCEL ver. 7.0 or later

15.1.5 The Industrial Demand Model

As for the rest of the demand-side models of PRIMES, the design aims at building a structural model to represent in detail the engineering of final energy demand formation in industry, while expressing the cost minimisation behaviour of industrial companies.

In the second version of PRIMES, the aim of the industrial sub-model is to project demand for fuels, electricity and steam. These fuels are purchased from other sub-models of PRIMES. The self-production of electricity and steam is not included in the industrial sub-model, but is a part of the large electricity and steam sub-model of PRIMES. This choice reflects the emerging market opportunities for independent generation of electricity and steam.

Modelling the structure of energy demand formation in industry is a complex task. This is due to the variety of technological cases in the industrial sectors and the heterogeneity of products, equipment, technologies, processes and energy uses.

Given the focus of PRIMES for policy analysis, the modelling objective has to emphasise on the potential for structural change in energy use in industry, a change that would involve new technologies, industrial processes, links to decentralised (self-supplied) energy conversion and the implied shifts in energy intensities.

The scope of the industrial demand sub-model of PRIMES is to represent simultaneously:

- the mix of different industrial processes (e.g. different energy intensity for scrap or recycling processes and for basic processing);

- the mix of technologies and fuels, including the use of self-produced by-products (fuels) and renewable energy forms;
- the links to self-supply of energy forms (e.g. cogeneration of electricity-steam, steam by boilers, use of by-products (fuels), heat recovery);
- the explicit and engineering-oriented representation of energy saving possibilities;
- the satisfaction of constraints through emission abatement, pollution permits and energy savings, and
- the rigidities of system change evolution because of existing capacities or dynamic technical progress.

Industrial behaviour is driven by the minimisation of total energy and environment costs subject to total useful energy needs, energy use capacities, technology availability and emission constraints. In optimising, the producer may act on the degree of basic processing, the fuel mix, capacity replacement, technological choices, the degree of internal energy conversion, energy savings, abatement technologies and pollution permits. The Figure 15-2 provides a scheme of the basic model structure.

The model evaluates consistently the potential of such new technologies, by considering simultaneously three types of mechanisms:

- economic optimality
- dynamics, i.e. constraints from existing capacity, and
- gradual market penetration and acceptance.

The non linear optimisation is performed period by period in a time forward direction. In a given period a set of lagged values are used that are updated dynamically by the single-period optimisation results.

Choices are constrained dynamically by the existing energy-use equipment. This may be renewed through investment and is retired on the basis of retirement rates.

The above decisions can be made on the grounds of available technologies in the different domains. Technology is considered to evolve in time, and is categorised in vintages (generations) presenting different characteristics.

Within the above context, the model represents the gradual penetration of new technologies and processes in several domains: industrial processing, end-use equipment, energy savings, pollution abatement. The model evaluates consistently the potential of such new technologies, by considering simultaneously three types of mechanisms:

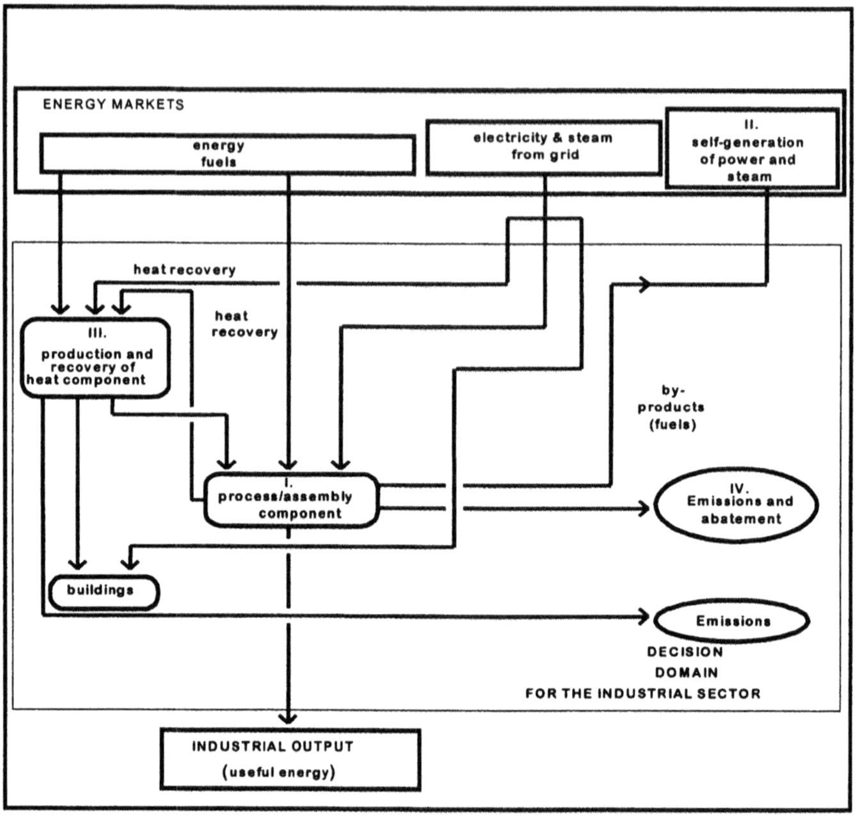

Figure 15-2: Structure of the industrial demand module (one industrial sector).

15.1.6 The Households Sub Model

15.1.6.1 Introduction

In the domestic sector, energy is consumed as input in processes that provide services to the households, such as space heating, water heating, cooking, cooling, lighting and other needs. The decision about the level of energy consumption is related to the need for services covered by energy, which are further related to changes in prices and income as is true for other consumption commodities.

Energy consumption has, however, several special features which need to be considered especially concerning the way they affect the dynamics of consumer response. In particular:

- The pattern of energy consumption is not usually controlled directly by the consumer, but is determined by the household technology (i.e. the type of fuel and equipment used for an energy service); the level of consumption is controlled, in the short run, by behavioural decisions in utilisation intensity;
- The household technology for energy consumption is largely embodied in the characteristics of dwellings and durable equipment. Consequently, responses to price shifts may involve long lags;
- Energy costs are normally billed to households periodically for several uses combined. Due to this fact, there may be no direct linkage between policy and cost, even for highly rational consumers; this also causes a delay of response when price shifts occur;
- Energy covers primary needs of households. The income elasticity is expected to be less than one, while substitutions by non energy commodities are rather limited. In industrialised countries the share of energy in total consumption is close to saturation, a fact that explains the observed asymmetry in price elasticities with respect to positive or negative shifts.

As a result of these special features of energy consumption in the domestic sector, the model has to include both technological and behavioural components. Technological components are necessary to capture the physical constraints on energy conservation and use, while behavioural components are necessary to explain consumer expectations and their influence on equipment choice as well as to explain the influence of energy prices on energy consumption intensity. The model is designed to provide energy consumption forecasts for each end-use by fuel aggregate. Four fuel aggregates are considered: solids, liquids, gas and electricity.

The fuel shares, for each end-use in which we have substitution between fuels, are assumed to represent fuel choice frequencies (which express the percentage of households that choose a specific fuel to serve an end-use). The probability that a given appliance (for space heating, water heating and cooking) is chosen to be installed in a dwelling is calculated as a function of a total perceived cost and of the maturity of equipment (so that inter-fuel substitution is constrained). The total perceived cost is a function of capital, maintenance and fuel (operating) cost of the equipment, as well as of the income of households. In the model design, we consider four household income categories each with a different elasticity. Especially, for cooking and water heating we assume that the total perceived cost also depends on the fuel choice made for space heating following the decision-tree approach mentioned above. This assumption leads to a *"nested logit model"* approach. The fuel shares obtained, are implemented for new dwellings and for the installation of new equipment due to normal replacement. As a result, updated fuel shares by end-use are computed, concerning both existing and new dwellings.

The domestic sector model is formulated as a system of non linear equations.

15.1.6.2 Model Overview

Four different end-uses are considered for the domestic sector, namely space heating, water heating, cooking and specific electricity

The general separability scheme assumed to model energy demand formation in the domestic sector is based on the distinction between:

- substitutable energy uses and
- non substitutable energy uses

For the former, a complex mechanism is formulated, consisting in evaluating the effective demand for fuel aggregates, such as solids, liquids, gas and electricity by end-use. Then, the demand for fuels is further driven by means of technical rates.

Specific electricity use is considered as non substitutable, for which an analysis in terms of electric appliances is formulated. The following figure presents the structure of the model.

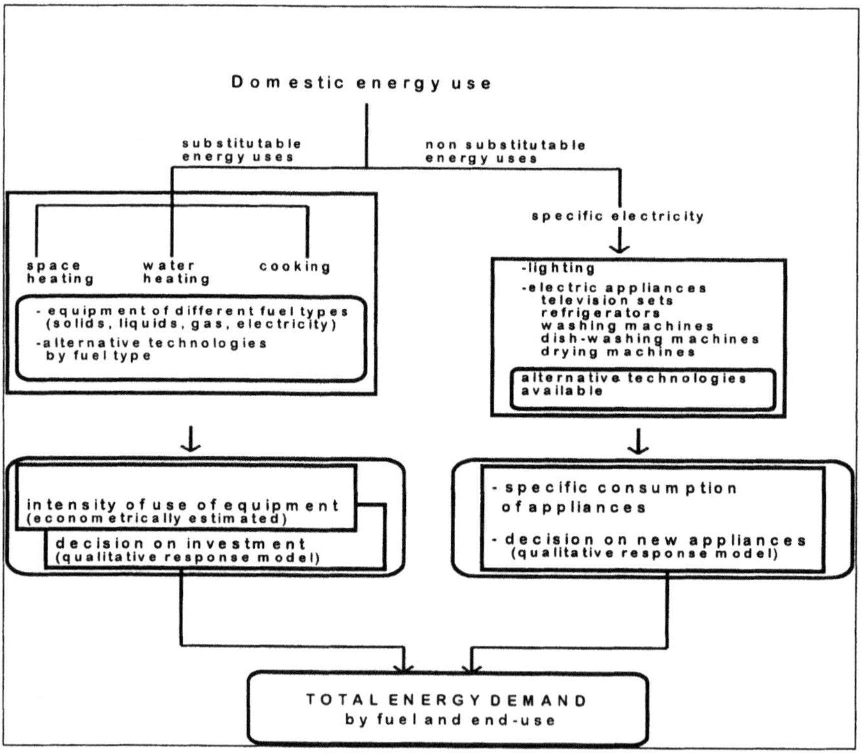

Figure 15-3: Model structure (domestic energy use).

15.1.7 The Tertiary Sector Sub Model

The purpose of the tertiary sub-model of PRIMES is to project final energy demand of the services sector, as a function of economic activity of the sector, which is exogenous, and the prices of the energy forms as transmitted to these sectors from energy supply. Possible interactions with the production of electricity and steam are modelled though links of quantities and prices with the corresponding component of PRIMES.

Tertiary is one of the fastest growing sectors in the last years. Energy demand in this sector also grows fast, as a consequence of the introduction of new energy uses, particularly regarding electricity, thermal comfort and cooling.

For modelling, the main difficulty for the tertiary sectors comes from the introduction of new energy uses (such as air cooling), the increase in use of electrical equipment (computers etc.), and the heterogeneity of equipment, technologies and energy uses.

Given the orientation of PRIMES to support long run policy analysis, the modelling objective has to emphasise on the potential for structural change in energy use in tertiary, a change that would involve new technologies, links to decentralised energy conversion and the implied shifts in energy intensities. A top-down approach based solely on econometrics could not capture such a complex evolution and would not provide a basis for policy appraisal in view of the environmental objectives.

Deliberately, to represent such mechanisms one must support economic modelling with engineering evidence, an objective that always has been challenging modelers. This is the model approach adopted by PRIMES, in view also of its future use in RTD strategy analysis and the need to deal explicitly with technologies.

The scope of the tertiary demand sub-model of PRIMES is to represent simultaneously:

- the mix of different energy uses;
- the mix of technologies and fuels, including the use of renewable energy forms;
- the links to district heating, steam production from boilers and cogeneration;
- the explicit and engineering-oriented representation of energy saving possibilities;
- the satisfaction of constraints through emission abatement, pollution permits and energy savings, and
- the rigidities of system change evolution because of existing capacities or dynamic technical progress.

Tertiary sectors behaviour is driven by the minimisation of total energy and environment costs subject to total useful energy needs, energy use capacities, technology availability and emission constraints. In optimising, the producer may act on the mix of different energy uses, the fuel mix, capacity replacement, technological choices, the degree of internal energy conversion, energy savings, abatement technologies and pollution permits.

For each tertiary sector one can distinguish between three components:

- the energy uses component;
- the link to electricity and/or steam production component, and
- the heat and cooling production component.

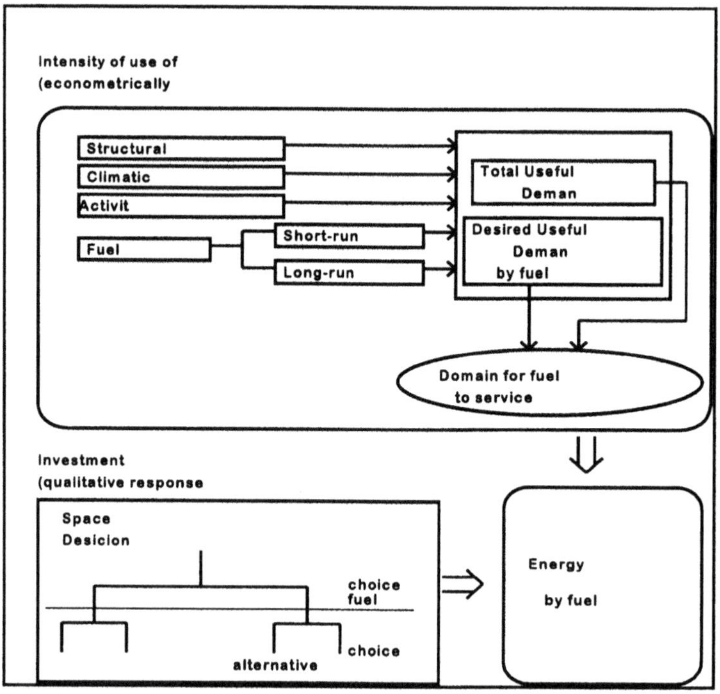

Figure 15-4: Space heating: the decision mechanism.

The consumer can purchase energy fuels, district heating and electricity from the energy markets (modelled in the other sub-models of PRIMES). The producer uses these fuels in three ways:

- directly in the energy uses component;

- to produce steam and/or electricity that are further used in the energy uses component;
- to produce heat and cooling that is further used in the energy uses component.

In all activities, the consumer generates atmospheric emissions that may be abated by means of equipment. Both at the level of the energy uses component and the production of heat and cooling the consumer may act on energy consumption by performing energy savings. In addition he can improve thermal integrity of buildings so as to decrease his needs for energy.

Of course, the degree of responsiveness of the sector heavily depends on capital turnover, which is relatively slow in this sector. For example, renovation of buildings and new constructions move slowly over time. In addition, energy demand growth in this sector is more due to new energy uses, rather than on old and less efficient equipment. These factors are taken into account in the dynamics of equipment penetration and replacement which underlies the model design. For example, a full vintage approach is adopted to represent the dynamics of equipment.

15.1.8 The Power and Steam Generation Sub Model of PRIMES

15.1.8.1 Introduction

The aim of the electricity and steam sub-model of PRIMES is to simulate the behaviour of agents that use fuels and other energy forms to produce, transmit and distribute electricity, industrial steam and district heating. This behaviour concerns the choice of equipment and the fuel mix to satisfy demand, the setting of selling prices and the purchase of fuels from the energy markets.

The model design is adapted to the very nature of the energy forms produced in this sub-model, related to the impossibility to use storage, the high degree of capital intensive equipment and the importance of technology choice for energy strategy.

The emergence of heat and power cogeneration possibilities and the prospects for increasing decentralisation of production led to the adoption of a unified modelling for power and steam production. On the contrary, the previous version of PRIMES has considered a separation between centralised electricity and the independent production of steam and electricity, the latter being modelled within the demand sub-models of PRIMES. That design has put more emphasis on the self-supply character of cogeneration, since such a situation has prevailed in the market for a long period of time. The emergence of efficient smaller scale technologies and the opening of the markets to competition created new prospects for cogeneration and independent production. The modelling needs then to tightly integrate producers of different nature, regarding for example economies of scale and market opportunities, into a single framework that will mimic the operation of the market.

The new version of PRIMES puts emphasis on the different nature of producers that will operate in the market and the interaction between electricity and steam markets, as enabled by cogeneration. For example, it is necessary to distinguish producers according to their scale, but also according to the captive markets they might address. A utility can exploit high economies of scale, but can hardly benefit from the market of steam, as steam cannot be self-consumed. On the contrary, an industrial independent producer will operate at smaller plant size, loosing competitiveness as far as the economies of scale are concerned, but obtaining benefits from a high base load demand for steam that he can supply. A company operating at the level of local authorities, may obtain benefits from niche markets (renewables, district heating), but it will face a highly fluctuating demand for heat and electricity.

The representation of different technologies that are now available or will be available in the future is a major focus of the model, as it is intended to also serve for strategic analyses on technology assessment. To support such analyses, the model uses a large list of alternative technologies and differentiates their technical-economic characteristics according to the plant size, the fuel types, the cogeneration techniques, the country and the type of producer. A model extension is also designed aiming at representing a non-linear cycle of the penetration of new technologies, for which learning through experience (and other industrial economic features) relates penetration with the technology performance.

The differences between the producer types play an important role in their ability to obtain interesting natural gas supply contracts. This issue seems to become very important in the future, as natural gas is emerging as the key fuel because of technology progress and environmental constraints. Again, a unified modelling approach is necessary to analyse the differentiated effects of natural gas for producers that differ as described above.

Both the market allocation from the producer perspective and the effects of natural gas supply conditions need a consideration of the time pattern of demand, production and fuel supply. In addition, the corresponding loads have to be considered in chronological terms, as serious limitations would arise if using load duration monotone curves, because of the need to analyse the synchronisation of the time patterns of electricity consumption, steam consumption and fuel supply (such as natural gas).

The consideration of intermittent energy sources, such as the renewables, also requires a representation of chronological curves, as the random availability of the source over time can be approximated. Nevertheless, the correct modelling of intermittent production also requires a representation of geographical characteristics of production and transmission and a modelling of congestion over the electricity networks. Obviously, such features are necessary to adequately represent the market for steam and heat.

Such features have not been yet introduced in PRIMES, as the model mainly aims to serve for integrated strategic analyses. The algebraic coding of the electricity and steam sub-model of PRIMES is enough generic and abstract, to provide a consistent framework for model expansion in the future, along the geographical or network congestion research lines.

The development of independent power and/or steam producers and their market forces heavily depend on the prevailing institutional regime in the market. In the past, market regulation, cross-subsidisation and the importance of returns to scale in power production has deprived small independent producers to enter the market. Exceptions have arisen in specific cases in which the scale of self-consumption or the existence of by-product fuels has permitted the survival of independent producers. The expectations for the future are different. New technologies allow for competitive production at a smaller scale, while the institutional regime in the market is increasingly opening to competition.

To represent these market dynamics, the model design preferred the representation of representative companies operating under a market competitive regime. For example, the exchanges of electricity and steam between the companies are performed under marginal cost pricing in the model. This choice has a limitation, as it cannot represent transitory phenomena of oligopolistic nature that might prevail in the market. However, a full competitive regime has been preferred as PRIMES puts emphasis on strategic analysis. Constraints regarding for example the degree of opening of the market can be introduced in the model through parameters regulating market allocation to producers.

In addition, constraints that would increase the inertia of the market are introduced in the form of contracts. These apply to both the exchanges between the companies and the provisions of fuels.

15.1.8.2 Literature on Power Modelling

There exists an important literature on electrical power economics and modelling. The different approaches have used a big variety of advanced mathematical techniques, including optimisation, probabilistic simulation and dynamic systems. Because of the complexity of electricity systems, the mathematical models have been specialised by focusing on some of the features of the system. With reference to their scope, the models can be classified in the following categories:

- models emphasising on the operation and dispatching of the electricity system; load flow and probabilistic techniques are commonly used
- models emphasising on reliability issues especially regarding the transmission network; stability analysis techniques are often used

- models emphasising on capacity expansion and plant selection; they usually follow dynamic programming and mixed-integer optimisation, as they represent discrete candidate plants
- models emphasising on the strategic economic issues and the interactions between the electricity system and the rest of energy systems and markets; a variety of economic modelling techniques are used

PRIMES' focus is mainly along the lines of this last category. However, given the importance of the sector, the model design gave importance to a minimum standard representation of the engineering aspects of the electricity system. In this sense it borrows elements from the engineering-oriented power economics modelling. In addition, it was aimed to design a model framework that would be easily expandable to construct sufficiently detailed model version that would be used specifically for the analysis of the electricity sector.

15.1.8.3 Decisions Represented in the Model

Decision-making by electric utilities (or steam producers) may be considered in three different, yet interrelated, problems:

i. the *strategic capacity expansion problem* which concerns the choice of new plants for construction, so as to meet future demand at a least long-run generation cost;

ii. the *operational plant selection and utilisation problem* which concerns the choice of existing plants to be committed in the system, so as to meet load at a least operation cost;

iii. the *cost evaluation and pricing policy* that has to be in conformity both with the long-term financial objectives of the company and with the aim to influence demand load.

In the electricity and steam sub-model of PRIMES, the representation of the above decision problems is in accordance with the optimal pattern of supply behaviour in a competitive equilibrium market. In particular, we formulate long run marginal cost principles for capacity expansion and short run marginal costing for dispatching and plant commitment. However, for price setting we formulate Ramsey pricing, which is close to average cost pricing, and we interpret this choice as representative of both the regulated monopoly and the monopolistic competition market regimes.

15.1.8.4 Engineering Features

The model puts emphasis on the following features that are dominant in electricity economics and engineering:

i. demand fluctuates over time in a year

ii. the optimality depends on exchanges between producing companies, via interconnections

A fundamental characteristic of electric or steam producers is that they cannot inventory their product in order to meet fluctuations in demand. Thus, an important feature to be captured in modelling is the implications of changing time-related patterns of demand on plant capacity selection and utilisation.

At the level of the electricity and steam sub-model, demand for electricity and steam is considered as exogenously given, varying widely between different times of the day and between different seasons. The representation of demand is based on the definition of a chronological load curve, which depicts the load (e.g. in GW) as a function of time in a year. Within an iteration of the overall PRIMES model, the demand sub-models provide estimates of demand (as a function of time in a year). They use the same representation of time as the electricity and steam sub-model. Changes in the demand-side, for example, induced by prices or other factors, influence the electricity and steam sub-model. The latter may, for example, change prices that may further affect demand.

The closed-loop interaction performing at the level of the overall PRIMES model represents the link between demand and supply of electricity and stream in an endogenous manner. This link is fully defined over time of a year, since all the sub-models are synchronised at the same representation of time. This synchronisation concerns not only electricity and steam, but also fuels that are inputs to power and steam production, such as natural gas for which the time pattern is important for their pricing. Of course, the synchronisation concerns both prices and quantities. As we will explain below, the synchronisation over time in a year includes also the supply of intermittent power sources, such as the renewables.

Other models adopt a representation through a load duration curve, which is simply a reordering of the loads in a monotone non-increasing way. The synchronisation feature as operates in the PRIMES model does not comply with a load duration curve.

The second important feature is associated to transmission and distribution of electricity and steam, which operate over networks (cables and pipelines). As explained in a previous section, PRIMES adopts a network representation but dissociates this from any geographical characterisation. In fact, nodes and arcs in PRIMES are virtual, without involving any physical characteristic other than losses and capacities. The adoption of a network representation endows the model with abstraction sufficient to expand the model so as to represent geographical or production characteristics in a more realistic way.

The interconnections between electricity companies are important, as electricity cannot be stored. Traditionally, interconnections between countries served to reduce the costs of system reliability for each country. They have also served to the establishment of contracts, concerning exchanges between countries that have reflected differences in the economics or capacities. The contracts have been always defined over time (time pattern), besides their contacted quantity. To this respect, the synchronisation feature of the model is useful to also represent contracts, which in fact are formulated as obligations for load exchanges.

As the electricity market is being increasingly liberalised in Europe, the interconnections also serve to support market operations, covering commercial transactions between a country, but also among companies within a country. In other terms, the development of the network physically defines upper bounds to the possibilities for market-related transactions of electricity. At a more technical level, congestion over the network also limits market development. Market competition regimes, such as the single buyer or third party access, following the jargon adopted in the EU, can be also represented through different topologies of the connections in a network. For example, a consumer may be allowed to connect to a utility in another country, to reflect a situation in a third party access regime.

In brief, the representation of the network in PRIMES is an abstraction that serves to support the representation of engineering information about transmissions, but also serves to formulate the possibilities for commercial transactions. The algebraic coding of the model, regarding the network, is general enough to allow for an expansion of the model either towards engineering representation of the network and/or a generalise formulation of commercial agreements.

15.1.8.5 The Temporal Framework

The model horizon is composed of a set of periods of equal length. The length of the periods may be one or several years. Activity variables related to generation, trade and sales are generally treated as time series, except some of the parameters that are assumed invariant with time.

During a year, the conditions under which the electricity or steam system works may vary from day to day. The model therefore considers a set of typical days in a year. Each typical day is characterised by a different set of operating conditions for the system.

Also, the operating conditions under may exhibit significant variations during a single day. This is taken into account by decomposing the day in a set of time segments, each time segment leading to different operating conditions.

Regarding the dynamics of the decision making, the model is flexible. Two anticipation regimes, activated through an optional switch, are formulated in the model:

The myopic time-forward anticipation, in which the decision-maker has information only about the past and the present time-period, while the model is solved dynamically along time-steps in a time-forward manner.

The perfect foresight over finite horizon, in which the decision-maker has full and correct information about the future, over a finite horizon, and the model runs simultaneously (intertemporally) the set of time periods from present up to the horizon.

The building of equipment in the electricity and steam system requires several years. This has important implications for planning and plant type choice. The model considers the financial costs associated to the construction period but ignores the fact that the plant types differ in construction time, which may influence plant selection in particularly uncertain circumstances. Furthermore, under the myopic anticipation regime, the model considers that the plants can be constructed and immediately used within the 5-years runtime period of the model. In this sense, the model operates as if the current 5-years period is perfectly known by the decision-maker.

15.1.8.6 The Spatial Framework

The model considers the electricity system of the member states of the European Union. To represent the interconnections between countries, the model considers links to Switzerland and the rest-of-the-world. However, their electricity systems are not represented. Regarding steam, the model considers that connections exist only within each country, separately.

15.1.8.7 The Mathematical Form of the Model

The model is designed to follow the general problem of optimisation of flows over a network.

Assume that the network of interconnections can be described as a set of nodes $n \in \{1,\ldots,N\}$ and a set of oriented arcs $tr = (n,m) \in \{N \times N\}$ defined on the Cartesian product of the set of nodes. The exact definition of the set of arcs determines the topology of the network that of course influences the solution. The flows (by fraction of time) over the arcs are unknown and denoted by x_{tr} while the capacities of some of the arcs are also unknown, denoted by K_{tr}. Then the problem consists of determining the values of the unknown variables so as to optimise the total cost of the network flows under feasibility of the network flows:

$$
\begin{aligned}
Min \quad & \sum_{tr} c_{tr} \cdot x_{tr} + \sum_{tr} k_{tr} \cdot K_{tr} \\
s.t. \quad & \sum_{tr\ inputs\ to\ n} x_{tr} = Eff_{tr} \cdot \sum_{tr\ outputs\ from\ n} x_{tr} \qquad \forall n \\
& x_{tr} \leq K_{tr} \qquad \forall tr \\
& \sum_{some\ of\ tr} x_{tr} \geq \text{Demand} \\
& \sum_{some\ of\ tr} x_{tr} \leq \text{Resources} \\
& \text{L} \leq \sum_{some\ of\ tr} x_{tr} \leq U
\end{aligned}
$$

The first constraint is the equilibrium on the nodes of the network and may involve an efficiency rate. The second constraint limits the flows by capacity, some of which may be expandable. The third constraint expresses the obligation to satisfy demand, while the fourth constraint may limit some of the flows to the available resources. Finally other constraints may restrict aggregations of the flows, as for example environmental regulation, cogeneration technical constraints, reserve margins, and so on. The objective function involves cost parameters.

If all parameters are linear and the unknown variables continuous, the problem corresponds to linear programming. This is the standard version of the model. By applying Kuhn-Tucker conditions, the problem can be transformed into a complementarity formulation, involving a system of equalities and inequalities to solve for the original unknown variables and the dual variables of the constraints.

In particular, in PRIMES, the network is separately defined for electricity and steam. The flows over the two independent networks are interrelated only at the level of the constraints that concern cogeneration of heat and power. Also, the unknown variables are defined as functions of time segments through which load and plant commitment is handled.

15.1.8.8 Model Design Principles

The electricity and steam sub-model of PRIMES is designed to be very general so as to provide a framework for studies specifically for the sector and not only for being a component of the overall model PRIMES. When building a model implementation, the user defines the content of the sets and gives numerical values to the exogenous parameters, at a level of detail that depends on the content of the sets. This definition can vary widely, so as to give rise to different types of electricity and team models. A variety of market situations can be represented.

When building the model implementation that is used as a component of PRIMES, the user has to consider time needed for running the electricity sub-model and the size. Usually, the component has to be more aggregated and simpler than other model implementations intended to support specific sectoral studies.

In the following, we present first the general model design principles, putting emphasis on the variety of possibilities of the model. Then, we present the current model implementation constructed to be a component of the overall PRIMES model.

The physical elements of the electricity and steam system that we consider are the production plants and the transmission and distribution networks.

15.1.8.9 The Production Technologies

Combustion of fossil fuels, or by transforming mechanical, chemical or nuclear energy, can perform production of electricity and/or steam. Regarding the output energy form, the plants are subdivided in three categories.

- Plants that produce only electricity
- Plants that can cogenerate electricity and steam
- Plants that can produce only steam (boilers).

For the combustion plants the distinction between electricity plants and cogeneration ones is necessary for the economic analysis of generation, as in engineering terms the distinction is not always clear. Combustion plants are also called thermal, a term that often includes nuclear plants and fuel cells, although the latter use a chemical process. Non thermal plants produce only electricity and are, in the model, subdivided in two categories, according to the possibility of dispatching:

- hydroelectric lakes, also termed reservoir plants; they are constrained by energy rather than power, as depending on the inflow of water in the reservoir;
- plants that depend on the availability of a randomly intermittent energy resource like wind, sun or maritime waves. Electricity produced by such plants is not dispatchable.

The model considers pumping as an exogenous shift made on the total demand load, implying of course higher production from base load plants and higher availability of hydro-power in peak hours.

15.1.8.10 Deterministic and Continuous Variables

The model generally follows a deterministic approach in the representation of the technical characteristics of the plants. More detailed engineering models of power systems treat some technical parameters of the plants as random variables. Usually, this is the case of the technical availability of thermal plants (for commitment), the availability of the energy resource in intermittent plants and the water inflow to lakes. These models consider probability distributions associated with these random variables and apply convolution techniques to aggregate the distributions and estimate system reliability and economic optimality.

In such models, the probabilistic approach combines with a discrete treatment of the power plants, and some times even the blocks of each plant. This is necessary because of the discrepancy of the technical parameters of each plant, fuel type or location. Also, the discrete approach reflects the non-linearity of the costs associated with the start-up, the operation and even the construction costs of the plants.

For PRIMES we consider that both the probabilistic modelling and the discrete treatment of plants are out of the scope of the model, mainly because of the great complexity that it would imply. If we included such features, the mathematical problem of PRIMES would be that of dynamic, non-linear, stochastic and mixed-integer programming, a problem that is extremely difficult to solve, especially when the capacity expansion problem is the main focus.

Therefore, PRIMES adopts deterministic approximations of the above random variables and considers the plant activity or capacity variables as continuous and not discrete. Also the model ignores non-linearity associated to start-up, block operation or location-specific costs of the plants.

However, PRIMES does consider that cost and performance of the plants can vary with the size of the plant. This relationship is artificially introduced in the model through the representation of a number of discrete sizes that are related to the cost and performance parameters, although the plant capacity variables are continuous. Of course, without introducing integer variables, it is also impossible to control that, for example, a minimum capacity is required to accumulate before producing from a plant-type, or that a plant with high start-up flexibility may be more preferable for peak load than a less flexible plant. The model, in fact, ignores such kinds of issues that normally heavily influence plant selection, in reality.

Regarding plant commitment availability, the model considers a single deterministic parameter, which expresses the percentage of time in a year in which the plant is available. There is no time pattern associated to maintenance schedules, nor to statistics of plant failure. However, if kept in deterministic terms, such technical information can be easily introduced in the model. Also, the annual availability rate is constant and does not depend on the rate of use of the plant, since such a dependency would introduce a non-linear relationship. Similarly the parameters on the operation costs do not depend on the shape of plant use in a typical data, as this would also require non-linearity. Such non-linear relationships, are however introduced in the mixed-complementarity formulation of the model, as it is will be explained below.

The random system reliability issue (often termed loss of load probability) is not represented, as mentioned above. Instead, a single reserve margin ratio is used as a constraint implying the building of capacity higher than peak load. As the plant selection depends on the ratio between capital and variable costs, it is expected that only peak devices will be adopted to fulfil the reserve margin requirement. This approach, of course, cannot address situations of over-capacity with base-load or renewable plants. The reserve margin ratio is exogenous and differs by country, to reflect the prevailing practice in terms of system reliability. The back-up issue, which is relevant for companies, even small producers, is also approximated through the reserve margin ratio, which then applies to a company.

The random availability of intermittent energy resources is approximated by means of an exogenous time-pattern of resource availability. Of course, this approach neglects situations in which additional back-up capacity is required to cope with probabilistic situations in which non-availability events occur jointly. Finally, the random availability of water in reservoir plants is also approximated through a deterministic annual inflow parameter.

15.1.8.11 ***Producers, Companies and Market Relationships***

Self-production of electricity or steam technically differs from the production of utilities because a self-producers traditionally does not consider trading of production surplus. Along the rapid development of trade in the electricity market, the self-production changes in character, as trading through interconnections alters the economics of the plants. The term "independent producers", which is now widely used to characterise self-production among other cases, does not make reference to the purpose of production, but to the scale of the company, in comparison to large utility companies. Because of this market evolution, PRIMES considers independent production only.

The economics of independent production usually are justified in situation when one or more of the following conditions hold:

- the producer is also a consumer of his output, the consumption costs less in terms of transmission and distribution, and there is sufficient development of a market for trading excess supply;
- the producer has access to a free or cheap energy source (waste, by-products, and specific locations for renewables);
- The liberalisation of the market and privatisation of utilities leads to smaller scale companies that are necessary to cope with the eventual diseconomies of scale of large utilities.

The design of the model is oriented to cover the first two situations. Differences in return to scale, related to the average size of plants that a producer is allowed to handle, are compared against eventual gains in transmission and distribution, while market allocation constraints may limit or favour the development of independent production. Similarly, among the variety of fuels considered, the model explicitly represents cheap or free resources to which some of the producers can only access. In this sense, the companies (producers), represented in the model differ in purpose: a utility makes business by selling electricity or steam, an industrial company is mainly addressing his own demand but can profit from market development, etc.

However, the model is not designed to represent adequately the development of multiple utilities in the same country or the case of independent producers, which transform from self-producers to small utilities. In compliance with the long-run orientation of the model, the design considers that these cases will be transitory, and that the in the long term competition can be adequately approximated by means of the "representative firm" assumption. The different representative firms (producers) reflect then differences in purpose, in size or market opportunities.

15.1.8.12 Cogeneration Technologies

The cogeneration technologies constitute a valid substitute of the traditional heat producing technologies, as for example the boilers. Their importance and their rapid emergence in the market justify their inclusion in the set of power technologies.

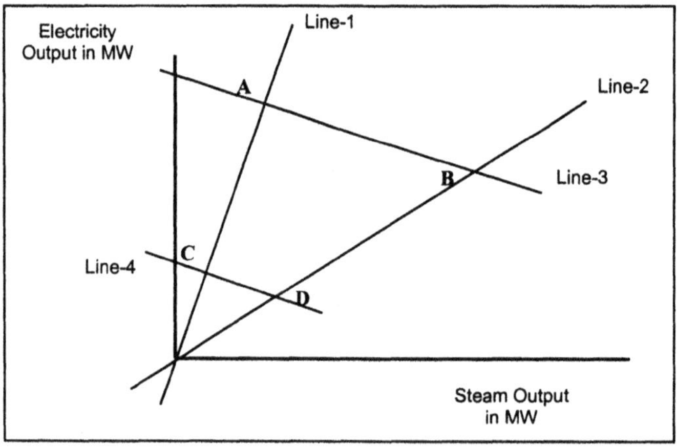

Figure 15-5: Feasible domain of cogeneration per unit of nominal electric power.

Cogeneration can be obtained through changes in the circuit of steam in thermal power plants. If it is intended to obtain big quantity of steam, then the change of the circuit is important, consisting in conducting steam to demand, instead of recycling through the cooling system of the plant. This kind of technology, usually termed "back pressure" technique, has significant negative implications for the electric efficiency rates of the plant and can achieve high steam to electricity ratios. If the quantity of steam, required to conduct to demand, is rather small, then steam can be just extracted at some point of the steam circuit. In this case, called "steam extraction", the steam to electricity ratio is mall, but also the implications on the electric efficiency ratios are also small. Finally, gas turbines, internal combustion engines and fuel cells can also obtain cogeneration.

All the above mentioned techniques of cogeneration are represented in PRIMES. To unify the modelling of cogeneration and simplify, the model considers that the feasible combination of electricity and heat output from a thermal plant are constrained within a surface delimited by four lines, as in the scheme.

The feasible domain of cogeneration is the area ABCD. Line 1 denotes the maximum electric power and minimum steam combinations. Line 2 denotes the minimum electric power and maximum steam combinations. Line 3 is an iso-fuel line (equal electric efficiency of the plant), per unit of maximum use of nominal electric power. Finally, line 4 is also an iso-fuel line, defined for the minimum electric output necessary to obtain a steam output. Of course, depending on the cogeneration techniques, the slope and the exact position of these lines can change, but the basic shape of the feasible domain for electric and steam output combinations remain the same. The above lines are introduced in the model as linear constraints, specifically calibrated for each cogeneration technique and type of plant. Line 4 (the minimum power) is incompatible with the continuous character of the activity variables, as assumed in the model. Therefore, this constraint is ignored.

15.1.8.13 Technical-Economic Characterisation of Plants

Power technologies are characterised by the type of fuel they can use, their efficiency in generating heat and/or power, their cogeneration technique (if applicable), their availability, their investment costs and their operating costs. In the linear optimisation version of the model, the corresponding parameters are constant. In the mixed complementarity version of the model, some of the parameters can be expressed as functions of endogenous variables, which obviously makes the model non-linear. The non-linearity functions are as follows:

- the availability rate is a decreasing function of the rate of use (output) of the plant in a year;
- the operation cost is a function of the dispersion of plant use across the time segments (to reflect start-up costs);
- the capital cost is a function of the accumulated stock of the technology (learning by experience or economies of scale in the producers of equipment). Similarly, the thermal efficiency rate is a function of the accumulated stock.

The model put emphasis on the representation of plant efficiency and performance as a function of plant size. Given the limitation from the assumption about continuity of the plant-related variables, the plant size is associated to the nature of the power and steam production companies. It is assumed, for example, that utilities can invest in large size plants and benefit from economies of scale, while industrial power and steam producers can invest only in small size plants.

The relationship of plant performance as a function of plant size is considered as varying with the type of technology and time. Through this assumption, the model attempts to capture a technology progress that would bridge the gap between the plant sizes, in terms of performances and costs. Such an example has been recently observed with the developments in gas turbine technologies and the combined cycle plants.

The following scheme illustrates how a certain type of technology progress reduces differences of plant performance across plant sizes.

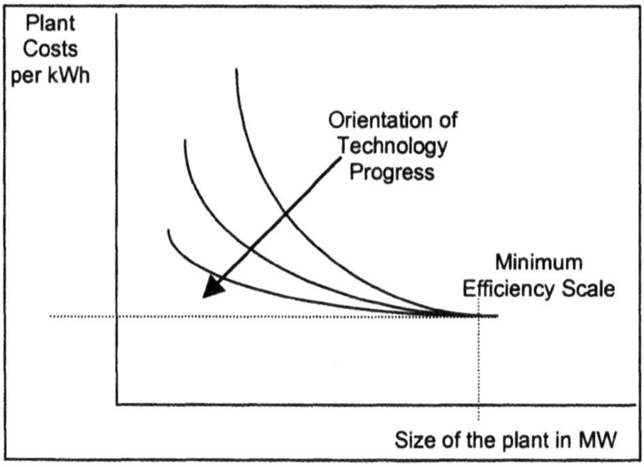

Figure 15-6: Relationship between pant performance and size.

15.1.8.14 Old and New Power and Heat Generation Plants

These are plants that use the aforementioned technologies to generate power and/or heat. We distinguish the old plants existing in the base year and the potential plants, i.e., those that might be built through investment.

The model does not consider investment in old plants but treats endogenous retrofitting of these plants. The extension of their lifetime may be judged as economical when scrapping is to be undertaken. Retrofitting involves capital costs and fixed costs that rapidly increase over time.

The technical-economic characteristics of the new power and steam plants are fixed over time. Future generations of plant technologies are represented as different technologies. Once a new plant is built remains as candidate for commitment in operation over its lifetime. New reservoir plants are not considered in the model, however all intermittent plants are fully endogenous.

15.1.8.15 The Grids and Pipelines

A transmission grid is a set of electric equipment that allows transporting electricity at high voltage. A distribution grid is a set of low voltage lines that connect the individual customers to the transmission grid. A distribution grid has a single owner. Steam pipelines and district heating distribution networks serve to connect producers and consumers for distances that are normally short. There is no interconnected network for steam.

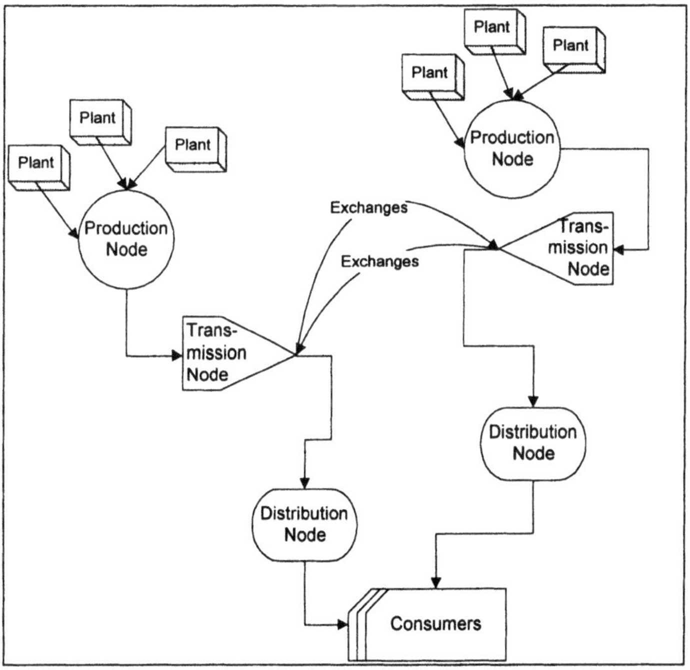

Figure 15-7: A typical power or steam network as presented in PRIMES.

The model neglects geographical characteristics of both electricity and steam networks. It considers a typical and simple network that has different topology for electricity and steam. The two networks are related only at a node representing cogeneration plants. The grids are characterised by two technical parameters: the connection capacity and the rate of losses (efficiency). The connection capacity is limited only is the case of transmission links, letting distribution links to have unlimited capacities.

The network comprises five types of nodes: plants, production nodes, transmission nodes, distribution nodes and consumers. The arcs link plants to production nodes, production nodes to distribution nodes, distribution nodes to consumers and transmission nodes to other transmission nodes, so as to allow trading between companies. Figure 15-7 illustrates the definition of the network.

15.1.8.16 The Polluting Emissions and Abatement

The model considers a set of polluting emissions. This set includes atmospheric emissions only. The pollutants are generated from combustion of fossil fuels and their emission is generally proportional to the quantity of fuel used. The emission factors are exogenous.

Installing emission abatement technologies may reduce the emission of the pollutants. The model assumes the simplification that these technologies operate for a single fuel type and can reduce only one pollutant. Hence one needs to set up separate abatement data for each fuel, plant and pollutant.

Because of the complexity of abatement technologies and their applicability to the various types of plants, the model adopts an implicit technology approach. It considers that the abatement is a service provided by an external supplier to the plant, which just pays a fee proportional to the achieved abatement. Given that the abatement technology exhibits decreasing returns to scale, the model represents the abatement service as a stepwise function in which abatement costs per unit increase non-linearly with the abatement level (in % of the corresponding fuel use).

15.1.8.17 The Actors in the Electricity and Steam Market

We shall consider three types of agents on the electricity market: the regulated companies, the independent generators and the consumers.

The consumers consume electricity and steam: government, private firms and households are consumers. They are aggregated in consumption sectors and are linked to the rest of the PRIMES model.

In all the models considered, the consumption sectors are price takers that maximise their utility. Their behaviour is represented by their electricity and steam demand curve, which is fixed for a given iteration of the overall PRIMES model. The electricity and steam sub-model sets the consumer electricity price which also include distribution costs, excises and taxes (including the VAT when it is not deductible).

The model considers several companies that are acting as suppliers of electricity and steam in the market. These companies are defined along the principle that their main aim is to sell electricity and steam to consumers, without having any self-demand (other than losses) to satisfy. Each company may own power plants, transmission and distribution grids. A company may have an exclusive right to serve some demand sectors. Because of their market power, these companies are subject to regulation.

Companies seek to achieve maximal profit under regulatory control. It is assumed that they follow the Ramsey-Boiteux regulated monopoly scheme. Given that they have to satisfy demand, this scheme implies that the regulated companies seek to minimise their long-term cost and tariff their output at average long run cost, on which a fixed mark-up may apply. They may perform the following actions:

- generate power
- transport electricity on transmission grids and maintain transmission grids
- distribute electricity on the distribution grids and maintain distribution grids

If a company only owns a share of a power plant, it may use up to a part of the total capacity of the plant. This part is proportional to the company's share in the ownership of the plant. The company may use its capacity independently of the actions of the other owners of the equipment.

The companies may also engage themselves in transactions with other economic agents.

These transactions are:

- sell and/or buy electricity from other companies
- sell and/or buy transportation services
- serve consumers

The companies determine the price and the quantities in the aforementioned transactions by exploiting their possible market power, subject to the regulatory constraints. They earn revenues from selling electricity and services. They pay the costs of the equipment that they own and their purchases from the other companies.

Several types of companies that are frequently met in practice are detailed below.

- The regulated generating companies are companies that own an important part of the power plant capacity in a country. They can be assimilated to those that have a dominant position in the common market. Because of that reason, the sales of these companies are subject to regulation.
- The manager of transmission grid owns and develops the transmission grid. It offers the grid for rent to the operator of the transmission infrastructure.

Because the development of the grid has a character of natural monopoly, the manager of the transmission grid is regulated.

- The operator of the transmission grid dispatches the generation plants. It rents the transmission grid from the owner of the infrastructure. It provides the transmission services. Because the dispatching of the generation plants has a character of natural monopoly, it is regulated.
- The manager of the distribution grid owns and develops the distribution infrastructure. It offers the distribution grid to the distributor. Because the development of the distribution grid has a character of natural monopoly, the manager of the distribution grid is regulated.
- The distributor delivers power to the customer. In order to do so it rents the distribution grid from the owner and insures that adequate supplies are available. The distributor may or may not be regulated.
- The utilities own the transmission grid, the distribution grids and the power plants in a region. The utilities have a character of natural monopoly and are therefore operated under regulatory control.

The independent generators are entities that produce and sell electricity and/or steam to the grid. They may be industrial firms that produce at least part of their needs in electricity or steam by the way of power technologies. They may also be firms developed at the level of local authorities or large tertiary sectors that also produce at least part of their needs in electricity or steam and may also distribute heat through district heating networks. They cannot be entities whose main goal is to produce electricity or steam for sale to the grid. It is assumed that these entities are not regulated.

In the general model design there is no reference to a country, as actor in the market of electricity or steam. In economic terms, the model does not consider that a company, grid or other item has an ownership relation with a country. However, in order to support regulatory policy analysis and country-specific reporting (like the energy balance sheets), the model considers the countries. A country is conceived as an institution related to a subset of the nodes of the network. A formal association is entered in the data of the model. This association allows for a representation of regulations, including:

- reserve margin constraint at the level of the country, to represent policy considerations regarding country independence
- upper and lower limits in the consumption of fuels by country, to represent special policy conditions in the provision of fuels in a country
- non fossil fuel obligations or obligations to use a certain percentage of renewable energy

- environmental regulations, as for example upper limits on certain total emissions in a country.

15.1.8.18 Relationships between Actors and the System Components

As mentioned, the physical components of the electricity system, their organisation and the actors on the market are represented as a network i.e. a set of nodes connected by arcs. A node represents a set of generation plants, a transmission grid, a distribution grid or a consumption sector. The arcs represent transmission capacities of flows between the nodes or exchanges among the actors.

Arcs represent the interconnections between two nodes. They should therefore not be considered as electric lines or sets of electric lines or steam pipelines, but as interfaces, or set of interfaces, between equipment. Also the arcs have no ownership.

The transportation of energy between two nodes requires a transportation capacity and it involves costs and losses. It is also limited by capacity. We consider one way arcs only. Therefore one or two arcs may connect two nodes. The latter case may only occur between two transmission grids. Since the arcs are directed, one may distinguish their origin and their end. One may also consider their origin node and their destination node.

The ownership relationships are presented below.

Transmission grids have a single owner. They are characterised by a transportation capacity, transportation cost

One or several nodes of the electric system represent the power plants in a country. Power plants may have several owners. At this stage, plants are characterised by the fuels they can use, their efficiency, their availability, their operating and investment costs. A power plant is connected to one and only one transmission grid.

Distribution grids have a single owner. A distribution grid is characterised by transportation capacity, transportation costs and transportation losses. Losses are due to the electric line resistance. A distribution grid is connected to one and only one transmission grid. At least one consumption sector is connected to a distribution grid.

One or several nodes of the electric system represent the consumption sectors in a country. These nodes have no ownership. A consumption sector is connected to one and only one distribution grid.

The companies are defined over the network through the ownership relationship. Except consumers, all other nodes are attributed to companies that may own plants, transmission grids and distribution nodes. The exact topology of the network and the definition of the companies as partitions of the network, determine their nature, according to the classifications given above.

A utility for example has no links to self-consumption and gives no priority to one of the consumers. The utility may have privileges in supplying exclusively a consumer, or even exclusively own a distribution system. An independent generator is also defined through the ownership of a number of nodes of the network. But the independent generator is constrained to give priority to supply self-consumption and normally has limited direct access to the consumers, while being able to transmit excess supply to utilities.

Consumers may be linked to several utilities or independent generators to allocate their needs to different suppliers. Wheeling, that is transmission through an intermediate grid, is possible, depending of course on the topology of the network. In this sense, the so-called third party access can be fully reflected in the topology of the model.

15.1.8.19 The Commercial Transactions

Several markets can be identified in the electricity and steam system. We first distinguish the market for power and the market for transportation services. In both markets we represent in a different way the transactions ruled by long term contractual agreements and the transactions that take place on the spot market, that is to say, where conditions for the sale of goods or services are arranged as a function of the current circumstances. Some transactions involve both transportation services and power.

The different types of contracts are further divided in new and existing contracts. An existing contract has already been settled and its clauses are fixed. A new contract has not been singed yet and some of its clauses may be negotiated. Such a representation allows a maximum flexibility in specifying the conditions that will prevail on the market in the future.

The physical movements of electricity, which cause costs and transmission and distribution losses, occur mainly through the nodes and not the arcs, which are only connection capacities. For this reason possibly unusual modelling options have been adopted for the representation of the commercial operations. The electricity delivery operations i.e. operations whereby the ownership of electricity changes are supposed to take place at the connection between two elements of the electricity system. Hence, we consider that the power deliveries take place on the arcs, that is to say, after losses and costs on the arc are incurred.

The transportation services are offered on a transmission or a distribution grid, between two connections of the grid. One can therefore represent a transit operation as flows on two adjacent arcs. We assume that the transportation services begin at the end of an arc and end at the end of another arc.

The power purchase contract is an agreement whereby a company commits to deliver power to another company at a specified arc. The power purchase contract may specify a power plant that will generate the power and power and a route for the deliveries.

The contract specifies:

- a price for the electricity or steam,
- the average amount of electricity or steam sold per year, typical day and time segment,
- a possible flexibility on these quantities,
- the delivery arc or a set of delivery arcs; if a set of delivery arcs is specified, the buyer may choose the distribution of the deliveries among the delivery point to its best interests,
- (possibly) the plant that will produce the power,
- (possible) the route or the set of routes to be taken by the deliveries. A contract path is specified as a set of arcs. The contract paths connect transmission grids owned by the seller, or power plants whose the seller owns a share, to a transmission or distribution grid owned by the buyer. The seller may only commit to Transport electricity on grids that he owns. Therefore, the contract paths should only cross nodes that the seller owns.

15.1.8.20 The Electricity Market

The contractual arrangements on the electricity market: exchange usually take the form of contracts, which may be considered as a special case of power purchase contract wherein neither a power plant nor a route is specified. Exchange contracts are not considered explicitly in the model design.

Besides the power purchase contracts, a spot market may take place at the different arcs. In such a market, power is assigned immediately on the arc where the transaction occurs i.e. it is a short term market. This kind of exchanges is taken into account in the model design.

The contractual arrangements on the transport market, as for example the transit contracts, can be considered as agreements whereby a company sells transportation services to another company. The contractual clauses may specify a route or a set of routes used for computing the charges of the transportation services. Such arrangements are explicitly represented when defining the content of the sets of a model implementation.

The sales to the consumption sectors may be constrained by regulations established in compensation of exclusive rights granted on some parts of the consumption sectors, or because of the dominant position of the seller. For example, the sales to consumption sector may be confined within requirement contracts, which are contracts whereby a company commits to deliver power to a consumption sector. Unlike the other types of contracts, the amount of electricity to deliver is determined by the consumption of the sector instead of being fixed (with a possible flexibility) by the contractual clauses.

The contract specifies:

- the share of the electricity consumption of the sector that the company must serve
- tariffs for the electricity
- the arc where the power has to be delivered

Short-term agreements (on the spot) may be concluded for power purchase between companies and consumption sectors

The companies perform also commercial transactions regarding the purchase of fuels from the market. The model design represents two types of fuel purchase transactions:

- the fuel contracts for which a yearly quantity, a duration, a degree of flexibility and a rate indicating dispersion of fuel load over time segments are determined;
- fuel purchases in the spot market, in which there are no quantity constraints at the level of the electricity and steam sub-model.

15.1.8.21 Model Implementation for PRIMES

In the following we present the model version as implemented to be a component of the overall PRIMES model, version 2.

The network includes plants, production nodes, transmission nodes, distribution nodes and consumers. The set of nodes (except consumers) is partitioned in three types of companies:

- a utility aiming at producing electricity and/or steam, transmit and distribute electricity and transmit steam to other companies (not distribute steam); the utility can exchange electricity with other companies through links of the transmission nodes;
- an industrial company aiming at producing electricity and/or steam, mainly to supply his own demand (for several industrial sectors); it can supply the excess power to the utility company, but it cannot distribute or exchange steam for purposes other than supplying its own needs (this assumption is made to reflect the special quality of industrial steam);

- a tertiary company aiming at producing electricity and/or steam; regarding electricity, the aim is to act as independent generator and sell electricity through the utility's grid or to directly supply tertiary electricity needs; for steam, the tertiary company acts as a utility for district heating, collecting production from the utility company and distributing district heating to the consumers (except industrial consumers); no steam connections among countries are allowed.

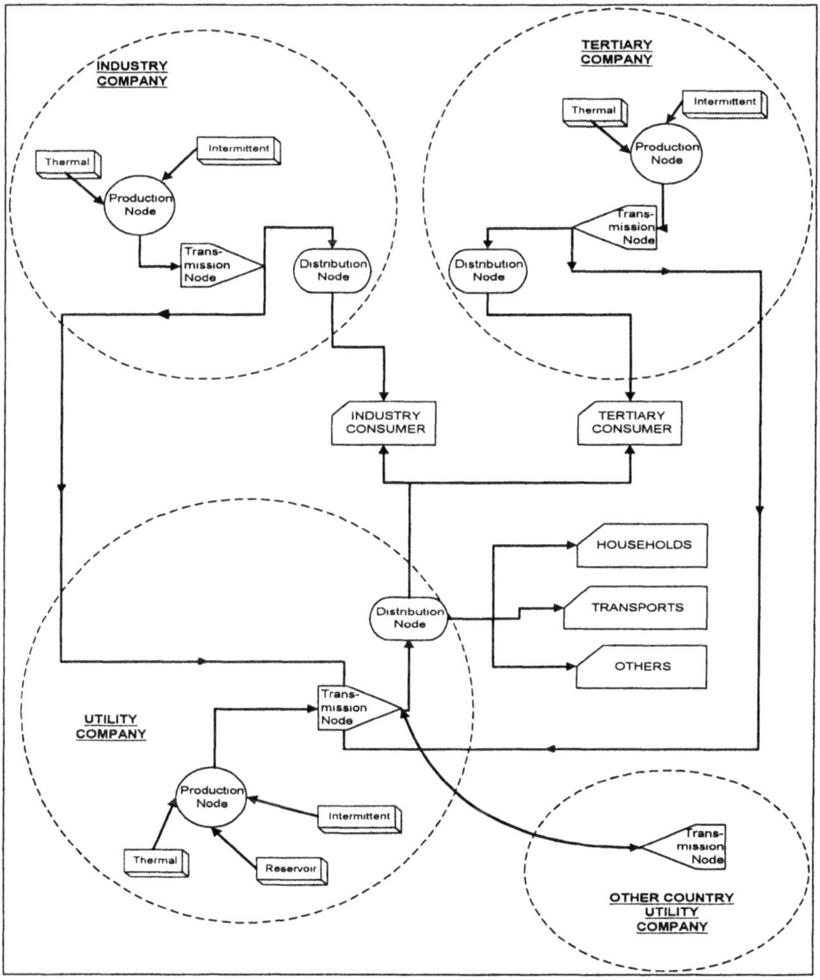

Figure 15-8: The electricity network in PRIMES Version 2.

All companies own production plants. It is not possible to share a plant among companies. There is no ownership of transmission and distribution grids, but the receiving company pays transport services at variable costs, which are also paid implicitly through the marginal cost (dual variable) of the corresponding capacity constraint. There exists one type of company per country. This assumption is equivalent to an assumption about the existence of a representative firm per company category. The consumers are not connected to utilities or other companies outside their country of origin. They can be indirectly supplied from other countries through the transmission nodes linking the countries, via the corresponding utilities.

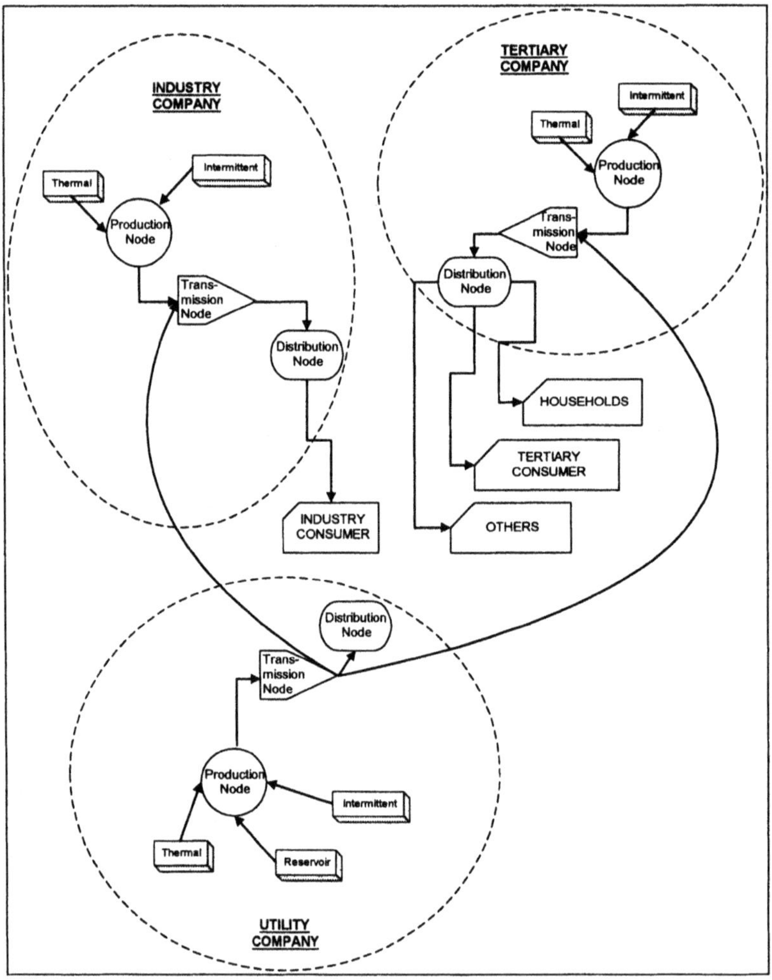

Figure 15-9: The steam network in PRIMES Version 2.

The companies can agree electricity or steam exchange contracts among each other and with the consumers. The contracts involve a quantity, a time pattern over typical days of a year, a duration, a flexibility degree and a route (sequence of nodes) that is used to fulfil the contract along the topology of the network.

Network connections are characterised through their: transmission capacity, transmission efficiency (through which losses are determined at the level of distribution and transmission nodes) and their operation cost (applying only to links between companies). The following schemes present the network as implemented in this version of the sub-model.

The model considers the following plant categories:

- new existing thermal plants of which the capacities are known in the base year
- new thermal plants that are candidate for investment
- existing hydroelectric reservoir plants of which the capacities are known in the base year
- existing intermittent renewable energy plants of which the capacities are known in the base year
- intermittent renewable energy plants that are candidate for investment

The variety of different plant types is incrementally specified in the model, by combining the elements from the following sets:

1. Generic plant technologies (7 existing thermal plant types, 27 types for new thermal plants, 1 reservoir type, 4 existing intermittent and 10 new intermittent)
2. Type of fuel combustion (if applicable), including multiple fuel capability (13 types)
3. Size of the plant (three sizes considered), 3 types.
4. Cogeneration technique (if applicable), 9 types.
5. Company type, 3 types
6. Country of origin, 15 countries

The technical-economic characteristics of the plants differ across the above items. Regarding new plants, technology progress is represented as technological generations that are considered as different plant types.

In total, the model considers:

- 148 different plant types per country for the existing thermal plants
- 678 different plant types per country for the new thermal plants
- 3 different plant types per country for the existing reservoir plants

- 30 different plant types per country for the existing intermittent plants
- 60 different plant types per country for the new intermittent plants

The model considers the following information to characterise a plant technology.

- Capital cost, including overnight cost (ECU'90/kW) and financial charges during construction.
- Variable cost (per kWh produced) and annual fixed costs (per kW). The fixed costs increase over time, as the plant becomes older.
- Thermal efficiency rate and multiple fuel capability, if applicable. Rate of electricity auto-consumption per plant.
- Plant availability rate and rate of utilisation for intermittent plants.
- Time-related characteristics of a plant, like technical lifetime and economic lifetime (used for capital amortisation).
- Technical parameters for the feasible combinations of electricity and steam output, if applicable.
- An old plant (in the category of existing plants) can be retrofitted at the moment of decommissioning (which is exogenous). The retrofitting extends the lifetime of the plant and involves capital and fixed cost payments.
- Intermittent plants are linked to renewable resources for which the time pattern of supply is given.
- Reservoir plants can operate at the limits of energy available from inflow water (exogenous) in the lakes, during each year.
- To reflect future technologies, it is assumed that new plant technologies will become available at different points in time (exogenous parameter).

The model considers several atmospheric pollutants and links them to the combustion of fuels through fixed emission factors. Emission can be abated per plant and fuel used. The costs and possibilities of abatement are represented by cost-supply curves specific to each plant-type, involving five abatement levels.

The model considers 10 fuel types and an association to 13 fuels that can be purchased in the market. Fuels are characterised by their purchase prices, taxes and availability limits.

Fuels can also be supplied through fuel contracts, which are specified per company. A contract involves a quantity, degree of flexibility and a price. Time of use limits on fuels is generally represented through constraints that restrict the dispersion of fuel use over time segments.

The companies can be constrained by policy, through three types of restrictions:

- upper emission limit, per pollutant
- obligation to use a certain percentage of non fossil fuels, or renewables, in the company's energy balance
- fuel obligations, expressed as lower bounds to fuel consumption per company
- upper bounds on investment in new plants
- capacity reserve margins.

The same constraints can apply at the country level, to reflect national policies.

Finally the model includes the possibility of having different discount rates per company and country.

15.1.9 The Oil Refinery Model of PRIMES

The oil-refinery model in the PRIMES model is a linear programming (LP) model, computing the optimal pattern of the refining units and the marginal costs of production, including exchanges between the European countries.

The objective function of the LP model is the total processing cost that has to be minimize according to a lot of various constraints and in order to satisfy a demand for oil products (variables and equations of the model are described in the following paragraphs).

Two types of long-run model are defined.

In the first case, the capacities of modelled refineries are variables of the problem and sole the right-hand-side coefficients of demand constraints are non zero. Duality theory leads under some assumptions stemmed from linear programming models marginal costs to correspond to average costs (Babusiaux, Champlon and Valais- 1982). Consequently, the supply prices of the oil products are computed from the marginal costs of the LP model.

In the second case, the existing capacities of plants are given and their costs are set to zero: the refining units investments are the results of the optimization.

The LP model is defined through the following variables and equations:

- variables of the models are:
    1. purchase of raw materials (crude oil, feedstocks)
    2. plant activities (pipe-still, cracker,...)
    3. blending activities (gasoline, middle distillate, heavy fuel oil, ...)
    4. fuel burnt (furnaces, powerplant, utilities)

v. import/export (including exchanges between areas) and

vi. investment

- the objective function consists in minimising the annual refining costs (including equivalent investment costs), according to a set of constraints:

i. balance of intermediate and finished products

ii. product quality control (octane, suphur, viscosity)

iii. refinery fuel burning balance

iv. avaibilities control (crude oil and feedstock supply) and

v. plant capacities control

- the refining scheme consisted in 12 classic refining units:

i. topping

ii. vacuum distillation

iii. catalytic reformer

iv. catalytic cracker

v. hydrogen unit

vi. hydrocracker

vii. hydrocracker

viii. isomerization unit

ix. alkylation unit

x. visbreaker

xi. coker

xii. hydrodesulfurization unit

As the present refinery model considers a representative agregated refinery scheme, it includes a deep conversion complex which can meet more stringent environmental constraint changes in the oil product specifications in Europe (lower sulphur content, reduction of olefins and aromatics content in gasoline, oxygenates in gasoline).

- This deep conversion complex includes:

i. a c5 deasphaltery unit

ii. a partial oxydation unit

iii. a dao hydrotreater and

iv. a deep hydrodesulfurization unit

Some other units and some other schemes are going to be introduced in the new version of the PRIMES model .

- The investigated schemes are:

  i. a flexicoker unit

  ii. an hydrotreating unit of atmospheric residuals

  iii. an hydroconversion unit of vacuum residuals and

  iv. an hydrotreating unit of atmospheric residuals with a catalytic cracking unit of hydrotreated residuals

Some agregations are done in order to limit the size of the model.

This the case of the crude oil supply. The size of a linear programming model is indeed approximatively proportionnal to the number of crude oil considerd in the model.

This reduction is essential to avoid giving to some crude oils an importance in the global supply they have not in the reality.

The oil refining model is a multi-refineries model which considers several refining areas and some exchanges of oil products between them. This approach is convenient to analyse the oil product flows between countries. For this purpose, the European Union (12-EU) was divided in six refining areas, each one being represented as a single refinery what enables simplifications in the modelling.

The six geographical areas, distinguished according to their characteristics (macroeconomic activities and geographical situation) are:

i. United Kingdom and Eire

ii. Germany and Denmark

iii. Belgium, Netherlands and Luxembourg

iv. France

v. Italy and Greece and

vi. Spain and Portugal

The extension of the model to the three new member states of the European Union (15-EU) will consider a new European refining area in the North (Sweden and Finland) and will extend the Germany-Denmark refining area to Austria.

15.1.10 The Transport Model

15.1.10.1 Introduction

The transport module of PRIMES has been developed to study mainly the penetration of new transport technologies and their effects on emissions, besides the evaluation of the energy consumption and emissions in the transport sector. The emphasis is on the use of car technologies and on the long term (2030). The model structure is kept deliberately simple as it is made to interact as demand module with supply modules (refineries, new fuel production) of PRIMES.

15.1.10.2 Scope of theModel

The model's scope is to represent all energy use per transport purpose in the EU.

Horizon:	1990 – 2030, year by year or by 5 years periods.
Spatial dimension:	EU, country by country.
Transport activities covered:	urban passenger transport;
	non urban passenger transport;
	freight transport.
Transport modes represented:	urban passengers: car, public transport, motorcycle;
	non urban passengers: car, bus, rail, air, navigation;
	freight transport: truck, rail, air, navigation.
Technologies represented:	6 to 10 alternative technologies for each mode (car, bus, truck);
	more limited number of alternatives for rail, air and navigation.
Air pollutants represented:	emission of NO_X, VOC, PM10, CO_2, SO_2.

15.1.10.3 General Overview of the Model Structure

The model covers three types of transport activity:

- urban passenger transport
- non urban passenger transport
- freight transport

and for each type, the model contains four levels as shown in Figure 15-10.

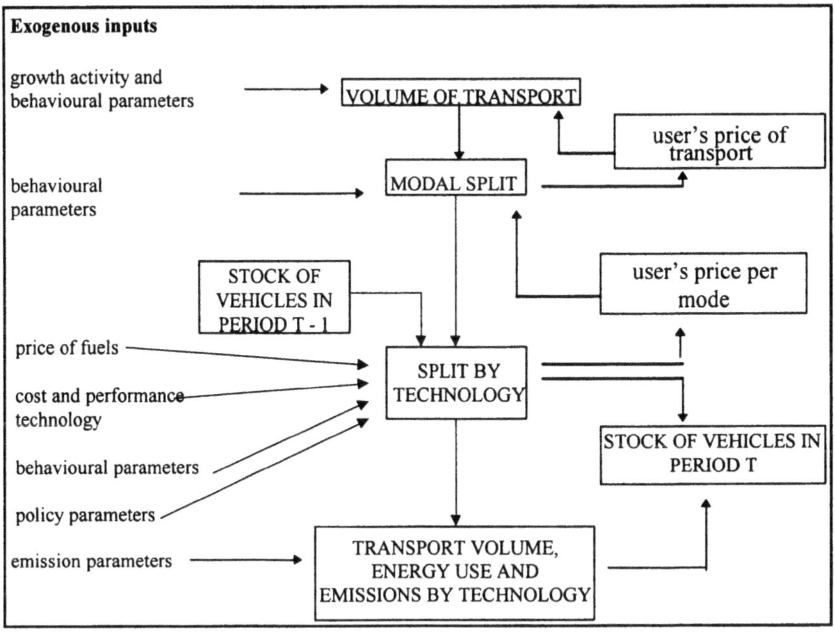

Figure 15-10: General structure of the model.

We see in Figure 15-10 that the overall demand for transport (passenger kilometers, ton kilometers) is determined by income/activity growth and by the overall price of transport. The overall price of transport is determined endogenously, as a function of the modal split and of the price per mode.

The split of the overall transport activity over the different modes is driven by the price per mode and by behavioural parameters. The price per mode depends on the choice of technology for new investment and on past investment for each transport mode. The technologies for new investment are chosen, based on the lowest expected usage costs.

The stock of vehicles inherited from the previous period is expanded in function of the transport needs per mode. The new stock composition determines the stock for the next period as well as the aggregate price per mode.

In the final stage, one computes transport volumes, consumption and emissions by technology as well as by the necessary aggregates.

15.1.10.4 User Price Concept

The choice of technology and of mode is driven by relative user prices. In this model, the user price concept used is close to the generalised cost concept in transportation economics.

Component:	function.
Fuel cost:	cost element.
Vehicle and maintenance cost:	cost element.
(Dis)comfort cost:	in order to represent differences in trunk space, refueling time, driveability among technologies.
Time cost:	in order to represent changes in average speed due to congestion or policy measures.

The generalised price concept is useful to represent other quality characteristics than out of pocket costs. In transportation economics, one often uses the time cost per km (equal to the value of time multiplied by inverse of speed) as an important component in the choice of travel mode. This concept is particularly useful to represent growing congestion phenomena and their impact on the modal choice (second level in Figure 15-10).

15.1.11 Primary Energy Supply, Transformations and Pricing Sub Model

The supply and pricing sub-model serves three main purposes.

- It ensures balance of demand and supply linking all sub-models with supply of primary energy and secondary energy (other than electricity and steam).
- It determines market allocation between domestic and foreign supply (imports).
- It determines the prices of primary and secondary energy forms (other than electricity and gas).
- It applies global environmental constraints and the pollution permit market.

Regarding secondary energy transformations and market allocation between domestic and imported sources, the supply-pricing sub-model behaviour is driven by minimisation of total energy costs. The model derives the optimal shares of domestic production and imports for primary fuels and the optimum shares of transformation output and imports of other fuels, so as to meet fuel demand and satisfy blending constraints. The solution of the above problem provides the shadow cost of meeting the demand constraints. The latter combined with the market pricing regimes, transformation and distribution costs, taxation policies and dual prices of environmental constraints provide fuel prices for each type of consumer. These prices re-enter the corresponding sub-models so that consumers can define their consumption behaviour.

15.1.11.1 Cost Supply Curves and Pricing

Regarding the overall model operation, the role of the supply model is mainly to transmit a positively sloped supply function to demand sub-models. This is achieved by combining the following:

- depletion of domestic resources as extracted quantities increase;
- positively sloped reaction of foreign suppliers of energy, as demand for imports increases;
- short-run pressures on domestic transmission, distribution and transformation capacities, as demand increases;
- pricing of energy forms incorporates those factors to further influence demand, so as to contribute to the overall market equilibrium.

Additional mechanisms are incorporated in the model regarding technology representation. They may influence energy market balances. Among others, the model incorporates modelling of economies of scale, learning by doing and asymmetric reaction of suppliers to demand.

In general, the pricing of energy forms follows the Ramsey-Boiteux principle. This leads to pricing close to average costs. The principle is interpreted as a regime of regulated monopoly for new technologies and fuels, but also as a result of long-run equilibrium of monopolistic competition in case of mature technologies.

The selling price that each consumer faces, is then derived by adding transportation and distribution costs, mark-ups and taxes where applicable. Again transportation and distribution costs involve non linear factors that modify prices when quantities demanded depart from the expected values. Congestion and other similar phenomena are thus covered.

15.1.11.2 Fuels from Transportation Processes

Transformation output processes (refineries, cokery etc.) are treated simultaneously and economic optimisation is performed as if they were within a single company. For each transformation process non linear cost factors are also included to represent penetration of new technologies or capacity use pressures on existing technologies.

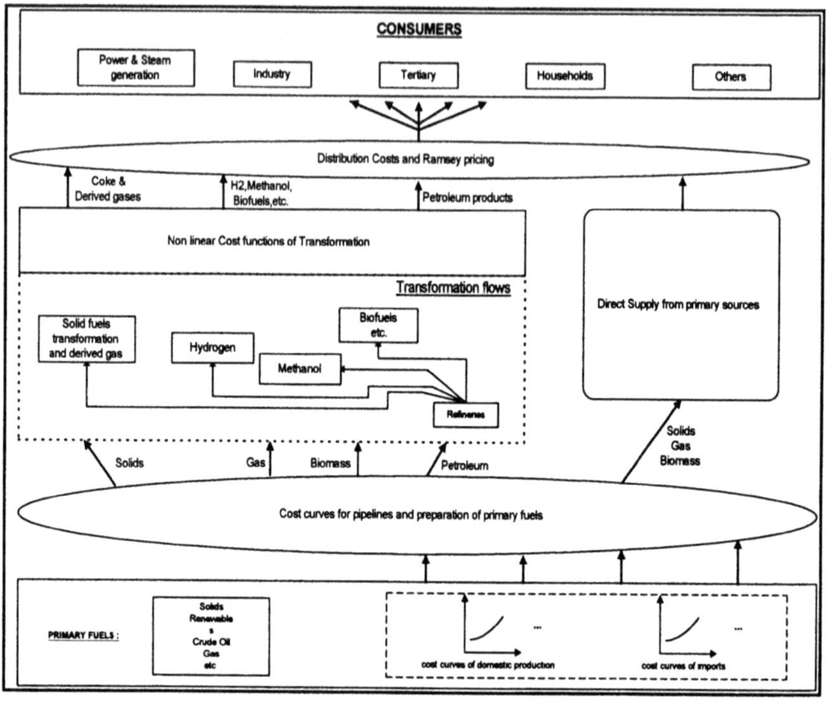

Figure 15-11: The structure of the supply-pricing sub-model

As shown in the figure above, two different levels can be distinguished in the model:

- supply and imports of primary and transformation processes output fuels
- computation of fuels transportation and distribution costs, performance of consumers fuel pricing

15.1.12 Global Environmental Constraints and Pollution Permits Market

Total energy system emissions must satisfy global environmental constraints introduced in the model in order to represent environmental regulations. The shadow price of the constraint which is uniform for each pollutant affects all fuel selling prices.

In the market for pollution permits, each sector computes the optimal amount of permits that it desires and the amount of permits that it wants to trade (sell or buy, depending on its initial endowment of permits). The supply and pricing sub-model ensures that the net sum of permits trading is zero (i.e. that for each unit of permit demanded, someone is willing to supply it). In other terms this equation can be written as the sum of permits used by all sectors must be less or equal to the amount of permits with which all sectors are endowed. The shadow cost of this constraint (again uniform for each pollutant), is again added to the selling price for each fuel. Note that the price of the pollution permits enters the objective function (in the NLP formulation) of cost minimisation for each sector, so that (in the NMCP formulation) derivatives with respect to the activity of each sector are modified accordingly.

15.1.13 General Structure of the Demand Side Sub Models

The demand-side sub-models of PRIMES V.2 have a uniform structure. Each sub-model represents a sector that is further decomposed into sub-sectors and then into energy uses. A technology operates at the level of an energy use and utilises energy forms (fuels). The following graphic illustrates the hierarchical decomposition of the demand-side models.

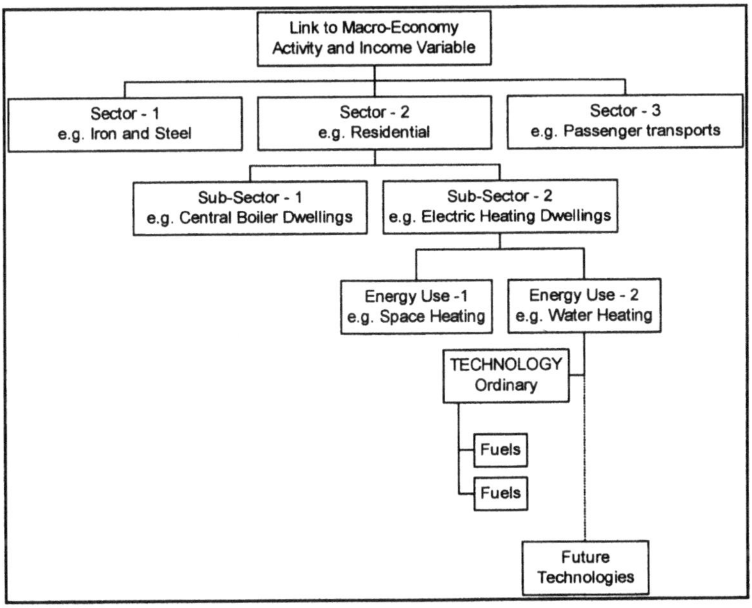

Figure 15-12: Decomposition structure of the demand-side models.

The data that are necessary to calibrate the model for a base year (1995) and a country (all EU member-states) can be divided in the following categories.

1. Macro-economic data that correspond to demographics national accounts, sectoral activity and income variables. These data usually apply to sectors.

2. Structure of energy consumption along the above-described tree in the base year and structure of activity variables (production, dwellings, passenger-kilometres, etc.). Some indicators regarding specific energy consumption are also needed for calibration.

3. Technical-economic data for technologies and sub-sectors (e.g. capital cost, unit efficiency, variable cost, lifetime, etc.).

The basic source of data for energy consumption by sector and fuel is Eurostat (detailed energy balance sheets). By using additional information (surveys of cogeneration operation and capacities and surveys on boilers), we modify the balance sheets in order to represent explicitly the production of steam.

According to PRIMES definitions, steam includes industrial steam and distributed heat (at small or large scale). In the balance sheets, Eurostat reports on steam production in the transformation input/output only if the producers sells that steam. If the steam, irrespectively of the way it is produced (e.g. a boiler or a CHP plant), is used for self-consumption only, Eurostat accounts for only the fuels used to produce that steam and includes these fuels in final energy consumption. Our modification consists in introducing that steam (for self-consumption) in the final energy consumption tables of the balance sheets and inserting the fuels used to produce that steam in the table of transformation input and output. This is necessary for the model to calibrate to a base year that properly accounts for the existing cogeneration activities (even if they are used for self-generation of steam).

The fuel types are as follows:

1. Solid fuels except lignite and peat
2. Lignite and Peat
3. Residual Fuel Oil
4. Diesel Oil
5. Liquefied Petroleum Gas
6. Kerosene
7. Gasoline
8. Naphtha
9. Other oil products
10. Bio-fuels
11. Natural and derived gas
12. Thermal Solar (active)
13. Geothermal low enthalpy

14. Steam (industrial and distributed heat)
15. Electricity
16. Biomass and Waste
17. Hydrogen

Data on fuel purchase prices, with and without taxes, are needed for large customers and small customers.

15.1.14 Industrial Sector

The industrial sector consists of nine sectors. For each sector different sub-sectors are defined. At the level of each sub-sector a number of different energy uses are represented. A technology at the level of an energy use may consume different types of fuels (one of which is steam generated from the power and steam sub-model of PRIMES, so only steam distribution and use costs are accounted for in the demand-side, together with a price for steam). The structure for the industrial sector is given below.

SECTORS	*SUB-SECTORS*	*ENERGY USES*
Iron and Steel		
	Iron and Steel integrated	
		Air compressors
		Low enthalpy heat
		Lighting
		Motor drives
		Rolled steel
		Sinter making
		Steam and high enthalpy heat
		Blast furnace
		Process furnaces
	Electric arc	
		Air compressors
		Low enthalpy heat
		Lighting
		Motor drives
		Electric arc
		Electric process
		Rolled steel
		Foundries
		Sinter making

		Steam and high enthalpy heat
		Process furnaces
Non ferrous metals production		
	Primary aluminium production	
		Air compressors
		Lighting
		Motor drives
		Electric furnace
		Electrolysis
		Process furnaces
	Secondary aluminium production	
		Air compressors
		Lighting
		Motor drives
		Electric furnace
		Electrolysis
		Process furnaces
	Copper production	
		Air compressors
		Lighting
		Motor drives
		Electric furnace
		Electric kilns
		Low enthalpy heat
		Steam and high enthalpy heat
		Process furnaces
	Zinc production	
		Air compressors
		Lighting
		Motor drives
		Electric furnace
		Electrolysis
		Process furnaces
	Lead production	
		Air compressors
		Lighting
		Motor drives
		Electric furnace

		Process furnaces
	Other non ferrous metals production	
		Air compressors
		Lighting
		Motor drives
		Electric furnace
		Low enthalpy heat
		Foundries
		Process furnaces
Chemicals production		
	Fertilizers	
		Air compressors
		Low enthalpy heat
		Lighting
		Motor drives
		Electric processes
		Steam and high enthalpy heat
		Thermal processes
	Petrochemical	
		Air compressors
		Low enthalpy heat
		Lighting
		Motor drives
		Electric processes
		Steam and high enthalpy heat
		Thermal processes
		Energy use as raw material
	Inorganic chemicals	
		Air compressors
		Low enthalpy heat
		Lighting
		Motor drives
		Electric processes
		Steam and high enthalpy heat
		Thermal processes
	Low enthalpy chemicals	
		Air compressors
		Low enthalpy heat
		Lighting

		Motor drives
		Electric processes
		Steam and high enthalpy heat
		Thermal processes
Building materials production		
	Cement dry	
		Air compressors
		Lighting
		Motor drives
		Low enthalpy heat
		Cement kilns
		Drying and separation
	Ceramics and bricks	
		Air compressors
		Lighting
		Motor drives
		Electric kilns
		Low enthalpy heat
		Tunnel kilns
		Drying and separation
	Glass basic production	
		Air compressors
		Lighting
		Motor drives
		Glass annealing electric
		Glass tanks electric
		Low enthalpy heat
		Glass annealing thermal
		Glass tanks thermal
	Glass recycled production	
		Air compressors
		Lighting
		Motor drives
		Glass annealing electric
		Glass tanks electric
		Low enthalpy heat
		Glass annealing thermal
		Glass tanks thermal
	Other building materials	

	production	
		Air compressors
		Lighting
		Motor drives
		Low enthalpy heat
		Material kilns
		Drying and separation
Paper and pulp production		
	Chemical paper	
		Lighting
		Motor drives
		Pulping electric
		Refining electric
		Steam and high enthalpy heat
		Low enthalpy heat
		Pulping steam
		Drying and separation
		Refining steam
	Mechanical pulp and paper	
		Lighting
		Motor drives
		Pulping electric
		Refining electric
		Steam and high enthalpy heat
		Low enthalpy heat
		Pulping steam
		Drying and separation
		Refining steam
Food, Drink and Tobacco production		
	Food, Drink and Tobacco goods	
		Air compressors
		Cooling and refrigeration
		Lighting
		Motor drives
		Drying and separation electric
		Steam and high enthalpy heat
		Low enthalpy heat

		Space heating
		Drying and separation thermal
		Specific heat
		Direct heat
Engineering		
	Engineering goods	
		Air compressors
		Lighting
		Motor drives
		Drying and separation electric
		Machinery
		Coating electric
		Foundries electric
		Steam and high enthalpy heat
		Low enthalpy heat
		Space heating
		Drying and separation thermal
		Coating thermal
		Foundries thermal
		Direct heat
Textiles production		
	Textiles goods	
		Air compressors
		Cooling and refrigeration
		Lighting
		Motor drives
		Drying and separation electric
		Machinery
		Steam and high enthalpy heat
		Low enthalpy heat
		Space heating
		Drying and separation thermal
		Direct heat
Other industrial sectors		
	Other industrial sectors goods	
		Air compressors
		Lighting
		Motor drives
		Drying and separation electric

		Machinery
		Steam and high enthalpy heat
		Low enthalpy heat
		Space heating
		Drying and separation thermal
		Specific heat
		Direct heat

Macroeconomic Data (for each sector, not sub-sector):
value added for the industrial sectors in Million ECU90.
Production Data (for each sub-sector):
index of industrial production – in physical units or an aggregator proxy.
Energy Consumption structure:
fuel consumption (steam is a fuel) per sector, sub-sector and energy use (in toe) for the base year (1995).
Specific energy consumption data:
energy consumption per unit of physical production (preferably at the level of sub-sector). It may be computed if all above data are available.
Useful energy indicator (in toe) at the level of each energy use:
it requires assumption on average energy efficiency of the energy use in a sub-sector.
Technical economic data for ordinary technology for each energy use:
capital purchase cost (ECU'90/toe-year) for a unit capacity of an energy use;
variable and fixed costs (ECU'90/toe) of an energy use;
efficiency rate per fuel type and energy use;
capital utilisation factor (average yearly load factor in %) of an energy use;
lifetime in years.
Statistical data for technology for each energy use:
average age in years in the base year (1990);
degree of overcapacity in the base year (%).
Technical-Economic data for sub-sectors:
non energy related capital investment cost per unit of production;
non energy and non-capital related operating costs per unit of production;
price of production of each sub-sector.

15.1.15 Tertiary Sector

The tertiary sector comprises of 4 sectors. At the level of the sub-sectors, the model structure defines groups of energy uses, which are further subdivided in energy uses defined according to the pattern of technology. The structure is as follows.

SECTORS	*SUB-SECTORS*	*ENERGY USES*
Agriculture		
	Lighting	

		Lighting
	Space heating	
		Electric
		Gas connected
		Boiler
		District heating
		Solids
		Greenhouses
	Electrical uses	
		Electrical uses
	Pumping	
		Pumping
	Motor energy	
		Motor energy
Offices and Services		
	Lighting	
		Lighting
	Space heating	
		Electric
		Gas connected
		Boiler
		District heating
		Solids
		Greenhouses
	Air conditioning	
		Electric
		Gas connected
		District heating
	Electrical uses	
		Electrical uses
	Water heating	
		Water heating
Trade		
	Lighting	
		Lighting
	Space heating	
		Electric
		Gas connected
		Boiler
		District heating

		Solids
		Greenhouses
	Air conditioning	
		Electric
		Gas connected
		District heating
	Steam uses	
		Steam uses
	Electrical uses	
		Electrical uses
	Water heating	
		Water heating
Public services		
	Lighting	
		Lighting
	Space heating	
		Electric
		Gas connected
		Boiler
		District heating
		Solids
		Greenhouses
	Air conditioning	
		Electric
		Gas connected
		District heating
	Steam uses	
		Steam uses
	Electrical uses	
		Electrical uses
	Water heating	
		Water heating

Macroeconomic Data (for each sector, not sub-sector):

value added for the tertiary sectors in Million ECU90;

employment per tertiary sector.

Structural Data (for each sub-sector):

index of surface per employee.

Energy Consumption structure:

fuel consumption (steam is a fuel) per sector, sub-sector and energy use (in toe) for the base year (1995);

useful energy indicator (in toe) at the level of each energy use;

it requires assumption on average energy efficiency of the energy use in a sub-sector;
average thermal integrity of buildings in each tertiary sector;
technical economic data for ordinary technology for each energy use;
capital purchase cost (ECU'90/toe-year) for a unit capacity of an energy use;
variable and fixed costs (ECU'90/toe) of an energy use;
efficiency rate per fuel type and energy use;
capital utilisation factor (average yearly load factor in %) of an energy use;
lifetime in years.

Statistical data for technology for each energy use:
average age in years in the base year (1990);
degree of overcapacity in the base year (%);
technical-Economic data for sub-sectors;
non energy related capital investment cost per unit of production of a tertiary sector;
non energy and non capital related operating costs per unit of production of a tertiary sector.

15.1.16 Residential Sector

The residential sector distinguishes five categories of dwelling. These are defined according to the main technology used for space heating. They may use secondary heating as well. At the level of the sub-sectors, the model structure defines the categories of dwellings, which are further subdivided in energy uses. The electric appliances for non heating and cooling are considered as a special sub-sector, which is independent of the type of dwelling. The structure is as follows.

SECTORS	*SUB-SECTORS*	*ENERGY USES*
Dwellings		
	Central boiler households that may also use gas connected to the central boiler (flats)	
		Space heating
		Cooking
		Water heating
		Air conditioning
	Households with mainly electric heating equipment (non partially heated)	
		Space heating
		Cooking
		Water heating
		Air conditioning
	Households with direct gas equipment for heating (direct gas for flats and gas for	

	individual houses)	
		Space heating
		Cooking
		Water heating
		Air conditioning
	Households connected to district heating	
		Space heating
		Cooking
		Water heating
		Air conditioning
	Partially heated dwellings and agricultural households	
		Space heating
		Cooking
		Water heating
		Air conditioning
Electric appliances		
		Washing machines
		Dish washers
		Dryers
		Lighting
		Refrigerators
		Television sets

Macroeconomic Data:
- population;
- number of households.

Structural Data (for each sub-sector):
- index of surface per dwelling category;
- number of households per dwelling category;
- number of electric appliances (per type) per household.

Energy Consumption structure:
- fuel consumption (steam is a fuel) per sector, sub-sector and energy use (in toe) for the base year (1995);
- useful energy indicator (in toe) at the level of each energy use;
- it requires assumption on average energy efficiency of the energy use in a sub-sector;
- average use per year of an electric appliance per household;
- average thermal integrity of buildings in each category of dwelling;
- technical economic data for ordinary technology for each energy use;
- capital purchase cost (ECU'90/toe-year) for a unit capacity of an energy use;
- variable and fixed costs (ECU'90/toe) of an energy use.

Efficiency rate per fuel type and energy use:

capital utilisation factor (average yearly load factor in %) of an energy use;
lifetime in years.
Statistical data for technology for each energy use:
average age in years in the base year (1990);
degree of overcapacity in the base year (%).

15.1.17 Transport Sector

The transport sector distinguishes passenger transport and goods transport as separate sectors. They are further subdivided in sub-sectors according to the transport mean (road, air, etc.). At the level of the sub-sectors, the model structure defines several technology types (car technology types, for example), which correspond to the level of energy use. The structure is as follows.

SECTORS	*SUB-SECTORS*	*ENERGY USES*
Passenger transports		
	Busses	
		Internal combustion engine
		Electric motor
		Fuel cell
		Gas turbine
	Motorcycles	
		Internal combustion engine
		Electric motor
	Private cars	
		Internal combustion engine
		Electric motor
		Fuel cell
		Gas turbine
	Passenger trains	
		Internal combustion engine
		Electric motor
		Fuel cell
	Air transports	
		Gas turbine
	Navigation passengers	
		Internal combustion engine
Goods transports		
	Trucks	
		Internal combustion engine

		Electric motor
		Fuel cell
		Gas turbine
	Trains	
		Internal combustion engine
		Electric motor
		Fuel cell
	Navigation	
		Internal combustion engine

Macroeconomic Data:
- gross domestic product in ECU'90 for the base year;
- volume of imports and exports.

Structural Data (for each sub-sector):
- passenger-kilometres and ton-kilometres per type of transport mean;

Energy Consumption structure:
- fuel consumption per transport mean and transport technology (in toe) for the base year (1995);
- useful energy indicator (in toe) at the level of each transport technology;
- it requires assumption on average energy efficiency of the energy use in a sub-sector;
- average use per year of a vehicle per type (kilometres per vehicle per year);
- average number of passengers per vehicle type;
- technical economic data for ordinary technology for each energy use;
- capital purchase cost (ECU'90/toe-year) for a unit capacity of a transport technology;
- variable and fixed costs (ECU'90/toe);
- efficiency rate per fuel type and technology;
- capital utilisation factor (average yearly load factor in %) of a technology;
- lifetime in years.

Statistical data for technology for each energy use:
- stock of vehicles per technology and transport mean;
- average age in years in the base year (1990);
- degree of overcapacity in the base year (%).

15.2 The POLES Model

The model aims at describing the structuring variables of national energy systems, in terms of their dynamics and international interactions. The objective is thus to produce coherent images of the world energy system in the long-term (2030), with a level of detail which provides a clear indication of technological and sectoral developments and which can contribute to informed decision-making in the energy system and R&D.

15.2.1 Main Characteristics of the POLES Model

The POLES model is a simulation model of the world energy system up to the year 2030 (for a detailed description of the model, see Criqui et al., 1996). It is a world energy system model in that economic and demographic trends in each of the large countries or regions of the world are considered exogenous, while changes in the variables which characterise energy consumption, transformation, production and prices are endogenous to the model. It is a recursive simulation model in that the dynamics are given, from the initial point then from year to year, through progressive adjustments to supply and demand variables on the one hand and price variables on the other.

Lastly, the projection horizon is deliberately limited to about forty years, so as to preserve "explicit" rather than "generic" technologies in the model. This period of time corresponds in fact to the time required between the initial emergence of a new technology and its widespread diffusion. In other words, it is assumed that technologies, which might have a significant quantitative impact in the year 2030, must be at least identified today.

15.2.2 Structure of the Model

The model is based on a hierarchical structure of modules, interconnected at the national, regional and world levels.

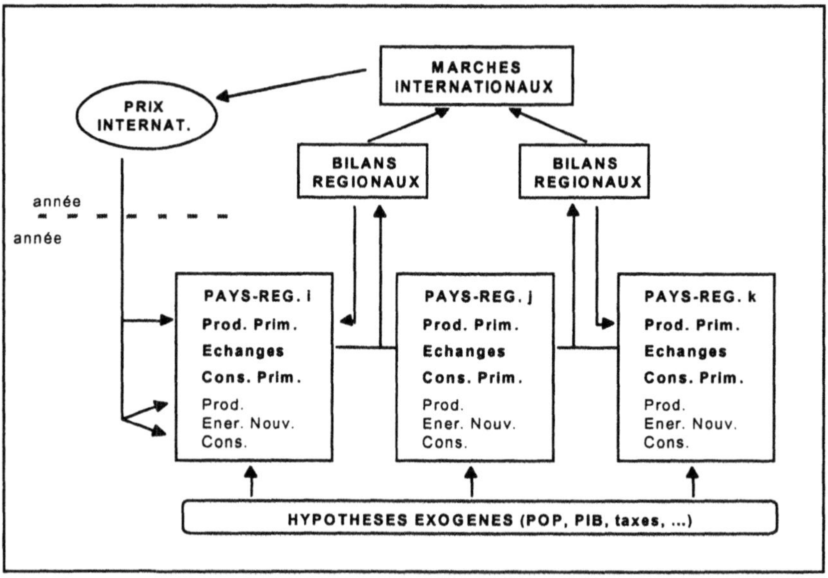

Figure 15-13: Structure and organisation of the POLES model.

The national dimension includes the modules of consumption, new energies, conversion to electric power and fossil energy production for each region. At the regional or world levels are found energy trade flows and the international price modules, for the three major energy sources involved in extensive international trade: oil, natural gas and coal.

At present the model distinguishes 26 geographical regions, which can be used as a basis for reconstituting various regional plans. One in particular is based on 11 regions and is used in most world energy studies (WEC-IIASA, 1995; IEA, 1996; DOE-EIA, 1997): North America, Latin America, Western Europe, Central Europe, CIS, North Africa and the Middle East, sub-Saharan Africa, Southern Asia, South-East Asia, mainland Asia, Pacific OECD countries. In addition, within each region, the most powerful countries - those of the G7 and the five major rapidly-developing economies (China, India, Korea, Brazil and Mexico) - are identified separately and are the subject of detailed models. As will be seen, this is particularly important for studying the costs of strategies for reducing greenhouse gas emissions, as well as the stakes involved in the international negotiations on climate change.

15.2.3 Simulation of Primary Energy Consumption

The simulation of primary energy consumption is carried out in steps, by analysing first the energy consumption according to sector and energy source, then the contribution of new and renewable energies and finally the contribution of the electricity production sector to the total primary consumption of the country or region in question.

Consumption by end-users is analysed in a detailed sectoral manner, by identifying three industrial subsectors, namely automobile transport and other transport modes, tertiary activities and residential consumption. In each subsector, changes in demand depend on specific activity indicators and on changes in consumer prices. The price-effect analysis is particularly detailed since it takes into account the short-term effects on behaviour and the delayed impact on investments. Similarly, the dissymmetry of demand elasticities is taken into account, in the case of a rise or fall in prices.

A dozen new or renewable energy technologies are identified (combined heat and power, wind power, solar energy...). The model identifies some aspects of the diffusion dynamics of new technologies, as demonstrated by the economic analysis of technical change. In particular, it uses the "cumulative learning effects" to explain the gradual reductions in costs (Ayres and Martinas, 1992) and draws on logistical profiles of diffusion within the identified potentials (Fischer & Pry's diffusion model, see Grübler, 1996).

Finally, the simulation of the dynamics and consumption characteristics of the electricity system distinguishes two stages: investment planning and the development of production facilities on the one hand, and the operation of production facilities, taking into account changes in demand, on the other hand. Investment programming starts with minimisation of total production costs, for anticipated investment and operating costs for a dozen cases of typical equipment, either existing or future (from the conventional coal-fired power station to new-concept nuclear reactors, and not forgetting combined-cycle gas turbines). Operation of the production facilities takes into account final demand for electricity and the associated load curve, as well as the variable costs and fuel costs of the different equipment available.

15.2.4 Development and Production of Fossil Fuels

The procedure used to assess future production of fossil energies is based in the case of oil and natural gas on a "discovery process" model (Adelman and Jacoby, 1979). It is assumed that the development of reserves depends on the oil prospection-exploration effort, according to following logic: reserves at time t are equal to total discoveries - themselves a function of the prospection-exploration effort - decreased by cumulative production at that date. Production is then derived either from the changes in reserves for the countries known as "price-takers", or, for the large exporting countries, from the constraints regarding closure of the world oil balance or regional gas balances (America, Eurafrica, Asia).

With this approach, natural constraints can be taken into account - ultimate resources, as considered in particular by the USGS (Masters et al., 1994) - but also the phenomena of economic renewal of reserves through prospection-exploration and technical progress - continuous in recent years - with regard to recovery rates. It is then possible to avoid the debate on resource endowment between "optimists" and "pessimists" (see Odell (1995) for the first thesis, Laherrère and Perrodon (1996) for the second). In the model, resource endowment and technical progress are integrated into a plan which is in line with that advocated by Adelman (1987): that of the "endless struggle of nature versus knowledge".

In addition to the logic of discovery and production of conventional oil, a further factor to consider is the dynamics of mobilising non conventional hydrocarbons (extra-heavy oils, tar sands, bituminous shale) or natural gas and coal liquefaction. The production of these non conventional liquid hydrocarbons is in fact dependent on their specific production and conversion costs in relation to the price of oil, which is itself a function of the rate at which resources are used.

15.2.5 International Energy Prices and Trade

The development of trade is based on the relationship between consumption and primary energy production by country or region. For oil, a single world market is considered (concept of *one great pool,* Adelman and Jacoby (1979)) and variations in oil prices depend on:

- a long-term variable corresponding to changes in the reserves/production ratio at the world level;
- a short-term variable corresponding to the utilisation rate of the production capacities of the Gulf States.

In the case of natural gas, the world market is made up of three major regional groups (America, Eurafrica, Asia). For each one, a different natural gas price is calculated according to:

- an index linked to oil prices;
- a specific term corresponding to the variation in the reserves/production ratio for the key producers of the region.

In the case of coal, international prices, again for three regional groups, depend on the coal production expansion rate in the major producing countries (USA, South Africa, Australia, China).

The model can thus give a complete image of the national and international energy systems, with a description - probably not exhaustive but nevertheless detailed - of the technologies used and can take into account the fundamental mechanisms involved in the adjustment of energy supply and demand to price variations.

List of Figures

List of Tables

References

Adelman, M.A., Jacoby, H.D. (1979), "Alternative Methods of Oil Supply Forecasting", in *Advances in the Economics of Energy and Resources*, JAI Press, Vol. 2, p. 1-38.

Adelman, M.A. (1987), "Economic Theory of Mineral Depletion with Special Reference to Oil and Gas", in International Association for Energy Economics (IAEE) 9th International Conference, Calgary, Rowse, J.G. (Ed.), p. 43-51.

Armand, F., Ayache, G. (1995), "Le Développment de l'éolien en France et dans le Monde: Quelles Perspectives pour l'éolin en France?", Brochure ADEMA, November.

Ayres R., Martinas K. (1992), "Experience and the Life Cycles: some Analytic Implications", in Technovation, Vol. 12 N° 7, p. 465-486.

Beer J., Worrell E., Blok K. (1995), "Long Term Energy Efficiency Improvement", in Zwerver, S., Van Rompaey, R.S.A.R., Kok, M.T.J and Berk, M.M. (Eds.), *Climate Change Research: Evaluation and Policy Implications*, Elsevier Science, The Netherlands, p. 1081–1086.

Berkhout, F. (1991), "Radioactive Wastes and the Nuclear Revival", Energy Policy, Vol.. 19, n°7, p. 615-620.

Blok, K., Farla, J., Hendriks, C., Turkenburg, W. (1991), "Carbon Dioxide Removal: a Review", Paper Pesented at the International Symposium on Environmentally Sound Energy Technologies and their Transfer to Developing Countries and European Economics in Transition (ESETT '91).

Bupp, I., Derian, J.C. (1978), "Light Water: how the Nuclear Dream Dissolved", New York, Basic Blackwell.

CCE-DG XII, (1995), " Extern, Externalities of Energy", Vol. 1 Summary, European Commission, Directorate General XII, EUR 16520 EN, p.179.

Committee on Science, Engineering and Public Policy (COSEPUP), (1991), "Policy Implications of Greenhouse Warming: Report of the Mitigation Panel", U.S. National Academy of Sciences, U.S. National Academy of Engineering and Institute of Medicine, U.S. National Academy Press, Washington, DC, USA.

Colombier, M., Dessus, B., Laponche, B. (1997), "Négociations Cimat: Quels Engagements?", in Global Chance N° 8, July, p. 57-64.

Coherence (1991), "Cost-Effectiveness Analysis of CO_2 Reduction Options", Synthesis Report, Report for the Commission of the European Communities, DGXII, May.

Coherence, (1991), "Cost-Effectiveness Analysis of CO_2 Reduction Options", Country Reports, Report for the Commission of the European Communities, DGXII, May.

Coherence, (1994), "Cost-Effectiveness Analysis of Energy Technologies and CO_2 Abatement Strategies: the Case of Five EC Countries Germany, France, Italy, The Netherlands, United Kingdom", Report for the European Commission, DGXII.

Cowan, R. (1990), "Nuclear Power Reactors: a Study in Technological Lock-In", The Journal of Economic History, Vol. 4, n° 3.

Criqui, P. et al. (1996), "POLES 2.2", European Commission, Directorate General XII, EUR 17538 EN, p. 99.

Criqui, P., Kouvaritakis, N., "What Sort of Environment in the XXI Century? Environment, Long-Term Control and Democracy" to Appear in the Proceedings of the Fontevraud Conference.

Department of Energy (DOE), (1993), "A Research Needs Assessment for the Capture, Utilization and Disposal of Carbon Dioxide from Fossil Fuel-Fired Plants", DOE, Washington, DC, USA.

Department of Energy/Energy Information Administration (DOE/EIA), (1996), "The Changing Structure of the Electric Power Industry: an Update", DOE/EIA, Washington, DC, USA, December.

Department of Energy/Energy Information Administration (DOE/EIA), (1997), "Electricity Prices in a Competitive Environment: Marginal Cost Pricing of Generation Services and Financial Status of Electric Utilities, A Preliminary Analysis Through 2015", DOE/EIA, Washington, DC, USA, August.

Department of Energy/Energy Information Administration (DOE/EIA), (1997), "International Energy Outlook", DOE/EIA, Washington, DC, USA, p.112.

Douglas, J. (1991), "Fuel Cells for Urban Power", Electric Power Research Institute (EPRI) Journal, September, p. 5-11.

European Commission (EC), (1995), "Techno-Economic Assessment Studies of Fossil Fuel and Fuel Wood Power Generation Technologies", Joule II-Programme R&D in Clean Coal Technology, Directorate-General for Science, Research and Development, Belgium, Brussels, p.125.

Electric Power Research Institute (EPRI), (1990), "New Push for Energy Efficiency", EPRI Journal 15(3): 4–17; and Barakat and Chamberlin Inc., 1990, Efficient Electricity Use: Estimates of Maximum Energy Savings, EPRI Report CU-6746, EPRI, Palo Alto, C.A., USA.

Environmental Protection Agency (EPA), (1990), "Policy Options for Stabilizing Global Climate", EPA, Washington, DC, USA.

European Wind Energy Association (EWEA), (1991), "Time for Action: Wind Energy in Europe".

Feron, P.H.M., Jansen, A. (1995), "Capture of Carbon Dioxide Using Membrane Gas Absorption and Reuse in the Horticultural Industry", Paper Presented at the 2nd International Conference of Carbon Dioxide Removal, 25–27 October 1994, Kyoto, Japan.

Finon, D., Menanteau Ph. (1989), "L'Echec des Surgénerateurs: Autpie d'un Grand Programme", Grenoble PUG.

Finon, D. (1995), "La Diversification des Modèles d'Organisation des Industries Electriques dans le Monde: une Mise en Perspective", Revue de l'Energie, Janvier-Février, p. 465.

Finon, D. (1996), "Les Nouvelles Fonctions du Régulateur et du Gouvernement dans les Industries Electriques Libéralisées: Les Leçons des Expériences Européennes", Revue de l'Energie, Avril-Mai, p. 477.

Fisher, D. (Ed.) (1990), "Options for Reducing Greenhouse Gas Emissions", Stockholm Environment Institute, Stockholm, Sweden.

Giovannini, B., Pain, D. (1990), "Scientific and Technical Arguments for the Optimal Use of Energy", Université de Genève, Geneva, Switzerland.

Gipe, P. (1995), "Wind Energy Comes of Age", John Wiley & Sons, Inc.

Gipe, P. (1996), "1996 Overview of Wind Generation", WWW, P. Gipe Inc.

Golomb, D., Herzog, H., Tester, J., White, D. and Zemba, S. (1989), "Feasibility, Modeling and Economics of Sequestering Power Plant CO_2 Emissions in the Deep Ocean", MIT-EL 89-003, Massachusetts Institute of Technology, Cambridge MA, USA.

Grubb, M., Edmonds, J. (1993), "The Cost of Limiting Fossil-Fuel CO_2 Emissions, A Survey and Analysis", in Annual Review of Energy and Environment, Vol. 18, p. 397-478.

Grubb, D., Meyer, N.I. (1993), "Wind Energy: Resources, Systems and Regional Strategies, in Johansson, T.B., Kelly, H., Reddy, A.K.N. and Williams, R.H. (Eds.), *Renewable Energy: Sources for Fuel and Electricity*.

Grubb, M., Chapuis, T., Duong, M.H. (1995), "The Economics of Changing Course. Implications of Adaptability and Inertia for Optimal Climate Policy", Energy Policy, 23, p. 417-432.

Grübler, A. (1996) "Time for a Change: on the Patterns of Diffusion of Innovation", in Daedalus, p. 19-42.

Hall, D.O. (1994), "Biomass Energy Options", Proceedings of the "Energy Technologies to Reduce CO_2 Emissions in Europe: Prospects, Competition, Synergy", 11-12 April 1994, Petten, The Netherlands, Organisation for Economic Co-operation and Development/ International Energy Agency (OECD/IEA), Paris, France, p.159–194.

Haugan, P.M., Drange, H. (1992), "Sequestration of CO_2 in the Deep Ocean by Shallow Injection", Nature, 357, p. 318–320.

Herman, R., Ardekani, S.A., Ausubel, J.H. (1989), "Dematerialization", in Ausubel, J.H. and Sladovich, H.S. (Eds.), *Technology and Environment*, U.S. National Academy Press, Washington, DC, USA.

Herzog, H.J., Adams, E.E. (1995), "Environmental Impacts of Ocean Disposal of CO_2", Paper Presented at the "International Energy Agency Greenhouse Gases: Mitigation Options" Conference, 22–25 August 1995, London, UK.

Hirschenhofer, J.H. (1997), "Fuel Cell Statuts: 1996", IEEE AES Systems Magazine, March, p. 23-28.

Hourcade, J.C. (1996), "Aspects Socio-Economiques, Consensus Inespéré ou Artefact Rhétorique ?", in Global Chance, N° 7 - July, p. 63-72.

Hunt, S., Shuttleworth, G. (1996), "Competition and Choice in Electricity", John Wiley & Sons.

International Energy Agency (IEA), (1991), "Energy Efficiency and the Environment", Organisation for Economic Co-operation and Development/ International Energy Agency (OECD/IEA), Paris, France.

International Energy Agency (IEA) Greenhouse Gas R&D Programme, (1994), "Carbon Dioxide Capture from Power Stations", Report Number SR2P, IEA, Gloucestershire, UK.

International Energy Agency (IEA), (1995), "World Energy Outlook", IEA, Paris, France.

International Energy Agency (IEA), (1996), "World Energy Outlook", Organisation for Economic Co-operation and Development/ International Energy Agency (OECD/IEA), Paris, France, p. 300.

International Energy Outlook, (1997), Energy Information Administration, Office of Integrated Analysis and Focasting, US Department of Energy, Washington, DC, USA.

International Energy Agency (IEA), (1997), "World Energy Outlook", IEA, Paris, France.

Intergovernmental Panel on Climate Change (IPCC), (1995), "Deuxième Rapport d'Evaluation du IPCC", OMM-PNUE, p. 55 & App.

Intergovernmental Panel on Climate Change (IPCC), (1996a), "Climate Change 1995, A Review of Mitigation Cost Studies", Cambridge University Press, Cambridge, UK.

Intergovernmental Panel on Climate Change (IPCC), (1996b), "Policies and Measures for Mitigating Climate Change 1996", Cambridge University Press, Cambridge, UK.

Ishibashi, M., Ota, H., Akutsu, N., Umeda, S., Tajika, M., Izumi, J., Yasutake, A., Kageyama, Y. (1995), "Technology for Removing Carbon Dioxide from Power Plant Flue Gas by the Physical Adsorption Method", Paper Presented at the "International Energy Agency Greenhouse Gases: Mitigation Options" Conference, 22–25 August 1995, London, UK.

Karras, K. (1992), "Wind Energy: what Does it Really Cost?", Paper Presented at the Windpower '92 Annual Conference of the American Wind Energy Association, Seattle, USA.

Kaya, Y., Fujii, Y., Matsuhashi, R., Yamaji, K., Shindo, Y., Saiki, H., Furugaki, I., Kobayashi, O. (1991), "Assessment of Technological Options for Mitigating Global Warming", Paper Presented at the IPCC EIS Group Meeting, 6–7 August 1991, Geneva, Switzerland.

Kemi, W. (1987), "Industrial Uses of Carbon Dioxide", in Aresta, M. and Forti, G. (Eds.), "Carbon Dioxide as a Source of Carbon", NATO ASI Series C, Vol. 206, Reidel, Dordrecht, The Netherlands.

Kliman, M. (1997), "Enhancing the Market Deployment of Energy Technology", Edited by the International Energy Agency (IEA).

Laherrère, J., Perrodon, A. (1996), "Contribution à l'Estimation des Réserves Ultimes", in Revue de l'Énergie, N° 398 , November-December, p. 53-56.

Levine, M.D., Geller, H., Koomey, J., Nadel, S. and Price, L. (1992), "Electricity End-Use Efficiency: Experience with Technologies, Markets, and Policies Throughout the World", Lawrence Berkeley Laboratory, Berkeley, C.A., USA.

Madsen, B.T. (1995), "Market Update", Wind Stats Newsletter, Vol. 8, n° 2.

Malinvaud, E. (1993), "Pour un 'New Deal' Européen", Le Monde, 16 November 1993.

Manne, A.S. and Richels, R.G. (1992), "Buying Greenhouse Insurance", The MIT Press, Massachusetts Institute of Technology, Cambridge MA, USA, p. 182.

Marchetti, C. (1976), "On Geoengineering and the CO_2 Problem", RM-76-17, International Institute for Applied Systems Analysis (IIASA), Laxenburg, Austria.

Masters, C.D. et al. (1987), "World Resources of Crude Oil, Natural Gas, Natural Bitumen and Shale Oil", Proceedings of the 12th World Petroleum Congress, 5:3–27, John Wiley, Chichester, UK.

Masters, C.D., Root, D.H. and Attanasi, E.D. (1991), "World Resources of Crude Oil and Natural Gas", Preprints of the 13th World Petroleum Congress 25: 1–14, John Wiley, Chichester, UK.

Masters, C.D. et al. (1994), "World Petroleum Assessment and Analysis", Proceedings of the 14th World Petroleum Congress, p. 529-541.

Mc Mullan, J. T., Williams, B.C., Campbell, P., McIlveen-Wright, D. and Bemtgen, J.M. (1995), "Techno-Economic Assessment Studies of Fossil Fuel and Fuel Wood Power Generation Technologies", European Commission, Directorate General XII, Brussels, Belgium.

Meier, A. (1991), "Supply Curves of Conserved Energy", in Proceedings of the International Energy Agency (IEA) International Conference on "Technology Responses to Global Environmental Challenges: Energy Collaboration for the 21st Century", 6–8 November 1991, Vol. 1, Inter Group Corporation, Kyoto, Japan.

Michaelis, L. (1994), "The Transport Sector", Proceedings of the "Energy Technologies to Reduce CO_2 Emissions in Europe: Prospects, Competition, Synergy", 11-12 April 1994, Petten, The Netherlands, Organisation for Economic Co-operation and Development/ International Energy Agency (OECD/IEA), Paris, France, p. 73–114.

Mimura, T., Shimojo, S., Suda, T., Iijima, M. and Mitsuoka, S. (1994), "Research and Development on Energy Saving Technology for Flue Gas Carbon Dioxide Recovery and Steam System in Power Plant", Paper Presented at the 2nd International Conference of Carbon Dioxide Removal, 25–27 October 1994, Kyoto, Japan.

Nakicenovic, N., Grübler, A., Inaba, A., Messner, A., Nilsson, S., Nishimura, Y., Rogner, H.-H., Schäfer, A., Schrattenholzer, L., Strubegger, M., Swisher, J., Victor, D. and Wilson, D. (1993), "Long-term Strategies for Mitigating Global Warming", Special Issue of Energy–The International Journal 18(5) p. 401–609.

Nakicenovic, N., (1995), "Carbon Dioxide Reduction Strategies", PIK/WI Workshop on "Cost-Benefit Analyses of Climate Change", 10-11 November 1995, Potsdam, Germany.

Newman, P. and Kenworthy, J. (1989), "Cities and Automobile Dependence: a Sourcebook", Gower Technical, Aldershot, UK.

Nishimura, F. (1991), "Technology Assessment for the Reduction of Carbon Emissions in the Electricity Sector", International Institute for Applied Systems Analysis (IIASA), Laxenburg, Austria.

Nishikawa, N., Morishita, M., Uchiyama, M., Yamaguchi, F., Otsubo, K., Kimuro, H. and Hiraoka, R. (1992), "Carbon Dioxide Clathrate Formation and its Properties in the Simulated Deep Ocean", Paper Presented at the 1st International Conference of Carbon Dioxide Removal, 4–6 March 1992, Amsterdam, The Netherlands.

Nishikawa, N., Ishibashi, M., Ohta, H., Akutsu, N., Matsumoto, H., Kamata, T. and Kitamura, H. (1995), "CO_2 Removal by Hollow-Fiber Gas-Liquid Contactor", Paper Presented at the 2nd International Conference of Carbon Dioxide Removal, 25–27 October 1994, Kyoto, Japan.

Odell, P. (1995), "Is Middle-East Domination of the International Oil Market Inevitable ?", 1995 Annual Meeting of the International Energy Workshop, International Institute for Applied Systems Analysis (IIASA), Laxenburg, Austria, p. 20.

Office of Technology Assessment (OTA), 1991, "Energy Technology Choices Shaping our Future", Washington, DC, USA.

Okken, P.A. and Ybema, J.R. (1994), "CO_2 Reduction through Materials Chain Management", Proceedings of the "Energy Technologies to Reduce CO_2 Emissions in Europe: Prospects, Competition, Synergy", 11-12 April 1994, Petten, The Netherlands, Organisation for Economic Co-operation and Development/ International Energy Agency (OECD/IEA), Paris, France, p. 303–324.

Perry, R.H. and Green, D. (1983), "Perry's Chemical Engineers Handbook", 6th edition, McGraw Hill, New York, NY, USA.

Saha, S. and Chakma, A. (1992), "Separation of CO_2 for Gas Mixtures with Liquid Membranes", Paper Presented at the 1st International Conference of Carbon Dioxide Removal, 4–6 March 1992, Amsterdam, The Netherlands.

Saji, A., Yoshida, H., Sakai, M., Kitamura, H. and Kamata, T. (1992), "Fixation of Carbon Dioxide by Clathrate Hydrate", Paper Presented at the 1st International Conference of Carbon Dioxide Removal, 4–6 March 1992, Amsterdam, The Netherlands.

Schüssler, U. and Kümmel, R. (1989), "Carbon Dioxide Removal from Fossil Fuel Power Plants by Refrigeration Under Pressure", in Jackson, W.D. and Hull, D.A. (Eds.), Proceedings of the 24th Intersociety Energy Conversion Engineering Conference, Washington, DC, USA.

Springmann, F. (1991), "Analysis of the Ecological Impact of Demonstration Projects in the Field of Rational Use of Energy: Development of Evaluation Criteria", Study on behalf of the Commission of the European Communities, Directorate General for Energy (DG XVII), Regio–Tec GmbH, Starnberg, Germany.

Steinberg, M. (1996), "The Carnol System for Methanol Production and CO_2 Mitigation from Coal Fired Power Plants and the Transportation Sector", Brookhaven National Laboratory, Upton, NY, USA.

Strubegger, M. and Reitgruber, I. (1995), "Statistical Analysis of Investment Costs for Power Generation Technologies", WP-95-109, International Institute for Applied Systems Analysis (IIASA), Laxenburg, Austria, p. 21.

TERES II (1996): "The European Renewable Energy Study Energy for Sustainable Development", ESD Ltd. Corsham, UK.

Train, K.E. (1985), "Discount Rates in Consumers' Energy-Related Decisions: a Review of the Literature", Energy 16(12), p. 1243–1253.

Train, K.E. (1991), "Optimal Regulation: the Economic Theory of Natural Monopoly", The MIT Press, Massachusetts Institute of Technology, Cambridge MA, USA.

Turkenburg, W.C. (1995), "Energy Demand and Supply Options to Mitigate Greenhouse Gas Emissions", in Zwerver, S., Van Rompaey, R.S.A.R., Kok, M.T.J. and Berk, M.M. (Eds.), *Climate Change Research: Evaluation and Policy Implications*, Elsevier Science, The Netherlands, p. 1013–1054.

Walker, J.A. (1996), "The Influence of Privatization on the Direction and Scale of R&D: Recent UK Experience in Competition and New Technology in the Electric Power Sector", International Energy Agency/Organisation for Economic Co-operation and Development (IEA/OECD), Paris, France.

World Energy Council and International Institute for Applied Systems Analysis (WEC-IIASA), 1995, "Global Energy Perspectives to 2050 and Beyond", The World Energy Council, London, UK, p.105 & Ann.

Wigley, T.M.L., Richels, R., Edmonds, J.A. (1996), "Economic and Environmental Choices in the Stabilisation of Atmospheric CO_2 Concentrations", Nature, 379, p. 240-243.

Worrell, E., Cuelenaere, R.F.A., Blok, K. (1992), "Energy Consumption of Industrial Processes in the European Commission", University of Utrecht, The Netherlands.

Wrixon, G.T., Rooney, A.M, Paltz, W. (1993), "Renewable Energy", Edited by Springer Verlag.

Yanagisawa, Y. (1995), "Separation, Conversion and Disposal of Carbon Dioxide as Mitigation Measures to Global Climate Change", in Proceedings of the "Role of Energy Technologies Towards Sustainable Development", 16–20 October 1995, Kansai, Japan.

GPSR Compliance

The European Union's (EU) General Product Safety Regulation (GPSR) is a set of rules that requires consumer products to be safe and our obligations to ensure this.

If you have any concerns about our products, you can contact us on ProductSafety@springernature.com

In case Publisher is established outside the EU, the EU authorized representative is:

Springer Nature Customer Service Center GmbH
Europaplatz 3
69115 Heidelberg, Germany

Batch number: 10725973

Printed by Printforce, the Netherlands